About the Authors

Jay Flynn has over ten years experience working with ArcInfo. He has worked as a GIS analyst for private firms and municipal government, and is now employed at Radford University, Virginia, where he teaches in the Geography Department and serves as manager of the department's GIS lab.

Teresa Pitts has worked in GIS for both the private and local government sectors, as well as for the U.S. Geological Survey and on U.S. Forest Service funded projects. She is currently employed as an instructor in the Geography Department at Radford University, Virginia.

Acknowledgments

The authors would like to acknowledge the people at ESRI for their help throughout this project, including technical review and especially their helpfulness in obtaining the software and permission to use data from their on-line web site *(www.esri.com)* for exercises.

We would also like to acknowledge Denise Dodd and Ben Logan for their proofreading and testing of the exercises.

Ms. Pitts would like to thank the staff of the Slow the Spread program, a U.S. Forest Service project conducted at Virginia Tech, for giving her the opportunity to expand her ArcInfo knowledge.

The authors would like to thank Cynthia Welch and Carol Leyba of OnWord Press for all their help in getting this book from manuscript form into print. And a special thank-you goes to Daril Bentley for his many hours of editing and proofreading. We could not have done it without you!

For Katherine, who brings joy to us every day!

Contents

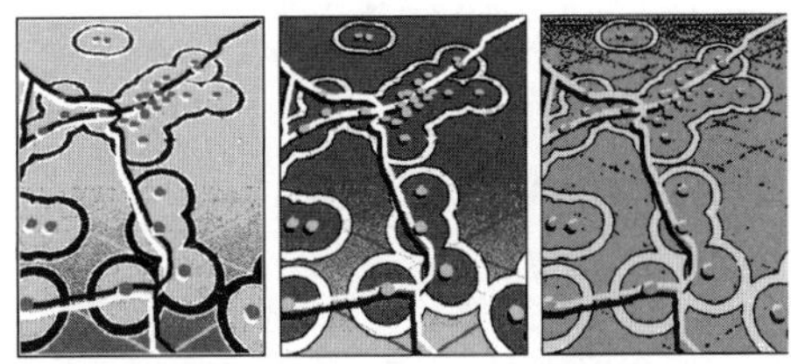

Introduction

Help is what you need and what this book offers. Designed for both new users and veteran GISers, *INSIDE ArcInfo* describes the new features found in Desktop ArcInfo, ESRI's new Windows NT software. Also described are the traditional core modules found in ArcInfo Workstation, ArcEdit, ArcPlot, and Arc. This book is organized to get you started quickly, whether you are an experienced user or new to ArcInfo or GIS.

Philosophy and Approach

Learning ArcInfo often seems like a daunting task to persons new to the software. This book seeks to make the process as simple as possible. Many hands-on exercises help you learn how to perform everyday tasks, as well as more complicated GIS operations. As you learn more about ArcInfo, you will find the sections of this book that provide quick descriptions of commands and menus found throughout ArcInfo an invaluable reference feature.

What This Book Covers

No book can cover all aspects of ArcInfo in depth. It is too vast and complex a program. Indeed, most organizations use only some of the modules of ArcInfo. The authors have chosen to concentrate on those modules that are included with out-of-the-box ArcInfo. The first part of the book focuses on the new drag-and-drop interface modules.

- ArcCatalog
- ArcMap
- ArcToolbox

The second part of the book examines the traditional modules.

- Arc
- ArcEdit
- ArcPlot
- INFO
- AML

Specialty modules such as Grid, Network, and so on are not covered in this book. Look to other OnWord Press publications to help you with them.

New in Version 8.0

With Version 8.0, ArcInfo is the custom, menu-driven software each organization wishes it had time to create. Not only do new drag-and-drop interfaces allow users to quickly compose map graphics, but also images and even shapefiles (ArcView formatted data layers) can be incorporated with the new ArcMap tool. The ArcCatalog tool makes it easy to find and manage your data: no more launching another file manager.

And remember all the times you had to look up Arc command syntaxes? No more! Popular Arc commands are now just a click away using the ArcToolbox, which can even be customized to suit your specific needs. ArcInfo Version 8.0 is a truly exciting development in the GIS industry. Its ease of use will open GIS capabilities to scores of new users.

Summary of New Features

ArcInfo 8.0 introduces Desktop ArcInfo, the new Microsoft Windows NT-based version of ArcInfo. ArcInfo Workstation, the traditional ArcInfo software already known to many users, is still maintained and supported by ESRI. The first half of this book describes the new Desktop ArcInfo software and includes many exercises to familiarize you with it. The following is a summary of the modules that make up Desktop ArcInfo.

- *ArcCatalog:* This module is used to organize your spatial data, preview coverages and shapefiles, list data tables, establish links between ArcInfo and third-party databases, and maintain your data. ArcCatalog is especially useful for keeping your data organized, including copying, renaming, or deleting coverages, shapefiles, and data tables.

- *ArcMap:* Use this module to create professional maps using a wide array of symbols and fonts. Included are a number of map templates to help you create maps very quickly. It features a graphical user interface (GUI) that makes map production a breeze. Great-looking maps can be viewed on screen or printed. On-screen features can be "hot linked" to photographs, documents, and even web pages on the Internet.
- *ArcToolbox:* A great benefit to all users, this module provides a menu system for accessing many of the commands in ArcInfo. Organized by function into tool sets, you can quickly find the commands you need without having to search for them. A few commands have step-by-step menus called wizards that make more complex commands more accessible to the new user.

Will My AMLs Still Run?

Yes! ArcInfo Version 8.0 retains all functions of Version 7. Any Arc Macro Language programs (commonly called AMLs) your organization relies on can still be used.

How to Use This Book

First, set up a workspace that is yours alone – a place to store your finished and unfinished exercises. You will want to access them easily. Second, as soon as you can, learn or review the basic concepts of GIS; this will enhance your understanding of the exercises and the software. Third, be consistent in your efforts: do not let weeks go by between learning sessions. And finally, relax: ArcInfo is a large, complex software package with thousands of commands; this book breaks it up into easy-to-tackle sections. You will find help and tips throughout the book.

This book allows you to chart your own course through this newest version of ArcInfo software. Each chapter can be worked through as a discrete unit, so you are able to learn what you need when you need it.

- *Totally new to GIS, ArcInfo, and panicking?* Don't worry. Head to the PowerStart exercise at the end of Chapter 1 now and have a map out in about an hour. At your leisure, return to Chapter 1 and proceed through the rest of the book.

- *Totally new to GIS but not panicking?* Begin at the beginning! Start with "What is GIS?" Learn the concepts behind GIS first to enhance understanding of the software and its varied commands. Then proceed to the PowerStart exercise following Chapter 1.
- *Know GIS but not ArcInfo?* Skip the first part of Chapter 1 and start at the section titled "ArcInfo and Topology," and then go to the PowerStart exercise to get you up to speed.
- *Already know ArcInfo Version 7?* Jump in at the PowerStart Exercise: it introduces the new drag-and-drop interface in Version 8.0. Then proceed to descriptions of the new modules.

Companion CD-ROM Installation

The companion CD-ROM contains a folder named *ai80exercises,* which holds the data used for all the exercises in the book. To use the data, you must copy it to your hard drive. The exercises will not work reliably if you attempt to run them from the CD-ROM!

Use Windows Explorer to copy the entire folder from the CD-ROM to your hard drive. Throughout this book, the hard drive is assumed to use the letter D. You may copy the exercise folder to any drive letter, but be sure to note your drive letter designator when working through the exercises.

There is also a file named *data_links.htm* on your PC. This file is a web document containing links to sources of spatial data available on the Internet. The links in this document are the same as those found in the appendix.

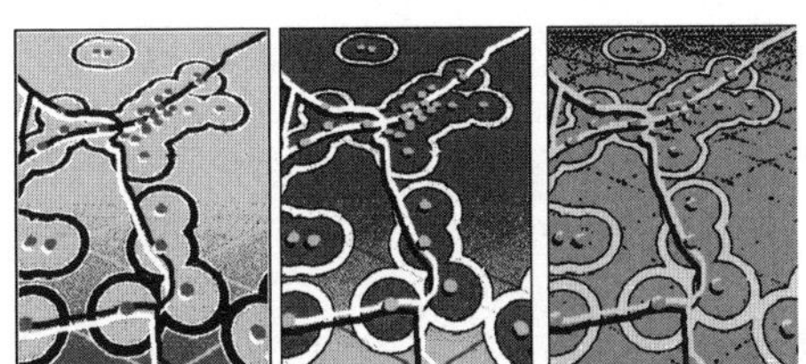

Chapter 1

An Introduction to GIS

What Is a Geographic Information System?

A geographic information system (GIS) combines computer mapping capabilities with a database, allowing users to store, analyze, and display data distributed across an area. GIS have been around since the early days of computer science. Originally programmed to run on large mainframes and to use teletype printers to create maps, GIS have evolved into a variety of software packages that run under several different operating systems.

ArcInfo, a software product from Environmental Systems Research Institute (ESRI, sometimes called "Ez-ree"), is the focus of this book. Its name refers to both its mapping ability (Arc) and its database connectivity (Info). This chapter introduces you to general GIS concepts and specific ArcInfo approaches to GIS.

Consisting of software, hardware, and various types of data, a GIS collects, creates, analyzes, and displays spatial data. Spatial data are data distributed across an area. An example familiar to most readers is a road map, which might show useable transportation routes (data) across a state (an area).

Another example is property (parcel) data: who owns what and where. In this case, the "where" may be a computer map attached to the "who" and "what" in a database. Three types of data are important to a GIS: points, arcs (lines), and polygons (areas). Table 1-1, which follows, includes examples of each type of the basic forms of GIS data.

Table 1-1: Point, Arc, and Polygon GIS Data Examples

Point	Arc	Polygon
Schools	Roads	U.S. states
Churches	Streams, rivers	Property parcels
Insect trap sites	Bus routes	National forests
Water wells	Wind flow	Climate regions

Sometimes GIS is referred to as a computer mapping program, but it is actually much more powerful. The power of GIS lies in its ability to take spatial data from one source (a digitized map), combine it with data from other sources (databases), and produce new data.

For example, a large storm is approaching and you need to alert property owners whose parcels lie in the river's floodplain to prepare for flash flooding. Using a GIS, you overlay the county's property map layer (layers are often called coverages) with the floodplain layer (perhaps derived from elevation or land use data) to produce a new map of just those properties within the floodplain. Doing this by hand would take hours or even days. With a GIS, the process is completed in minutes.

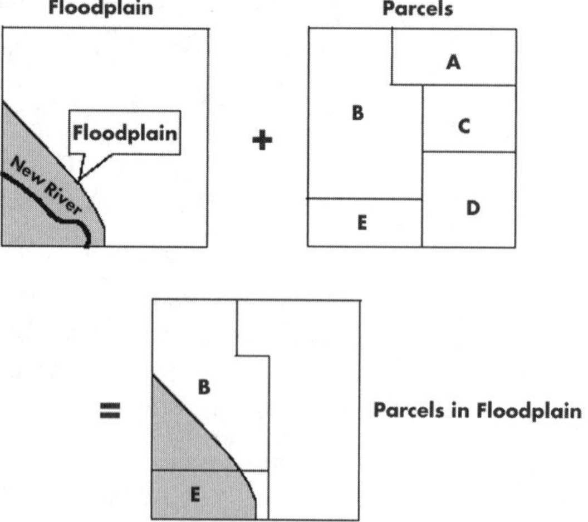

A GIS can quickly combine property map data with floodplain data to create new information on which properties lie within a floodplain.

GIS Versus CAD

There are many computer-aided design (CAD) software programs and hardware packages on the market, some of which are used for mapping. Although some CAD programs duplicate some of the functionality of a GIS, none incorporate the full range of GIS capability. GIS typically have a database component most CAD systems do not. A GIS can analyze and merge two or more data sets to create an entirely new map. Most CAD systems lack this analysis and merge functionality.

Perhaps the most important difference between a CAD system and a GIS is how each stores data. Generally, a CAD system uses lines to draw its features, but cannot tell you about the relationship between those features (their distance apart, the quickest route from one to the other, and so on). A GIS, however, has a built-in intelligence called topology that tracks the relationships between features. Topology is integral to the way ArcInfo stores and organizes data, which it does as something called a coverage. The next two sections discuss coverages and topology.

Organizing Data in ArcInfo

ArcInfo stores its graphic data as coverages. Think of each coverage as a layer: your GIS may have one layer for roads, one for county boundaries, one for land use, and so on. All of these layers may be combined to make a map of the county, but each is stored and accessed separately. This idea is shown in the following illustration.

The roads, county boundaries, and land-use coverages are stored as separate data layers. They can be combined for analysis and display.

County Boundary

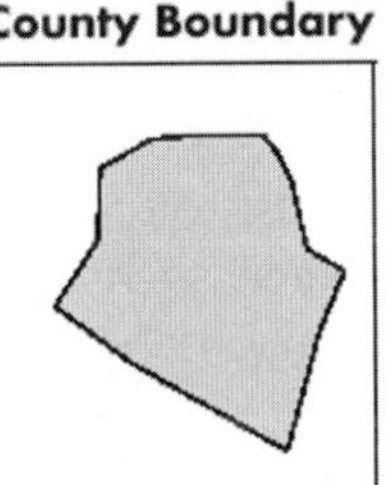

Roads

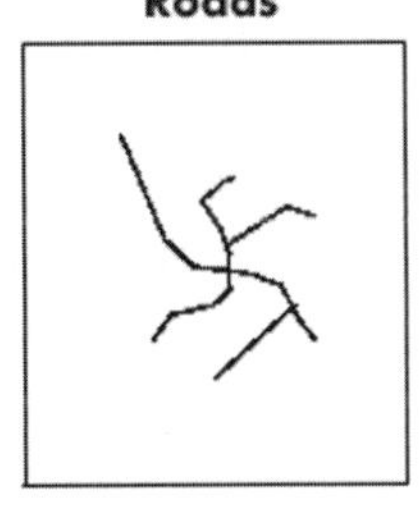

Land Use

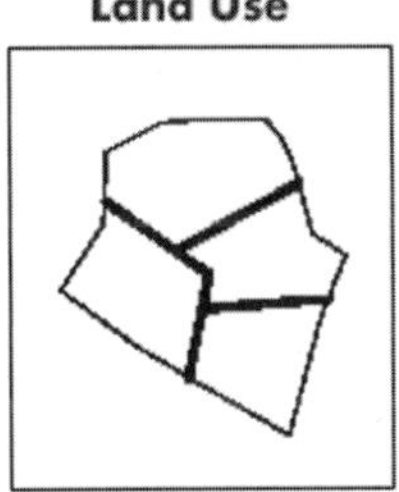

Three Layers Combined

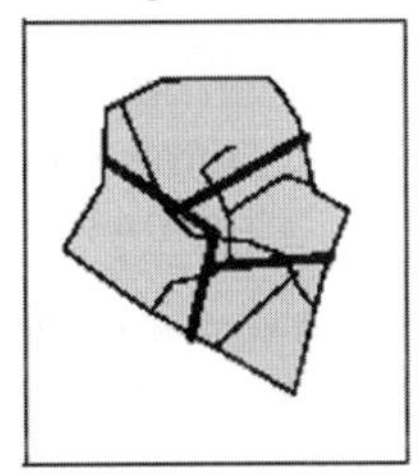

Each coverage can contain several features classes. A feature class represents a type of geographic data. Each feature can have associated attribute tables that hold descriptive data about it. Points, arcs, and polygons are types of feature classes. The following are descriptions of ArcInfo feature classes.

- *Label point:* A label point has two functions: (1) within a polygon, it links the polygon to database information about that polygon, and/or (2) it represents a point feature such as a water meter, a telephone pole, and so on.
- *Node:* Nodes exist at the beginning and end of each arc. Each has its own unique identification number. Other attributes can be assigned to nodes.
- *Arc:* An arc consists of a "from" node, a series of vertices, and a "to" or end node. Vertices and nodes are stored as X-Y coordinates. ArcInfo assigns each arc its own unique identification number. Other attributes can be assigned to arcs.
- *Polygon:* A polygon is defined by the arcs surrounding it. Each polygon has a label point inside its boundaries. This label point links it to database attribute tables.
- *Tic:* Tics are registration points used to align a coverage with specific points on Earth. They are also used to register a paper map on a digitizing tablet.
- *Annotation:* Annotation is descriptive information used for display purposes. Annotation includes labeled street names, river names, and so on.
- *Section:* Sections are a means of representing part of an arc.
- *Route:* A route consists of one or more sections. You can define data specific to routes, such as travel cost along a network, type of pavement present, and so on. Using sections and routes requires an extra ArcInfo module (Network) and is beyond the scope of this book.
- *Region:* Regions are useful for collecting a lot of small area features into bigger units. For instance, regions could be used to create a region feature of zoning categories. This reduces the number of polygons needed to be processed. Regions provide flexibility in managing and modeling data.

Although you may work with all of the feature types previously listed, you will probably work more often with the polygon, arc, and label point features. Some feature types—tics, for example—

may be created the first time you create a coverage, and then never used again. Table 1-2, which follows, presents examples of three of the most commonly used feature classes.

Table 1-2: Examples of Three Common Feature Classes

Polygons	*Arcs*	*Label Points*
Parcels	Rivers	Telephone poles
Census blocks	Roads	Utility lines
City boundaries	Airplane flight paths	Manholes
National forests	Railroads	Crime locations
Parks	Hiking paths	Streetlights

As previously mentioned, coverages can contain more than one feature class. For instance, a coverage of property parcels may contain both polygon and region features. You may work exclusively with polygon features for data entry and editing but use region features for aggregating and displaying.

What are X-Y coordinates?

X-Y coordinates locate a point in 2D space. The X coordinate represents the horizontal distance of a point from the origin, and the Y coordinate represents the vertical distance of a point away from the origin. The origin is the starting point of the distance units counting system. Traditionally it begins with 0,0 and is located at the lower left-hand corner of the graph (you probably remember this from high school geometry class).

In GIS the X-Y coordinates are in the same units as the data layer's projection coordinates. For example, if your data layer units are in latitude and longitude, the X-Y coordinate pair units will also be in latitude and longitude.

Topology and ArcInfo

In ArcInfo, topology refers to the way in which line data are stored and referenced. Topology records the spatial relationship between the arcs and polygons in a coverage. Each arc has a "from" node, a "to" node, a left and a right polygon, and a unique ID associated with it. Groups of arcs that form closed shapes (polygons) are associated with a unique label point. Storing data in this manner allows the system to determine which polygons neighbor which, which arcs are part of which polygons, how far apart arcs and polygon centers are from each other, and so on.

Keep in mind that topology is not created by the software automatically. The user has to execute a command to create it. ArcInfo's commands for creating topology are Build and Clean. These commands are discussed in depth in Chapter 8. Topology, along with the built-in database, provides the power, and the advantage, of a GIS.

Arc 531 has corresponding entries for its "from" and "to" nodes and left and right polygons in the coverage's .AAT table.

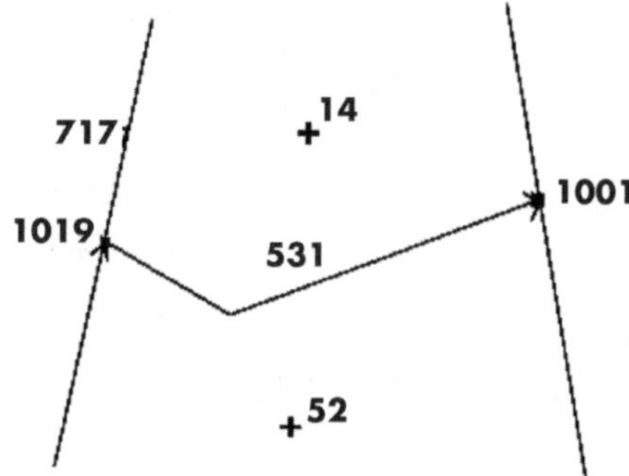

The items in table 1-3, which follows, correspond to the previous illustration. Each arc has a "from" and a "to" node, as well as a unique ID. Items storing "right" and "left" polygon IDs give the arc directionality, the direction in which data may flow along that arc. This is an important component for such things as water stream data and 911 emergency route data.

Table 1-3: Coverage Arc Attributes

From#	To#	Lpoly	Rpoly	Coverage#	Coverage_ID
1019	1001	14	52	531	1

The X-Y coordinate pairs that constitute each arc are hidden from view but are linked to the *.AAT* by the *coverage#* value. This value is assigned by the ArcInfo software. For a refresher on X-Y coordinates, see the previous sidebar.

Every arc and polygon coverage will have an *.AAT* file. You can add additional attributes to an *.AAT* file, but ArcInfo predefines the following *.AAT* attributes, which cannot be deleted or changed.

☛ **WARNING:** *Always add new attribute items after the* <coverage>-ID *item. Adding new items before it can destroy your coverage!*

- *fnode#*: "from" node; software-defined value for the beginning node
- *tnode#*: "to" node; software-defined value for the end node
- *lpoly#*: Polygon to the left of the arc
- *rpoly#*: Polygon to the right of the arc

The following attributes will only have values if polygon topology has been created.

- *length*: Length of the arc
- *<coverage>#*: Software-defined unique identifier for this arc; links this file and the file containing the actual X-Y coordinates
- *<coverage>-ID*: Software-defined identifier for this arc; users can change this value

Getting Data into ArcInfo

There are as many ways of entering data into ArcInfo as there are data types. Data from paper or other flat sources may be scanned or digitized, coverages can be created from tabular (database) files, coordinates can be keyboard entered, and CAD and other types of files can be converted for entry. Even remote-sensing images (such as aerial photography) can be displayed alongside mapped data. You can create data in house or purchase data from vendors, or tap into the U.S. government's vast sources of GIS data.

NOTE: *For more information on data sources, see Appendix A, or the web page of free GIS data sources on the companion CD-ROM.*

Digitizing and Scanning Data

Perhaps the most common source of data exists in house. Organizations typically convert their existing data into a form useable by ArcInfo. Paper maps remain a common source of data for a GIS. Two devices, digitizers and scanners, will help you get this type of data into ArcInfo.

Digitizers are flat boards containing a fine grid of wires. The boards are usually about two inches thick and are available in a variety of sizes. A puck, consisting of an eyepiece with a crosshair and several buttons, connects to the board via a cable.

The person operating the digitizer uses the puck's crosshairs to trace the desired features. Points along these features are registered in relation to the wire grid. In other words, the wire grid acts like a sheet of graph paper and a feature's coordinates are then stored in a file. That file is used to recreate the features in ArcInfo. Digitizing is fairly inexpensive and relatively simple, but can be prone to human error and should be carefully quality controlled.

An alternative to using a digitizer is to use a scanner. A scanner is a device that enters the entire source map into ArcInfo at one time. The map is secured to the scanner, and a photoelectric eye scans it, moving back and forth until it has completely scanned the map. Scanning often takes just a fraction of the time a digitizer would to accomplish the same task.

The problem with scanning, however, is that it scans the entire map. Someone must now work with the scanned map to extract just the features needed. That person might scan the paper map for roads, streams, and municipal boundaries only. This can be a time-consuming process that requires good editing skills and quality control checks.

Creating Data Layers from Files

Using specific ArcInfo commands (such as GENERATE, discussed in Chapter 4), coverages can be created from tabular data. Generally the file must be in the form of ID, X, Y. That is, each coordinate pair needs a unique ID number. Other attributes associated with each pair are later "joined" to create the coverage. This process is described in detail in Chapter 9.

You can also enter coordinates and feature IDs in ArcEdit, using the keyboard. This can be time consuming, however, if you have many points to enter. Instead of specifying a digitizer or mouse as the input device, identify the keyboard. See "Data Entry" in the ArcEdit on-line help for more information.

ArcInfo contains ready-made conversion tools for most popular files. AutoCAD's *.DXF* files, for example, are easily converted to coverages. The file conversion tools are described in Chapter 4.

Entering coordinates by hand is another option, although this is best done with small files. It can be time consuming and prone to human error.

Analyzing Data

As previously mentioned, a GIS goes beyond simply mapping: it can analyze and even create new data by combining data layers and accessing databases. For example, you can combine a road coverage with a national forest boundary coverage and create a new data layer of roads within the national forest. From this data layer a table can be generated, which might be used to assign oversight responsibilities to county or federal agencies for road repairs. The following are examples of other types of analyses GIS users conduct.

- Finding persons who live within a mile of a hazardous chemical spill
- Determining if a new housing development is within a stream's floodplain
- Locating the best position for a new cellular phone tower
- Determining high-income neighborhoods for siting a gourmet grocery

ArcInfo incorportes a full suite of tools that allows you to easily conduct wide varieties of spatial analysis. You can even combine them to form shortcuts for frequently repeated analyses. Version 8.0 extends the power of these tools by providing an easy-to-use GUI called ArcDesktop. Most analysis commands run in the Arc module. The commands are discussed in depth in chapters 4 and 8. The following are types of spatial analysis you can perform in ArcInfo.

- *Extract:* Removes desired features from one coverage to create a new one
- *Overlay:* Combines coverages to add, erase, or update features
- *Proximity:* Determines the distances between or around coverage features
- *Statistics:* Offers some standard statistical analyses
- *Surface:* Creates and analyzes 3D surfaces

Specialized ArcInfo modules such as GRID and NETWORK offer extensive raster data analyses, route analyses, and other capabilities. However, the specialty modules are beyond the scope of this book.

Displaying Data

Many options exist for displaying data with ArcInfo, ranging from just looking at it on the screen to printing paper maps to creating WWW-ready graphics. You can even print a descriptive tabular report. In most cases, however, you will want to create a map graphic that displays your work. Creating effective maps is discussed in chapters 3 and 7. ArcMap is the GUI-driven map composition module that makes designing a map graphic very easy. This GUI-driven module is new to Version 8.0.

ArcInfo 8.0 has a full complement of cartographic commands, fonts, and symbols to help create professional maps. Symbols specific to many types of industries (such as oil and gas exploration charting, highway and transportation markers, and municipal symbols) are built into ArcInfo.

For printing or plotting maps, ArcInfo supports everything from black-and-white laser printers to large-format inkjet plotters to professional photo-ready printing presses. Creating plotfiles (graphic files of map compositions) makes repeated printing or plotting of popular maps easy.

Saving map graphics as a file allows you to e-mail, ftp, or post them to a World Wide Web page. Sometimes file size and browser capabilities can be factors with these methods. For example, when sending a plotfile via e-mail, be sure the receiver's mailbox can accommodate the file's size. In addition, colors in graphics files posted to the World Wide Web can vary between browsers, platforms, and monitors.

Summary

This chapter provided an introduction to GIS. You have learned the concepts of GIS and how GIS differ from CAD programs. The unique way in which ArcInfo organizes spatial data was also discussed, along with some methods of creating, analyzing, and displaying GIS data using ArcInfo Version 8.0.

A GIS combines graphic computer mapping capabilities with powerful database tools. GIS users have available to them many methods of entering, analyzing, displaying data. Having learned about the capabilities of GIS, and ArcInfo in particular, get going! Turn on your computer and head to the PowerStart exercise that follows. In an hour or so you will know the basics of using the ArcCatalog, ArcMap, and ArcToolbox modules in ArcInfo.

PowerStart Exercise

This exercise acquaints you with ArcInfo's three newest modules: ArcCatalog, ArcMap, and ArcToolbox. Together they comprise ArcDesktop. Using the data sets provided, you will learn the following.

- How to examine and choose available data sets
- How to conduct a simple analysis
- How to display your analysis as a map

This exercise examines the process of a simple market analysis for a pretend company called KidSweets, which owns a chain of neighborhood candy stores with special appeal to kids. Using U.S. Bureau of the Census data, you will determine potential sites for a new retail store.

Phase 1: Preliminary Steps

It is 9:00 a.m. on a Monday at KidSweets corporate headquarters and you are just reaching for that first cup of coffee. Your boss rounds the corner in an obvious hurry and her eyes light up when she sees you. "Hey!" she says. "You're the technical type. I'm meeting with the big guy at 10:30 this morning and he's interested in expanding our market into Virginia. Use this new software and find locations for possible KidSweets stores in, say, Charlottesville. Thanks!" With a wave of her hand, she continues down the hall.

Traditionally KidSweets kid-appeal stores are located in neighborhoods within a quarter mile of a school. The KidSweets corporation also owns a chain of high-end stores selling gourmet chocolates, coffees, and ice creams. These stores, called The Sweet Life, are traditionally located in neighborhoods that have a property value of $125,000 and are within a half mile of schools. You decide you will begin the task by examining the data.

Examine the Data

First, set up a workspace named *ai80exercises* on your hard drive. Copy the folder named *ai80exercises/PowerStart* from the companion CD-ROM to that workspace. Now examine the data available to you. ArcCatalog is the tool to use for viewing data.

1. Start ArcCatalog: click on Start ➥ Programs ➥ ArcInfo ➥ ArcCatalog.
2. Click on the D drive and navigate to your workspace (assumed to be *D:\ai80exercises*). Find the folder named *PowerStart* and open it. It contains several data sets.

Use ArcCatalog to preview the data sets. Select the *Blockgroup* data layer, which is a data layer of census block groups within Charlottesville, Virginia. It contains income information by census block. Click on the gray Preview tab to see the map displayed. The *Blockgroup* data layer is shown in the following illustration.

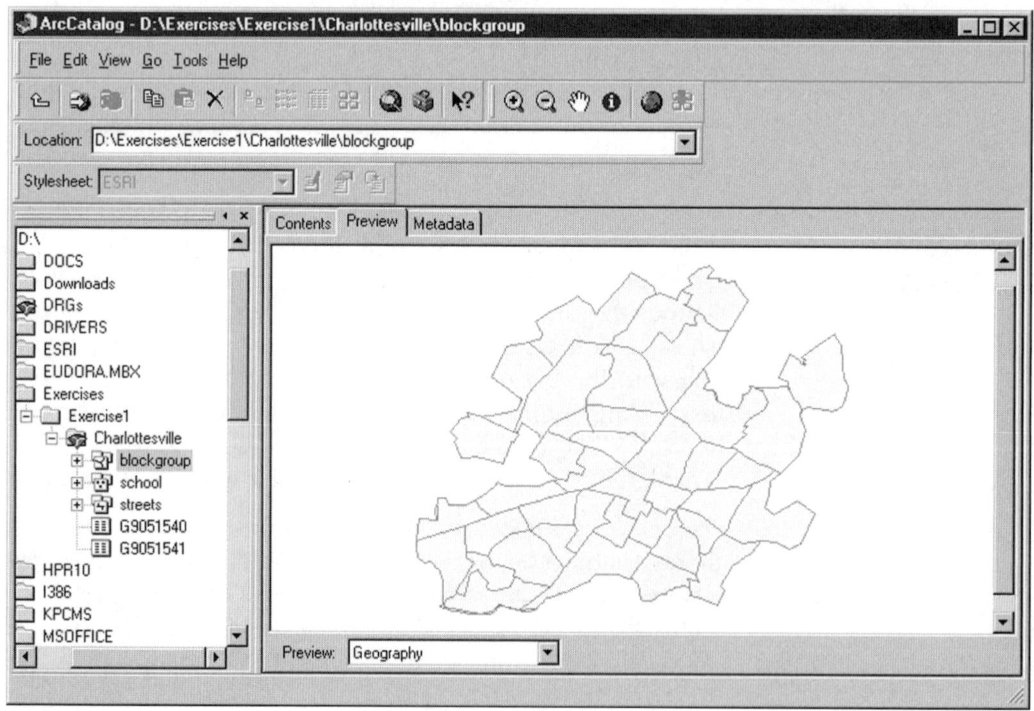

BLOCKGROUP data layer.

Now move that data layer into ArcMap by performing the following steps.

1. Open ArcMap by clicking on the ArcMap icon (the small globe with the magnifying glass) on the toolbar.

2. In ArcCatalog, drag the *Blockgroup* icon into the right-hand window of ArcMap.

Using ArcCatalog, you have examined a data layer and added it to the mapmaking "canvas" called ArcMap. Using the same steps as for *blockgroup*, examine and add the data layers *Streets* and *Schools*.

Once you have added the *Blockgroup*, *Streets*, and *Schools* layers to your ArcMap window, you will notice that the order in which they appear on the "canvas" is the same order in which they are listed in the table of contents window (the left-hand window). In other words, the first data layer listed is on "top" and the last data layer in the list is on the "bottom,"

Test this by clicking on the *Blockgroup* layer (in the table of contents window) and dragging it to the top. The *Streets* and *Schools* layers disappear because they are now "under" the *Blockgroup* layer. Now drag the *Blockgroup* layer back to the bottom of the list.

Change the Data Layer Symbol for Schools

ArcMap assigns default symbols to data layers. You will notice that the *Schools* data layer contains data as point locations; it has the symbol for schools showing as a dot. Change the dot symbol to something more appropriate by performing the following steps.

1. In ArcMap, look in the table of contents window on the left side of your composition. Notice that the default symbol for *Schools*, *Streets*, and *Blockgroup* is shown beneath the title of each feature. Double click on the dot symbol that represents schools.

The symbol selector window pops up; with this window you can change the shape, color, size, and angle of the symbol.

2. For now, scroll down the symbol choices until you find the one for schools, as shown in the first of the following illustrations. Select it. ArcMap will immediately reflect the new symbol choice, as shown in the second of the following illustrations.

Scroll down until you find the school symbol.

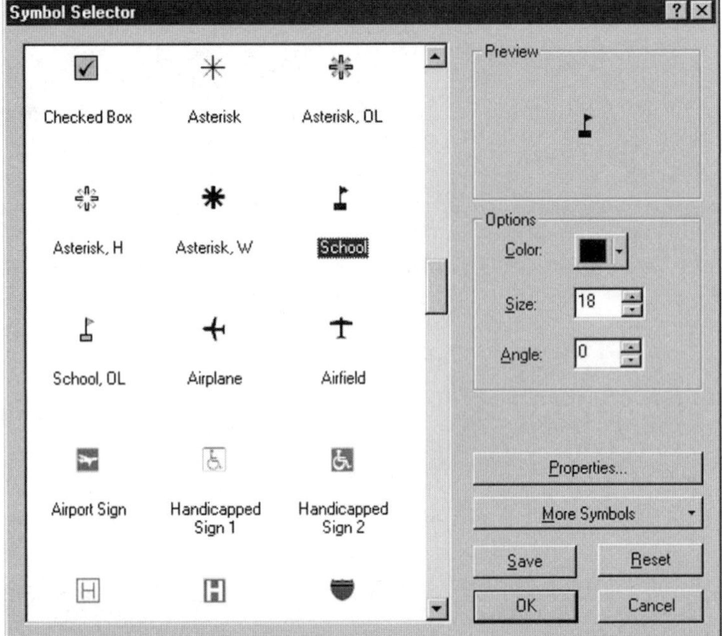

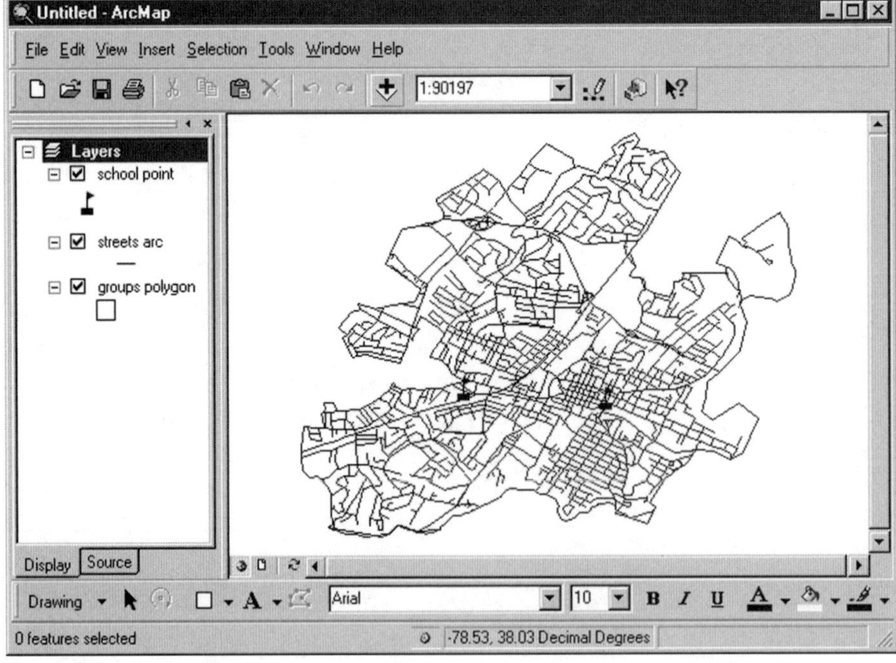

New school symbols as they appear in ArcMap.

✓ **TIP:** *When creating a new map composition, ArcInfo chooses colors and symbols at random. Be sure to change them to something that matches your data.*

Determine Likely Locations for Stores

Your candy store chain has two types of stores: the kid-appeal stores that serve primarily walk-in customers and cater to impulse purchasing, and the high-end stores that are more like a coffee shop, selling gourmet chocolates, fine coffees, and ice cream. Siting criteria is fairly simple. Kid-appeal stores are located within 1/4 mile of a school, and high-end stores are within 1/2 mile of a school and in an area with a median home price of $125,000.

Phase 2: KidSweets Store Siting

To determine potential sites for kid-appeal stores, you need only determine what streets are within 1/4 mile of each school. Perform the following steps.

1. In ArcMap's gray menu bar, click on Selection to pull down its menu, and select Select By Location. The Select By Location dialog box, shown in the following illustration, appears.

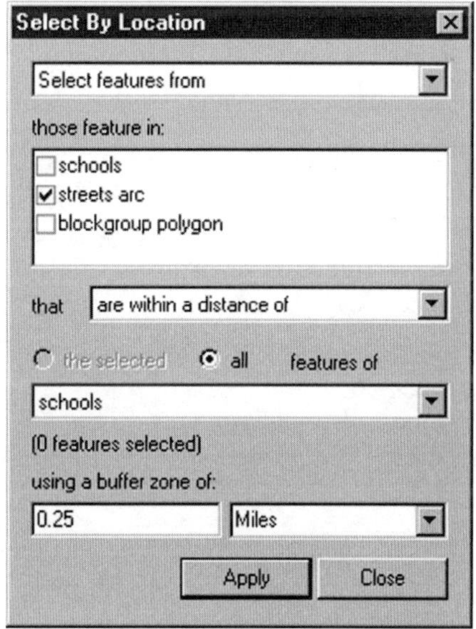

Select By Location dialog box.

Continue by filling in the blanks of this dialog box. You want streets within a distance of all schools using a buffer of 0.25 miles.

2. Uncheck *Schools* and *Blockgroup* by clicking on their boxes. This leaves *Streets* checked.
3. Select "within a distance of" and "schools."
4. Type in *0.25* and select "miles" for your buffer zone.
5. Click on the Apply button. The screen now shows potential locations for KidSweets stores.
6. Again, click on Selection and select Zoom To Selected Features. This gives you a close-up of the streets within a quarter mile of the two schools, as shown in the following illustration.

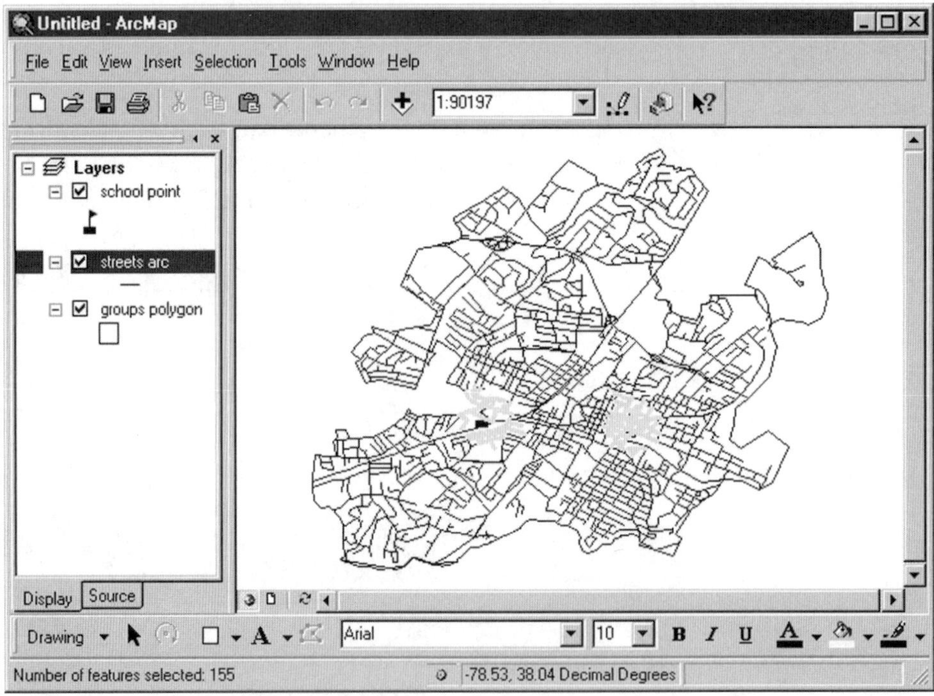

Streets within 1/4 mile of schools.

Finally, save and print this map for the boss by performing the following steps.

1. Click on File and select Save.
2. Enter a name for your map and click on the Save button.

3. Select File and Print (or click on the printer icon) to print your map.

Phase 3: Advanced Store Siting

Now for siting the high-end stores. The Sweet Life stores must be within 1/2 mile of a school and in an area with an average home value of $125,000. You first need to clear what ArcInfo calls the "currently selected set," which in this case is all roads within a quarter mile of schools. Do this by performing the following step.

1. Pull down the Select menu and select Clear Selected Features. The Clear Selected Features option is shown in the following illustration.

Clear Selected Features option in the Select menu.

Combine Data Layers

Next, determine which roads from the *streets* data layer match the selection criteria. This involves comparing the features from one data layer (*streets*) with area values from another data layer (the median house price in *blockgroups*). In this exercise, you will actually combine the two data layers to form a new data layer named *streetgroups*. To do this, you will use an overlay tool from ArcToolbox. Perform the following steps.

1. Start ArcToolbox: click on Start ➥ Programs ➥ ArcInfo ➥ ArcToolbox.
2. Double click on Analysis Tools ➥ Overlay and the Overlay Wizard icon (the magic wand). The Overlay tools are shown in the following illustration.

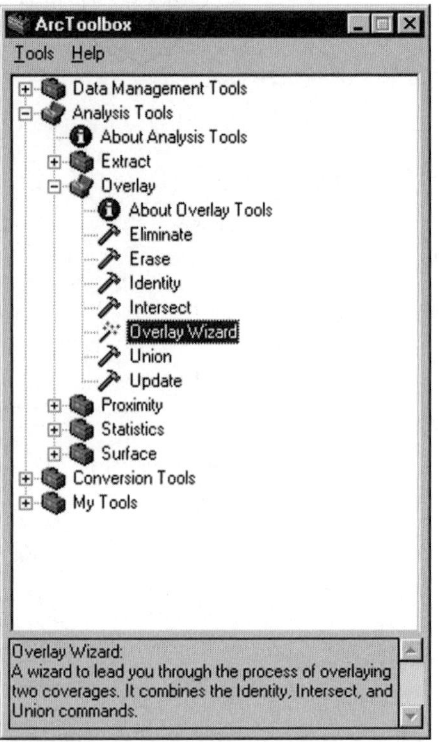

Overlay tools in ArcToolbox.

The Overlay wizard presents several screens that require you to enter information. First, the wizard asks which of three types of overlay you want to perform (all three of which are covered in detail in a later chapter of this book). Then the wizard asks you to specify input and overlay coverages. Here, coverages are the same as data layers. Finally, the wizard asks you to choose the attribute format and to assign a new name to the resulting coverage. To enter information in these screens, perform the following steps.

1. Select Intersect in the first screen of the Overlay wizard, shown in the following illustration.

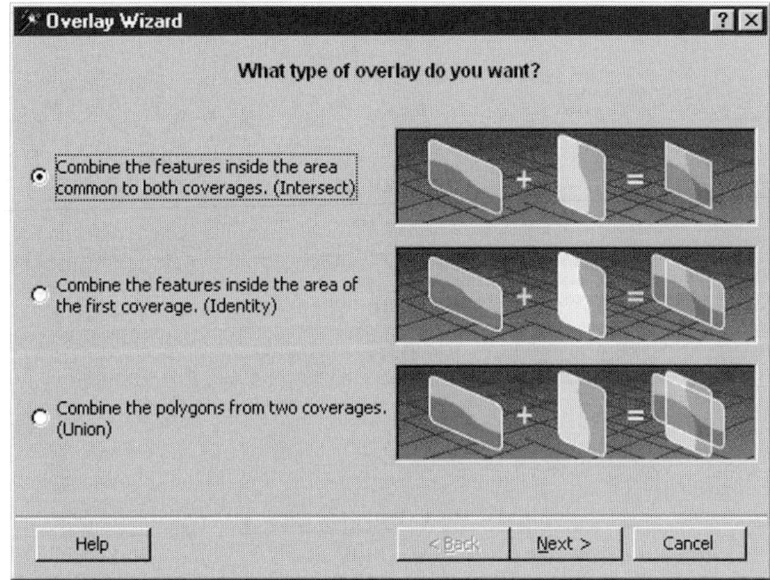

First screen in Overlay wizard.

2. Click on Next.

3. Specify STREETS as the input coverage, as shown in the following illustration. Be sure to give the full path name.

✓ **TIP 1:** *Rather than type in lengthy path names, open ArcCatalog and navigate to your PowerStart workspace. Now you can drag and drop coverage names from your ArcCatalog window to your Overlay wizard window.*

✓ **TIP 2:** *Your path and data layer (coverage) names cannot contain blank spaces. ArcInfo will not accept path names with blank spaces.*

Specify streets *as the input coverage.*

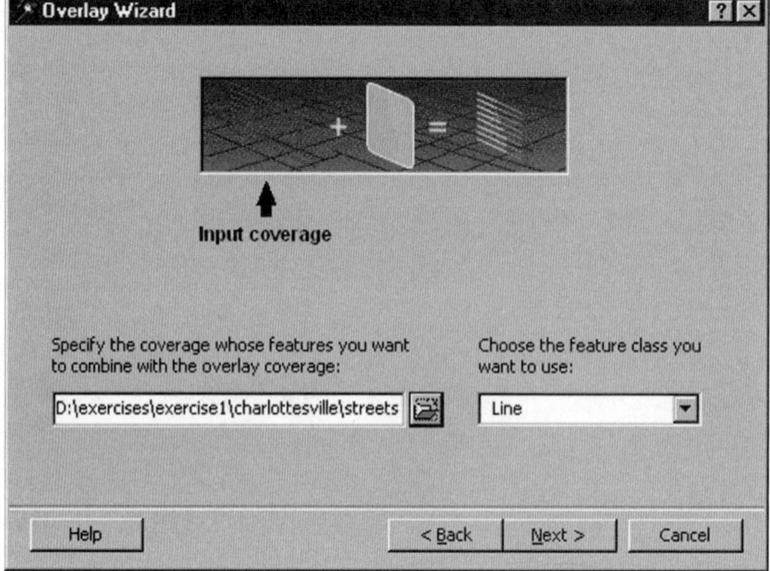

4. Click on Next.
5. For the overlay coverage, specify *blockgroup* (again, giving the full path name), as shown in the following illustration.

Specify blockgroup *as the overlay coverage.*

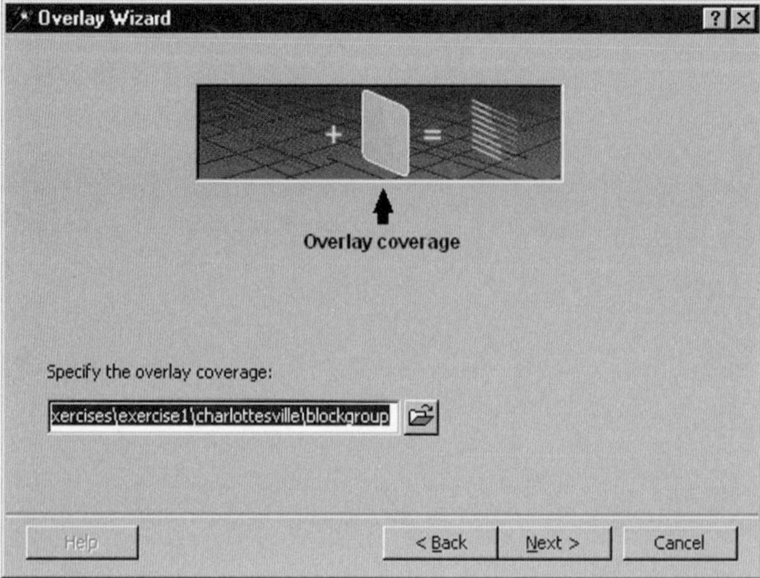

6. Click on Next.
7. Select the "Keep all attributes" option (see the following illustration). Do not worry about the other option for this exercise. Attribute properties are discussed in later chapters of this book.

Be sure to keep all attributes from both coverages!

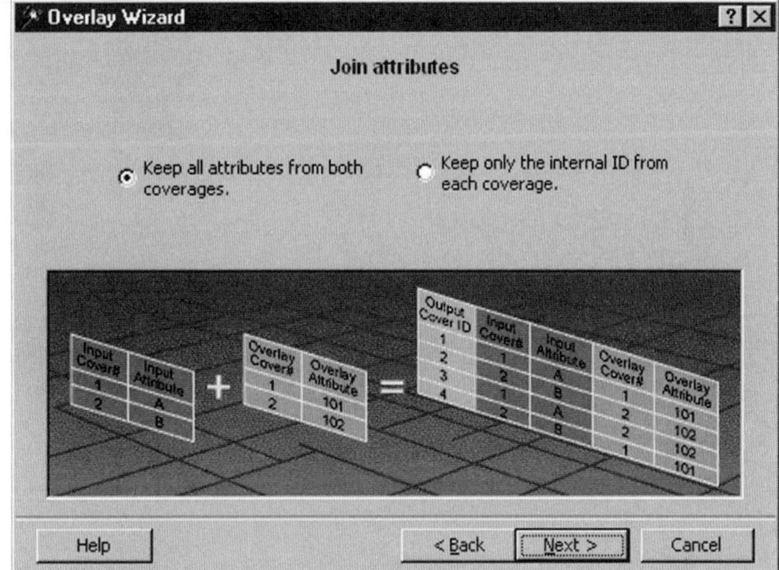

8. Click on Next.
9. Type in *groupstreets* as the new name for your resulting coverage, as shown in the following illustration. Do not worry about the "fuzzy tolerance" options. Again, that topic is covered in later chapters.

Enter groupstreets *as your output coverage.*

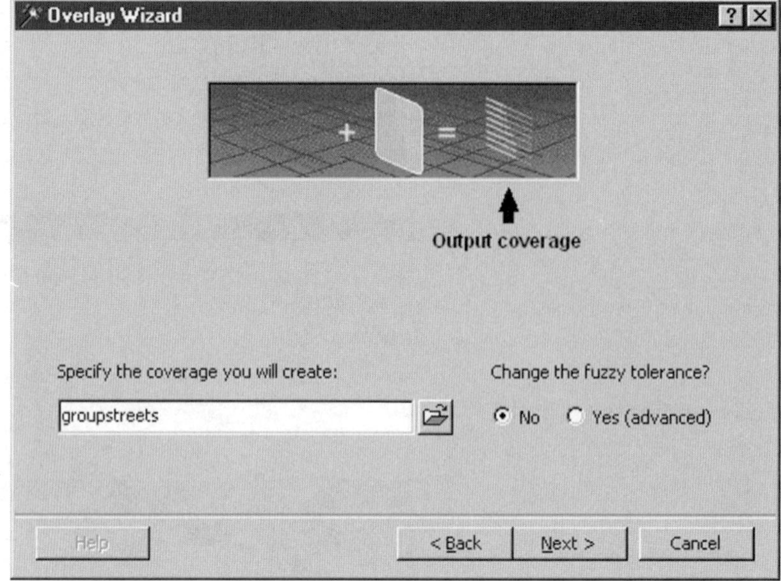

10. Click on Next.

The Overlay wizard will display a screen containing all of your responses and ask if you want to continue. Look over the information you have typed in or otherwise entered and verify that it is correct. The Save To AML option allows you to create a short Arc Macro Language program of these commands. This is particularly useful for repetitive operations and is discussed later in this book. For now, ignore this option.

11. Select the Finish button.

The wizard executes the Intersect command using the options specified. It may take several seconds.

The *groupstreets* data layer contains all streets from the *streets* coverage, as well as all attribute data from the *blockgroups* data layer. In other words, each street now has the census block information attached to it. Now reopen your ArcMap window to view your new creation. Replace *streets* with *groupstreets* in your map composition by performing the following steps.

1. Remove *streets* by clicking the right-hand mouse button on the data layer name in the table of contents (left-hand) window. From the pop-up menu, select the Remove option. Do not worry about deleting data; the *streets* coverage still exists, it is just no longer a part of your map composition.
2. Click on the Add Data (the black "+" on the yellow triangle) button on the gray toolbar. A window will show available data layers in your *PowerStart* folder, including the new *groupstreets* data layer. Highlight *groupstreets* with the mouse and click on the Add button. The view in your ArcMap window will look about the same as before.

You should now have *schools*, *groupstreets*, and *blockgroups* in your map composition. Be sure all three are listed in the table of contents window; their boxes should have check marks activated. Be sure they are arranged so that all three layers are visible. At this point, you will not be using *blockgroups* for any analysis, but keeping it in the map composition provides city boundaries and a soft background color to your map composition.

Select for Locational Criteria

Continue your locational analysis for the high-end stores. First, select all roads within a half mile of schools. Use the previous method of clicking on Selection and Select By Location previously described. Perform the following steps.

1. Click on Selection ➥ Select By Location.
2. In the pop-up box, select "Select features from" for the first box.
3. Uncheck *schools* and *blockgroup*. Leave only *groupstreets* checked.
4. Select "Are within a distance of," "All," and *schools* for the next section.
5. Type in *0.5* and select Miles.
6. Click on Apply.
7. Zoom in to those chosen streets by clicking on Selection ➥ Zoom To Selected Features.

The selected set should look like that shown in the following illustration.

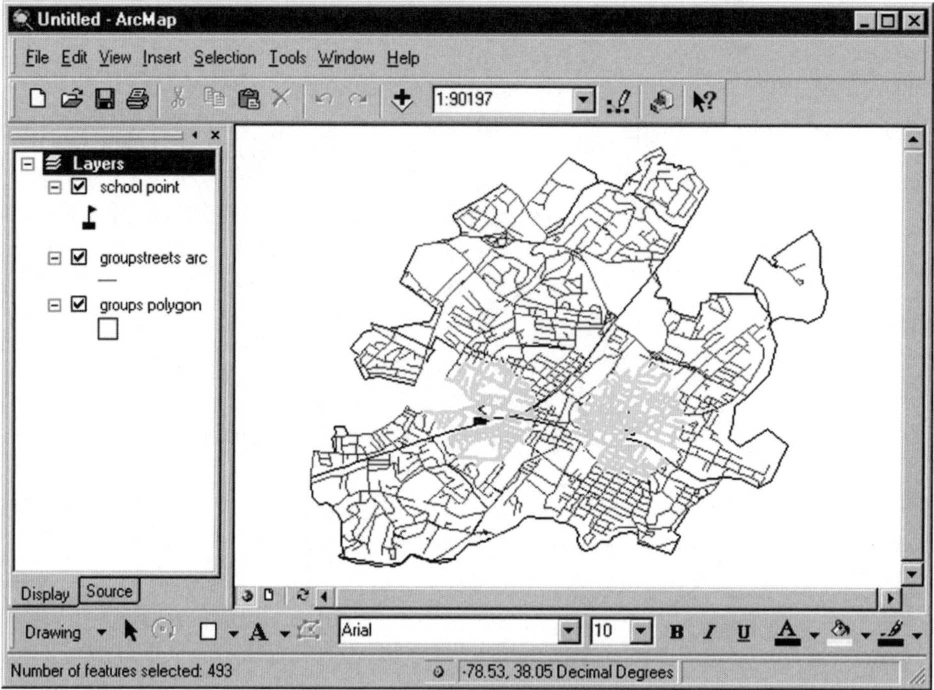

Streets within 1/2 mile of schools.

From this set you will reselect those streets containing homes with a median value of $125,000. To do this, perform the following steps.

1. Click on Selection ➥ Select By Attribute.
2. For the Layer option, select *groupstreets.*
3. Scroll down, close to the bottom, to find VAL_MEDI, shown in the following illustration. This is the eight-character name for median home values. The Fields box lists all available attributes. Remember that the list is quite long because you added the census block attributes to the streets attributes.

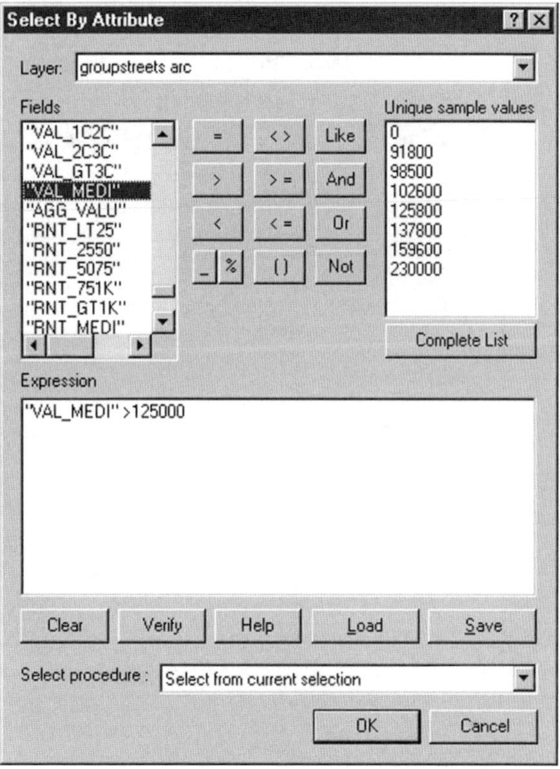

Select By Attribute menu showing "VAL_MEDI" highlighted.

4. On the right is a box with the title "Unique sample values." This shows you the values for the highlighted attribute. Below the box is a button named Complete List. Click on this button to get a list of all values for that attribute.

5. In the Expression box, form the expression "VAL_MEDI" > 125000. First, click in the Expression box. Then click on VAL_MEDI. VAL_MEDI should automatically appear in the box. Select the ">" button to transfer that symbol into the box. Type in 125000. Alternatively, you could choose a value from the "Unique sample values/Complete List" box.

6. Select "Select from current selection" in the "Select procedure" box. This results in a reselection of streets (with a median home value greater than $125,00) from the streets within a half mile of a school.

Do not worry about the Clear, Verify, and other buttons. They are discussed later in this book. Your new set of selected streets should look like that shown in the following illustration.

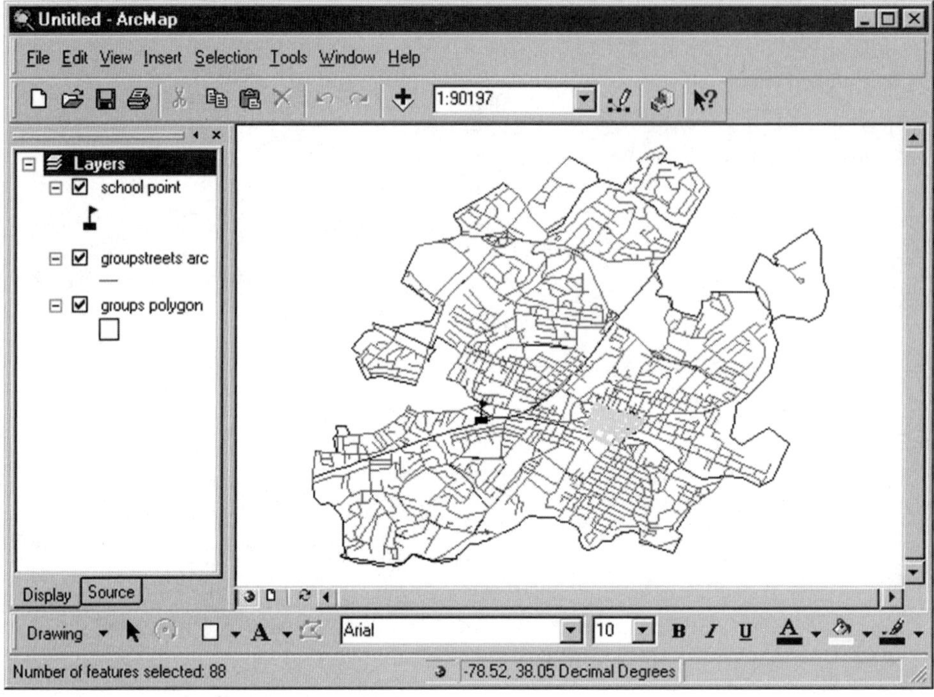

Streets meeting criteria for "The Sweet Life" high-end stores.

That is it! In less than an hour you have completed two locational analyses. Print this map and, along with the first one, hand it to your boss. Do not forget to ask her for a raise!

By completing this PowerStart exercise, you are now familiar with the three new drag-and-drop modules of ArcInfo. Using ArcCatalog, you have examined and moved data layers. With ArcMap, you created two map compositions, one showing the results of an analysis completed with an ArcToolbox wizard. Using these basic skills with your own data and needs, you are now able to create your own analyses and map compositions.

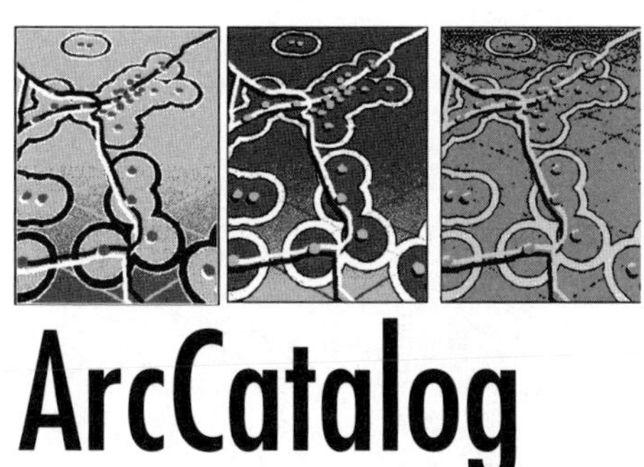

Chapter 2

ArcCatalog

Managing Data

Version 8.0 of ArcInfo is a powerhouse capable of combining many types of data. ArcCatalog helps you organize, track, and find the data layers and components you need for your analyses. With ArcCatalog, you know what you have, and where to find it.

Using GIS technology, you can very quickly accumulate more data than you can efficiently track. Your data might be of many types, including ArcInfo coverages, ArcView shapefiles, CAD files, database files, map compositions you have already created using ArcInfo, and even digital and nondigital photographs.

Although ArcCatalog employs the simplest menu system of the three new components constituting Desktop ArcInfo, it may be the most difficult for both new and seasoned users of ArcInfo to get used to. This chapter helps you start quickly with ArcCatalog, which you will soon find an invaluable tool for data organization.

Once you are familiar with ArcCatalog, you will be able to create shortcuts that will obviate the need to remember those long path names that have plagued users. You will learn how to quickly organize and browse your data, and to drag and drop what you need for use in ArcMap or ArcToolbox. You will even be able to add comments about your data layers using the metadata feature. This chapter serves as an introduction to these ArcCatalog capabilities, covering the following.

- Types of data ArcCatalog can access
- The menu interface
- Exercise 2-1: Exploring Data Using ArcCatalog
- Managing your data
- Customizing the catalog
- Exercise 2-2: Customizing a Toolbar
- Reference section: ArcCatalog commands

Types of Data ArcCatalog Can Access

With Version 8.0, ESRI has increased the number and types of spatial data with which ArcInfo can work. In ArcCatalog, data types and their locations are referred to as *items*. In other words, the data unit name and its path name constitute one item. The major data types with which ArcCatalog works are discussed in the sections that follow.

NOTE: *The term* item *as used by ArcCatalog refers to a data layer and its path name. Do not confuse this with the way* item *is used in a database sense.*

ArcMap Maps and Layers

Maps and layers are documents created using ArcMap. These documents may contain several elements in addition to the data item. A map of, for example, Alaskan forests might contain a title, a legend, a North arrow, a source statement, and the data item (perhaps a coverage) of forest locations. Together, these elements constitute the map. ArcCatalog can create direct links to maps, which allows you to quickly access frequently used maps.

Layers are essentially shortcuts to the location of geographic data on your computer. Layers may exist independently or may be part of an ArcMap map. They can be created from tables of geographic data, coverages, or information accessed from an SQL database.

Layers are an important addition to ArcInfo, as many data sources can be combined into a group layer. For instance, separate coverages of secondary roads and highways can be grouped into a transportation group layer. By adding the group layer to the catalog, users do not have to know the names or locations of the separate coverages.

Symbology (for example, highways displayed in red and secondary roads in black) may be easily added to layers using interactive query and palette functions in ArcMap (see Chapter 3). Veteran ArcInfo users will much appreciate this alternative to traditional "reselect" or look-up table symbology assignment.

✓ **TIP:** *Add symbology to layers to create a trademark "look and feel" for your organization's map products. Clients will appreciate symbology standardization, and effective maps are less dependent on the person creating them.*

Folders

When starting ArcCatalog for the first time, you will see something called *folder connections.* These are icons that represent each drive on your computer. Another icon, called *database connections,* allows you to create and store connections to external databases. You can add folder connections to access specific directories on your hard drive, other shared resources on the network, or even data held on a CD. Folder connections are flexible: you may add or delete them at any time. Setting up folders for regularly used data sources will be one of your top time savers in ArcInfo.

➻ **NOTE:** *Clicking on a folder opens the folder to reveal its content. However, ArcCatalog shows only data sources (such as coverages and shapefiles) and folders that contain data sources. Other files, such as word processing documents, will not be shown.*

Shapefiles and Associated dBase Files

Shapefiles are the standard data set used by ESRI's ArcView software. Shapefiles have associated dBase files that are viewable by ArcCatalog. By default, other types of ArcView files, such as projects (with the *.apr* suffix), are not viewable by default in ArcCatalog.

You can make ArcView projects viewable in ArcCatalog by adding them to the list of viewable file types. You do this by selecting Tools ➥ Options ➥ File Types and adding the *.apr* file extension to the list of recognized file types. ArcView projects will then appear in ArcCatalog. Clicking on a project in ArcCatalog will display the project in ArcView, however, because ArcInfo does not read ArcView project files.

Coverages

ArcInfo coverages consist of point, line, or polygon features. ArcCatalog identifies each feature type and makes it viewable. Selecting the polygon feature, for example, would let you see the content of the polygon attribute table.

Recall from Chapter 1 that a polygon coverage consists of more than one feature type, specifically line and polygon features. This means that both will be viewable in ArcCatalog.

TIN and Raster Files

ArcInfo also makes use of spatial data stored as raster or TIN (Triangulated Irregular Network) data sets. Examples of raster data include satellite imagery and USGS digital elevation models (DEMs). TIN data consist of X, Y, and Z points (usually elevation) and a set of lines that connect these points into irregularly shaped triangles, which, for example, are used to model terrain.

TINs are also useful for 3D analysis and viewing. Separate modules for creating and manipulating raster and TIN data are available for purchase with ArcInfo, but a detailed discussion of them is beyond the scope of this book. Suffice it to say that these data types are viewable just like other data types within ArcCatalog.

CAD Files

CAD files (computer-assisted drawing/design) can also be viewed within ArcCatalog. Such popular programs as AutoCAD and MicroStation generate CAD files. Often CAD files will contain more than one layer. In these cases, ArcCatalog will present a drawing as two files, one representing the entire drawing and the other representing point, polygon, line, and annotation classes. When viewing a CAD drawing, ArcCatalog allows you to select which layers it will display.

Geodatabases

Geodatabases are new in ArcInfo 8.0. Simply put, a geodatabase is a relational database that contains spatial information. Traditional ArcInfo information such as coverages and related data files (such as Info files) can be represented as a geodatabase. What makes geodatabases so useful is their ability to store relationships between features. By creating something called a relationship class, you can specify the relationship between features. For instance, you can create a relationship between a coverage of states and counties. Then, when selecting a county you would also automatically see all information about its state.

Coordinate Systems

A Coordinate Systems folder is normally hidden from view. You make it visible by selecting Options ➥ Customize menu. This folder contains information about the coordinate systems (map projections) used by your data sources. It can be useful when creating new spatial data sets, as you need only point to an existing projection in this folder to correctly register your new data.

NOTE: *See Chapter 7 for more information on map projections.*

XML Files

XML (Extensible Markup Language) is a programming language similar to HTML. It is used by ArcCatalog to display and edit the metadata for all ArcInfo data sources. Programming in XML is outside the scope of this book, but XML programmers may find uses for their own files in ArcCatalog.

The Menu Interface

The ArcCatalog screen, shown in the following illustration, consists of two main components: the Tree panel and the Views panel. The Tree panel lists the Catalog content (the data), and it is here that you select an item. The Views panel displays the content of the selected item. There are also customizable toolbars that act as shortcuts to various operations in ArcCatalog.

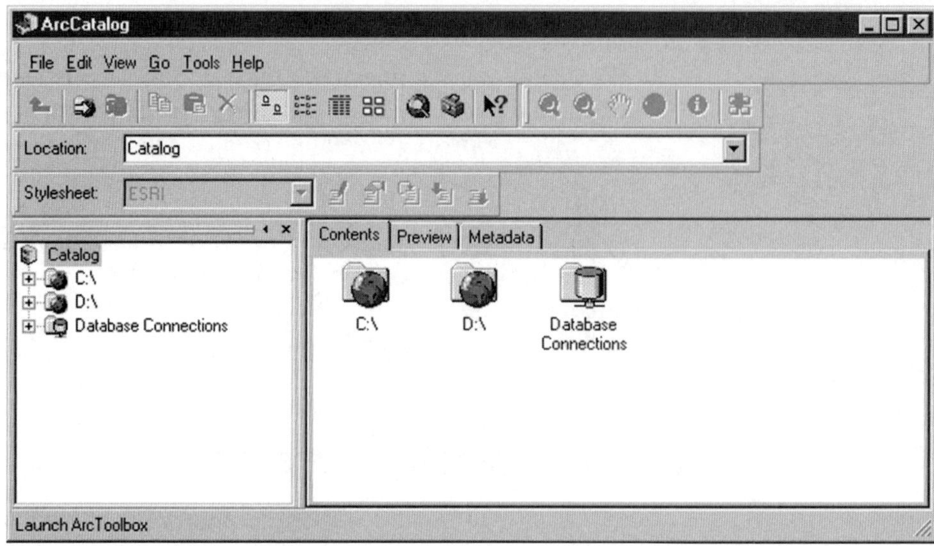

ArcCatalog screen.

The Tree panel is normally the left-hand window. This window can be resized, switched with the Views window, or dragged out of the ArcCatalog window entirely. Notice the small left-pointing arrow and the X in the upper right corner of the Tree panel. Clicking on the arrow minimizes the Tree panel, leaving the View panel to fill the screen. Restore the Tree panel by clicking on the small arrow on top of the dotted, left-hand side. Clicking on the X closes the Tree panel. Restore a closed Tree panel by selecting View ➥ Catalog Tree.

The Views panel contains three tabs along its top: Contents, Preview, and Metadata. Each tab displays something different about the highlighted selection in the Tree panel. The Contents window contains information about what is within the highlighted selection. For instance, if you have a polygon coverage highlighted in the Tree panel, the Contents tab will show the feature types that comprise such a coverage (in this case, arc, label, polygon, and tic folders).

Clicking on the Preview tab presents a selection of graphic file types. The Metadata tab is used to view descriptions of the highlighted data type. Exercise 2-1, which follows, shows you how to use these features to explore your data.

Exercise 2-1: Exploring Data Using ArcCatalog

This exercise takes you step by step through the basics of using ArcCatalog to explore your data. Be sure that the data for exercise 2-1 on the companion CD-ROM has been loaded to your hard drive.

NOTE: *For this exercise, assume that the data has been loaded to a folder named* D:\ai80exercises\chap_2. *If you have loaded it to your C drive (or a different drive letter), be sure to substitute your drive letter for D.*

Step 1: Exploring the Content of Exercise 2-1

You can explore the content of a folder by highlighting it in the Tree (the left-hand window) of the main ArcCatalog window. Use the Tree panel to navigate to *D:\ai80Exercises\chap_2*. Highlight the *Exercise_2* folder. Your screen should resemble that shown in the following illustration.

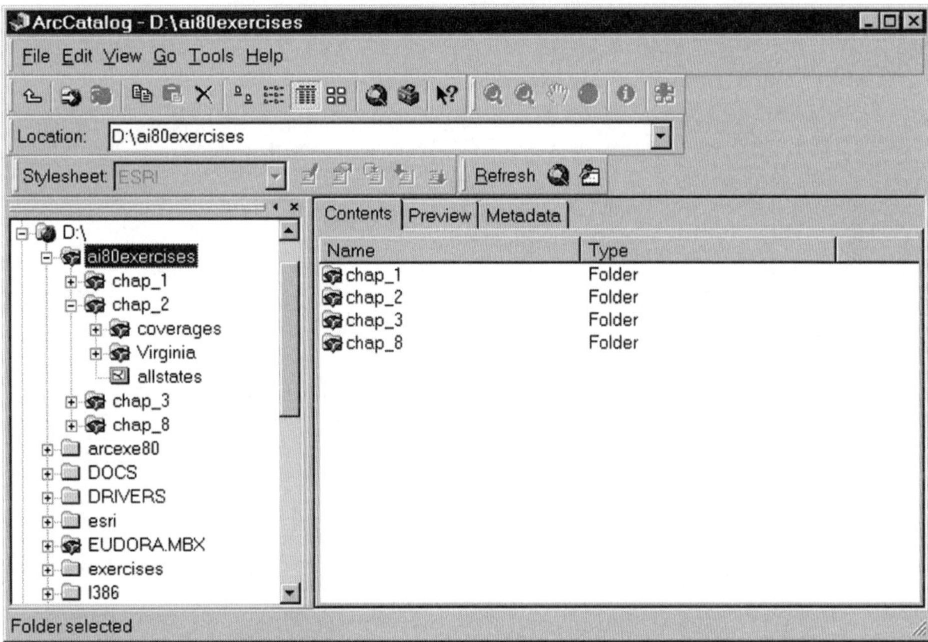

Starting screen.

Notice the two folders named *coverages* and *Virginia*. ArcCatalog portrays these folders with a small blue symbol over the traditional yellow folder, which simply means that ArcCatalog believes there are spatial data sets within these folders. A shapefile, called *Allstates*, should also be present in the folder for exercise 2-1.

Select the *Allstates* shapefile. Notice that the Contents tab is selected in the View panel and that nothing is displayed. Why? Remember that the content ArcCatalog looks for are other folders, or feature types such as polygons and lines. Because shapefiles store data differently, no content is listed. Although shapefiles originate in ArcView, another ESRI GIS software package, they are viewable, and useable, in ArcInfo.

Step 2: Examining Information About a Shapefile

With the shapefile *Allstates* highlighted, click on the Preview tab in the View panel. You should see a screen similar to that shown in the following illustration.

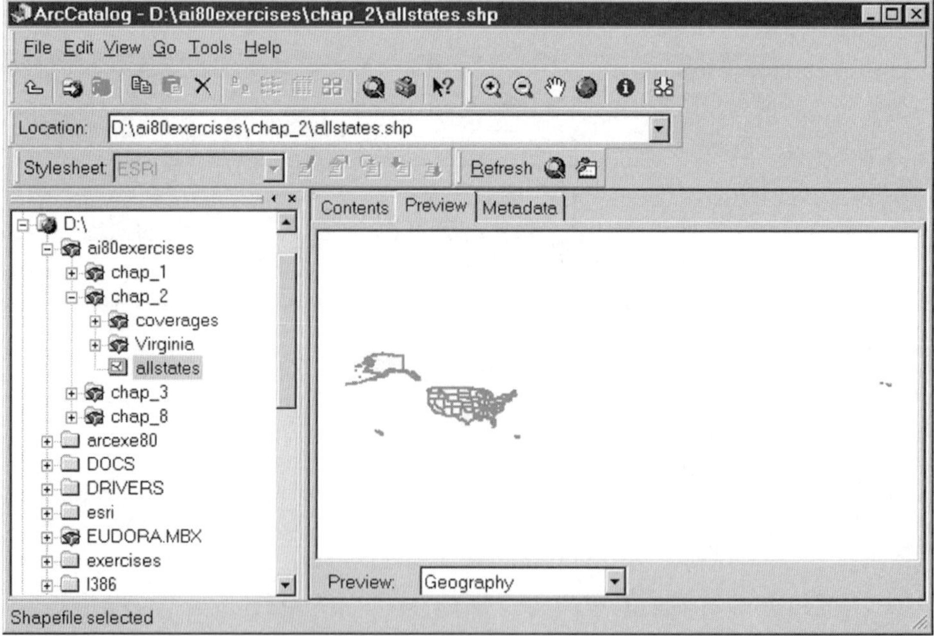

Screen resulting from use of the View panel's Preview tab.

ArcCatalog draws the graphic content (the "geography") of the shapefile in the Preview panel. Notice the scrollable window in the bottom section of the Preview panel. Click on this window, shown in the following illustration, and change its view from "geography" to "table."

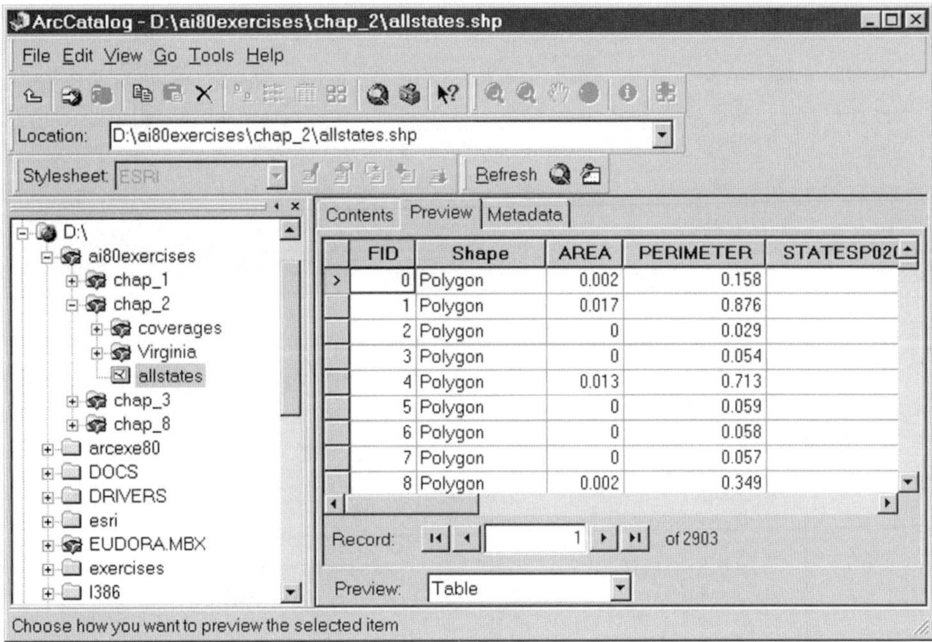

Preview panel scrollable window.

Now the shapefile's associated tabular data is displayed. You can quickly examine data to determine if they suit your needs for a particular project. You can also toggle between "geography" and "table." Using the Preview panel is a convenient means of quickly previewing data items. It is a lot like browsing books in the library: you can look at the pictures or read the content.

Step 3: Examining Metadata

Click on the Metadata tab in the View panel. A new screen with three tabs (Description, Spatial, and Attributes) appears, shown in the following illustration. These tabs display information about the shapefile. This is essentially what metadata is: data about data.

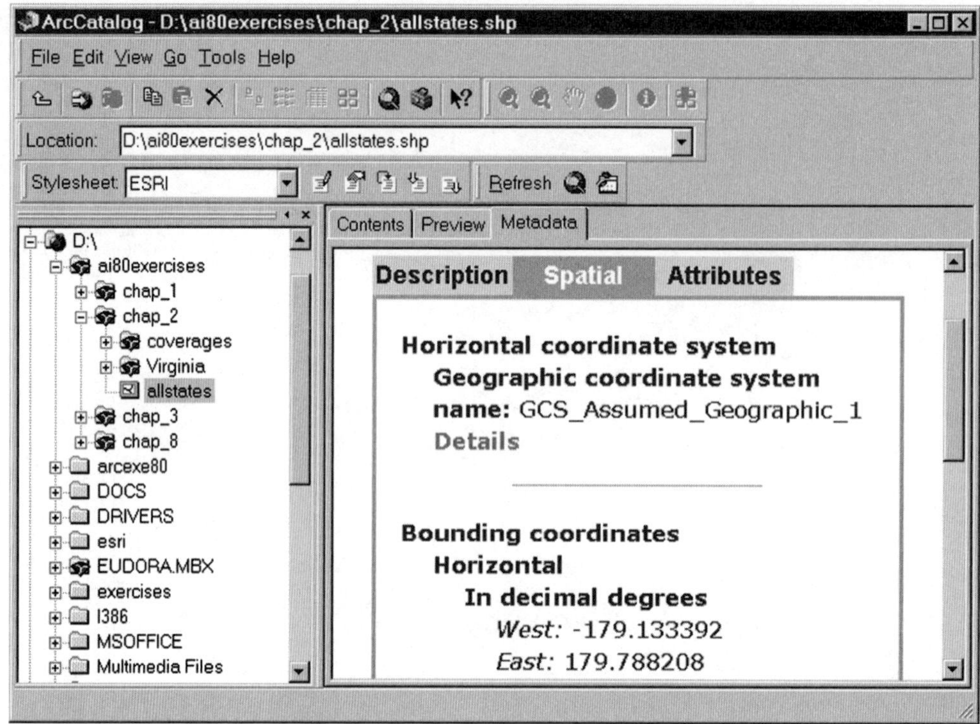

Options under the Metadata tab.

Select the Description tab, shown in the following illustration, to access an area provided for user-added text that describes the shapefile. This can be edited using the metadata editor: click on the "paper and pencil" icon located on the metadata toolbar just above the View window. Information useful to your organization and clients can be added here.

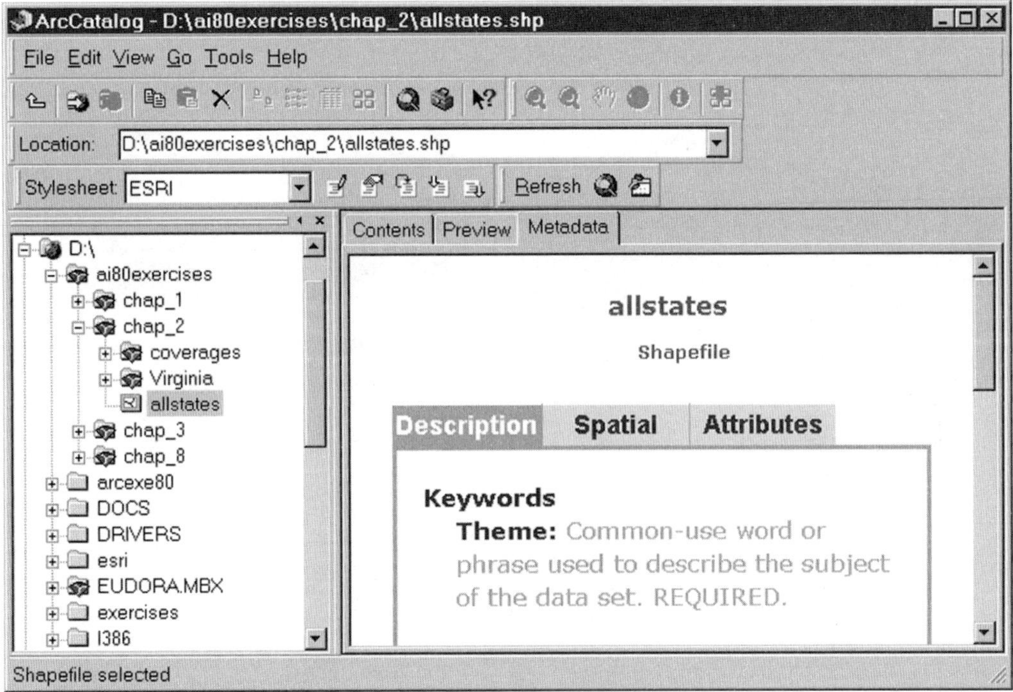

Description tab.

✓ **TIP:** *Take advantage of the metadata feature to add information such as date of creation, purpose, person responsible, data sources, and so on to your data items. As data age and staff changes, this can help prevent mistakes resulting from using old or inappropriate data items in current analyses. It can also be a nice "value-added" feature for work done for your clients.*

Select the Spatial tab to see information about the shapefile's projection and its geographic coordinates. This data cannot be edited: it can be viewed only. If you want to change its projection, for example, you must use a different ArcInfo tool.

Select the Attributes tab, shown in the following illustration. This lists the shapefile's attributes. Again, the attributes may be viewed only, not edited. Changing attributes requires different ArcInfo tools.

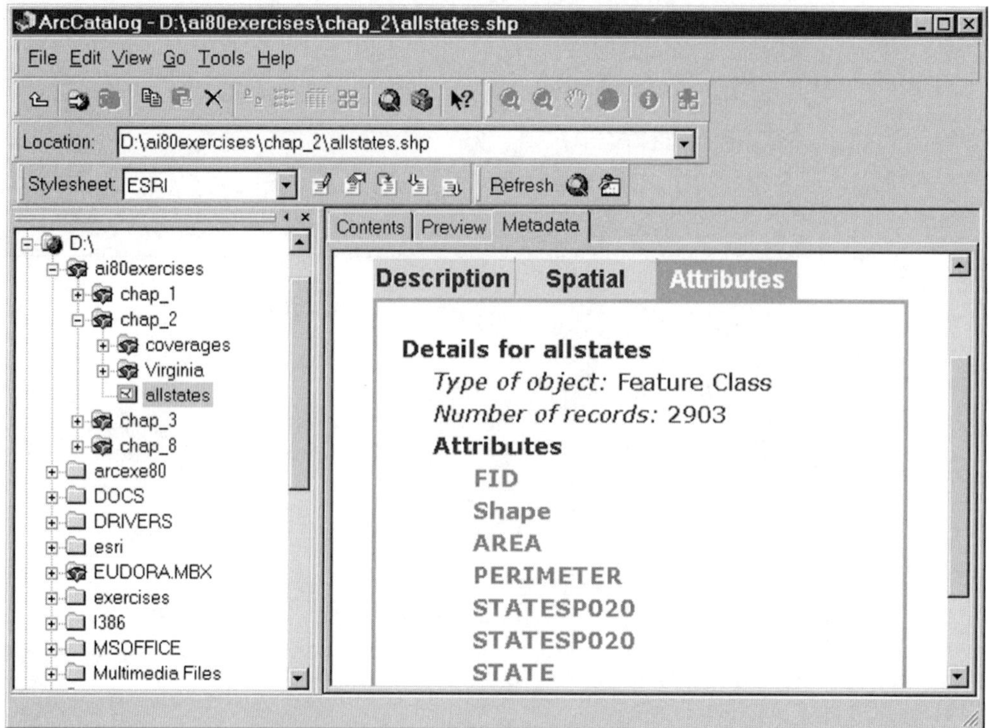

Attributes tab.

Step 4: Examining a Coverage

In the Tree panel, access the coverages folder and select the coverage named FED_LANDS. Click on the Contents tab in the View panel. You should see a screen like that shown in the following illustration.

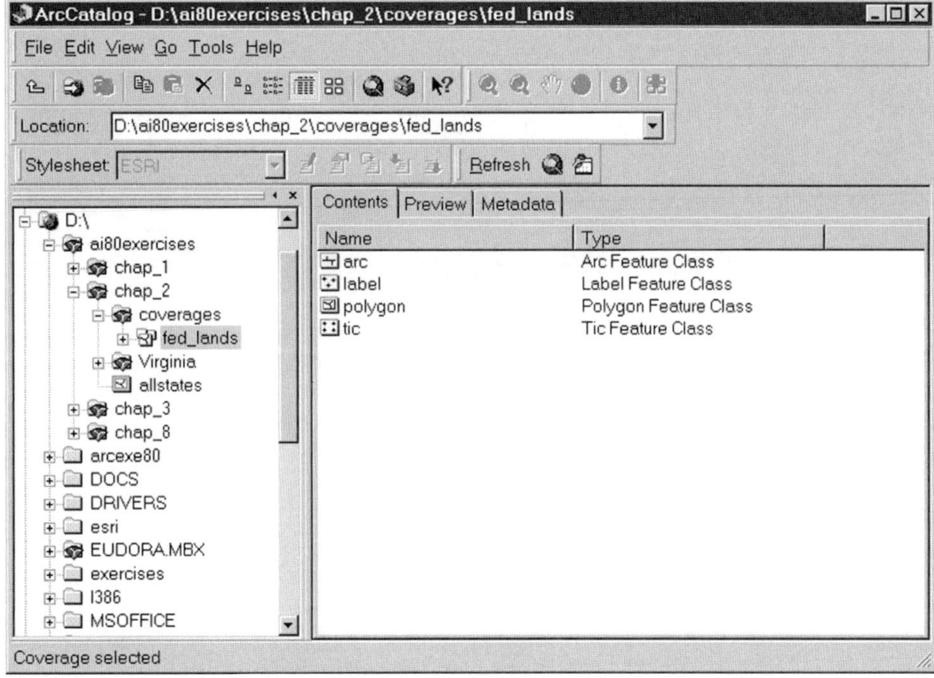

Contents tab screen.

Now the View panel shows four entities: arc, label, polygon, and tic. Each of these is a feature within this ArcInfo coverage. You can explore an individual feature type or look at the coverage in its entirety. Look first at the entire coverage by clicking on the Preview tab in the View panel. The screen should look as it does in the following illustration.

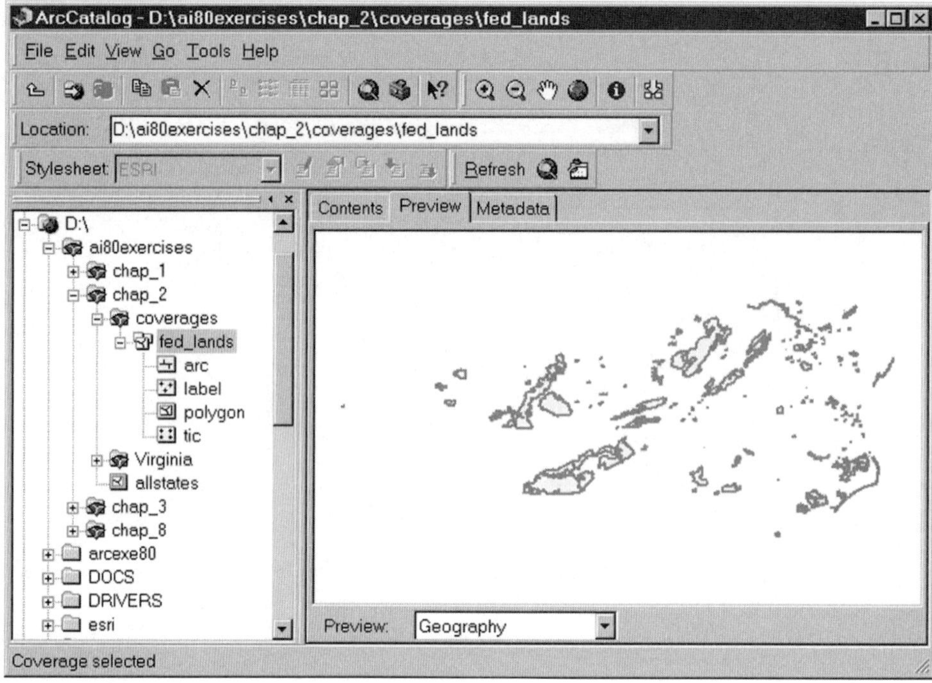

Preview tab screen.

A graphic of federally owned land in the state of Virginia is displayed. As with a shapefile, you may select the Table option in the preview's scrollable window. This shows you the polygon attribute table (*.PAT*) associated with this coverage. Now select the Metadata tab to explore the attributes, projection, and geographic bounds.

Next, look at just one of the features within the coverage. Select the label feature in the Tree panel. Select the Preview tab in the View panel. The screen now shows only the center points of the polygons, as shown in the following illustration. Because coverage features can be extracted and used to create new coverages (for example, extracting the points from one coverage for use in another), viewing just one feature type can be useful. Similarly, you may select the tic, label, or polygon features. ArcCatalog will display just those features you select.

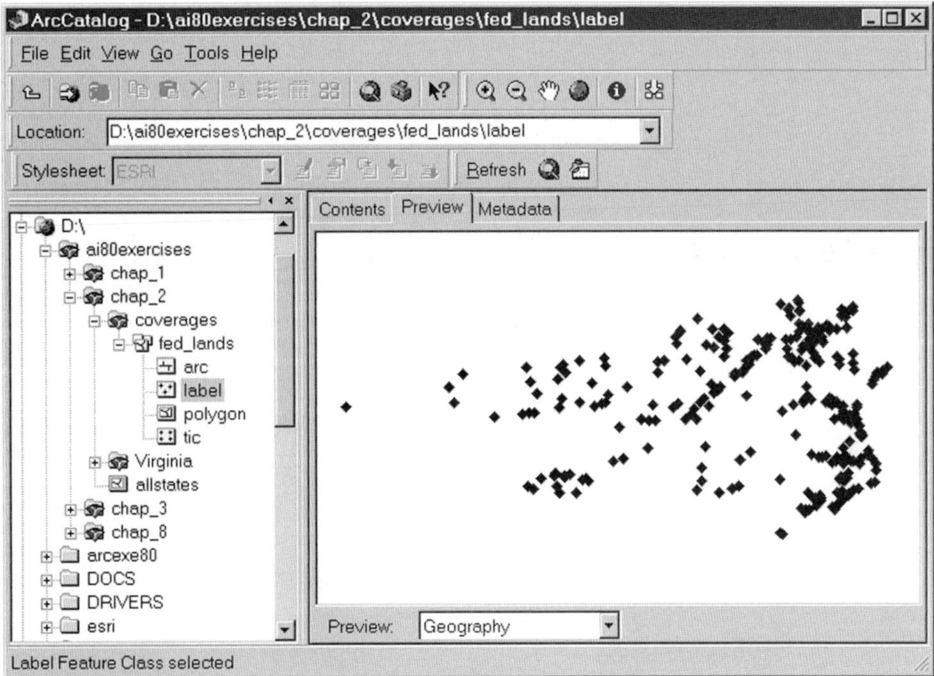

Screen showing center points of polygons.

By now you should be comfortable navigating and exploring your data using ArcCatalog. This exercise covered viewing the geographic and tabular data associated with data items. It also introduced you to the function and usefulness of metadata, and demonstrated how to view components of a data item. You can now use ArcCatalog to effectively navigate, explore, and manage your data.

Managing Data

As you have seen, ArcCatalog provides an effective means of browsing spatial data: no more long path names or switching to the operating system's search functions. ArcCatalog also provides excellent tools for organizing data. The sections that follow discuss using ArcCatalog's tools to create new folders, workspaces, and coverages, and to move, rename, copy, and delete items. Examples provided draw upon data items used in exercise 2-1.

Exercise 2-1 folder menu.

Creating a Folder

A folder is really like a directory: it can hold data items or other folders. To create a folder, go back to the Tree panel and highlight the *exercise_2-1* folder. Then, with the *exercise_2-1* folder highlighted, right-click the mouse. The menu shown in the illustration at left will appear.

Select the New option, and then Folder from the list of choices, shown in the following illustration. Type in the new folder name, *my_data.*

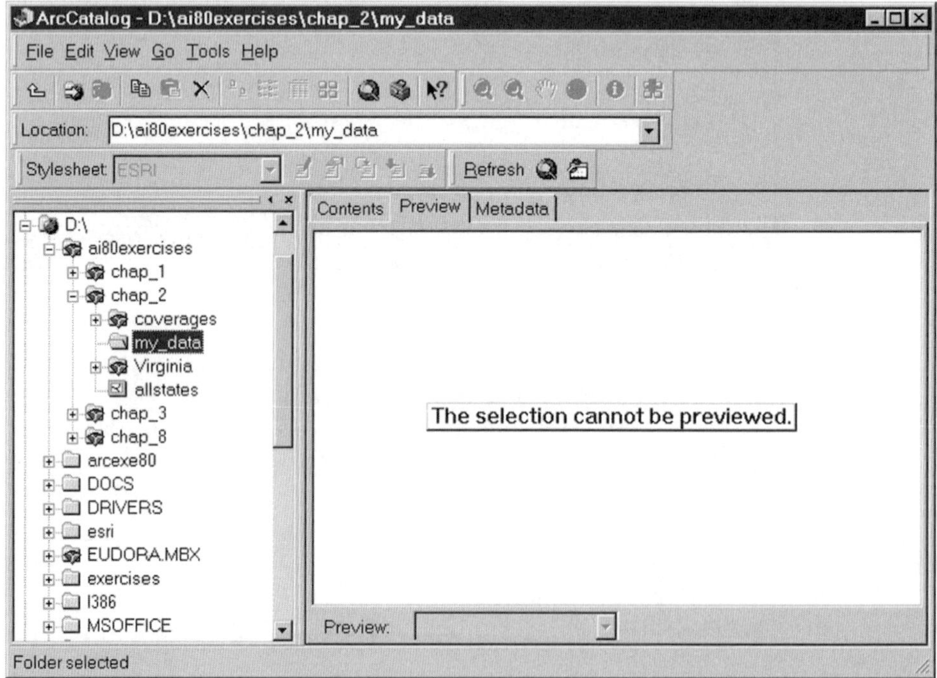

Selections under the New option.

Your *exercise_2* folder should now have a subfolder named *my_data.*

Moving Data Sets to New Folders

Now move the *Allstates* shapefile into your new folder. Highlight the *Allstates* shapefile in the Tree panel and, while holding down the left mouse button, drag the shapefile to the *my_data* folder. Once there, release the mouse button.

The *Allstates* shapefile is now in its new folder. Notice that all files associated with the *Allstates* shapefile moved with it. Practice moving other data sets into the folder.

Copying Data Sets to a New Folder

In the last section, you physically moved the *Allstates* data set from one folder to another. This time, copy it. Copying creates an exact duplicate of the data set and puts it in a new location.

To make a copy of *Allstates*, access the *my_data* folder and highlight it. Now right click and select Copy from the pop-up menu. ArcCatalog now knows that *Allstates* is selected for copying. Release the mouse button and select the *exercise_2* folder. Right click on this folder name and select Paste. The result is shown in the following illustration.

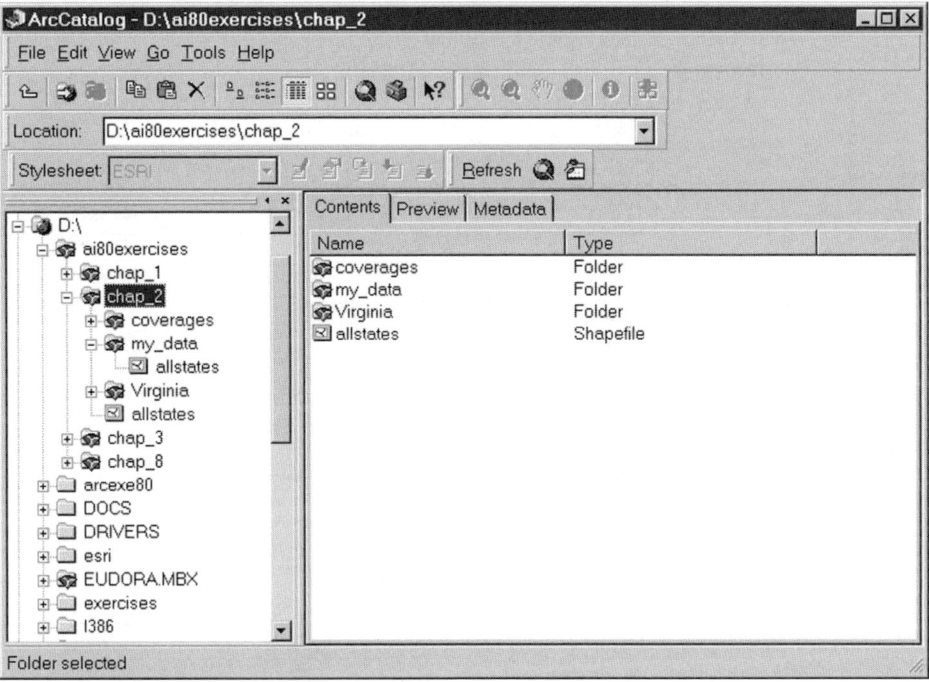

Result of the Paste operation.

Now there are two copies of *Allstates*: one in the *exercise_2* folder and one in the *my_data* folder. To copy several items at once, hold down the <Shift> key and select them all. Then use the Copy and Paste commands. You can also copy entire folders or groups of folders.

✓ **TIP:** *Organize your work as you go. For example, if your current project has several data items, grouping them in one folder makes backups and information exchanges easy.*

Renaming Data Sets

Renaming a data set is similar to copying one. Practice by renaming the *fed_lands* coverage (in the *coverages* folder) with the more descriptive name *va_fed_land*. Again, use the Tree panel to open the *coverages* folder, and highlight the *fed_lands* coverage. Click the right mouse button and select Rename. The result is shown in the following illustration.

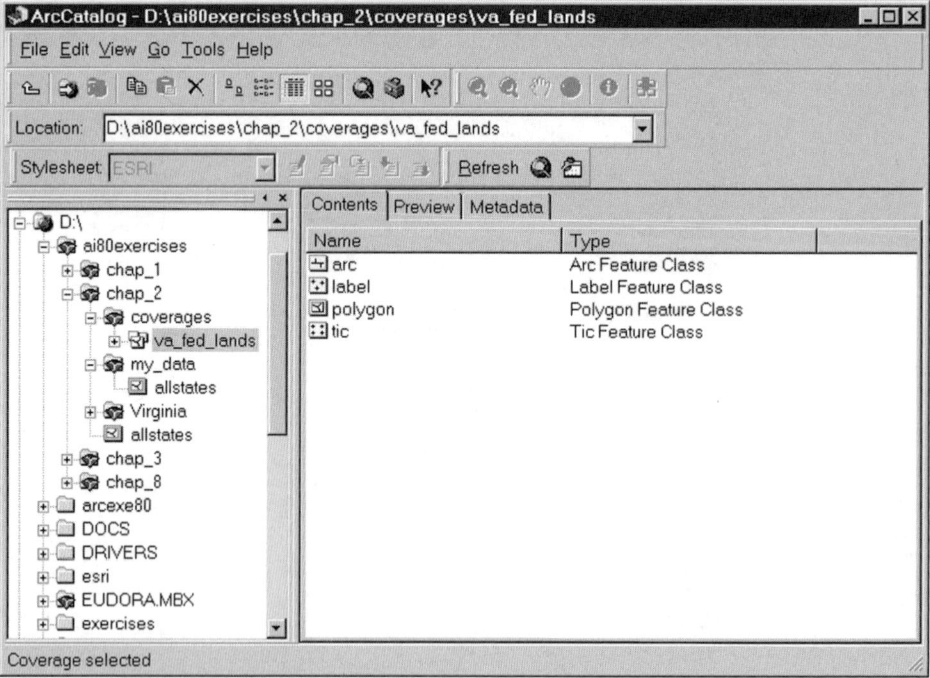

Result of selecting Rename.

Type in *va_fed_land* as the new name. The new name takes effect immediately.

☛ **WARNING:** *Never change the name of a coverage or shapefile from Windows Explorer! For shapefiles and especially for coverages, many associated files are hidden from the user. Data sets can be destroyed if the hidden file names are not changed. Use the ArcCatalog commands (or the appropriate Arc commands, discussed later in this book) to ensure that those hidden files also have their names changed.*

Deleting Folders and Data Sets

Again, deleting a folder or data item is similar to copying and renaming one. Practice by deleting the shapefile you just moved into the *my_data* folder. First, open the *my_data* folder. Highlight the *allstates* shapefile. Right click with the mouse and select Delete. All files associated with *allstates* will be erased.

☛ **WARNING:** *Be careful when deleting! No "Undo" command exists in ArcCatalog, so anything deleted is gone for good.*

To delete a folder, perform the operation previously described. This time, however, select the *my_data* folder and delete it by right clicking with the mouse.

☛ **WARNING:** *Anytime you delete a folder, you also delete all data residing within it. Therefore, be careful and keep your backups current!*

Creating ArcInfo Workspaces

Workspaces differ from working directories. A workspace is unique to ArcInfo, and when created includes an Info (the database component) subdirectory. Workspaces can be first created and then have data moved into them, or you can make a workspace from a directory by copying a coverage into it. Using ArcCatalog to create a workspace automatically sets up an Info directory for you.

To create a workspace, select the folder in which you want a workspace. Then select File ➥ New ➥ ArcInfo Workspace. You will be prompted to enter a name for the workspace, which now resides in the selected folder.

Creating ArcInfo Coverages

Creating a new ArcInfo coverage is similar to creating a new workspace. Again, select the folder in which you want to create a new coverage. As practice, create a new coverage named *Test* in the *coverages* folder.

Next, select File ➥ New ➥ Coverage. The menu that appears takes you step by step through the process. A menu box appears, shown in the following illustration, prompting you for the name of the coverage. Type *Test*. Click on the Next button.

New Coverage dialog.

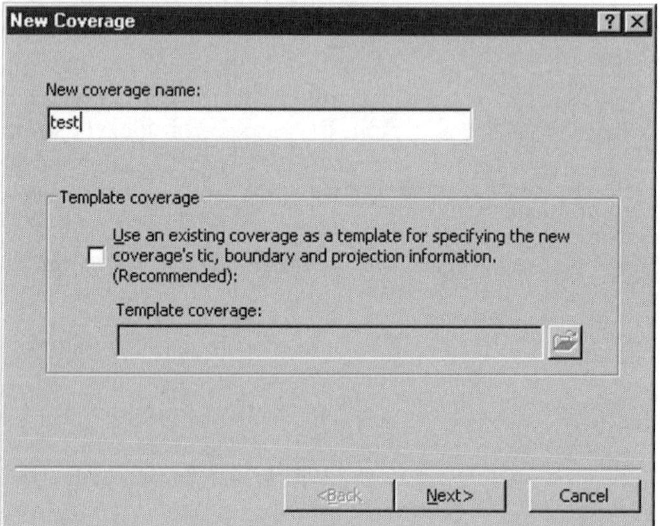

Define a projection for the coverage (see the folowing illustration). Projections are mathematical formulas used to translate the coordinates from a 3D Earth to a 2D piece of paper. Projections are covered in more detail in Chapter 4. For now, know that the data items must have the same projection in order to be viewed or used together.

Menu box for name of coverage.

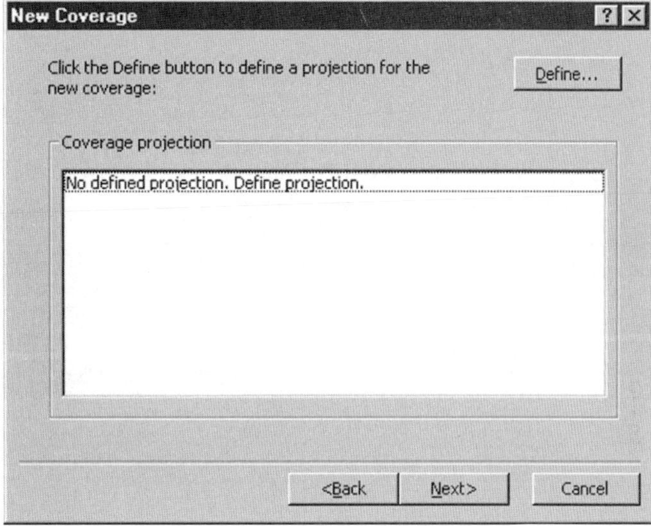

You can choose a projection interactively or specify the name of an existing coverage whose projection you want to copy. Selecting the Interactive option starts the Define Projection wizard, discussed in Chapter 4. Selecting an existing coverage prompts you for the name of a coverage from which a projection file is copied to this new coverage. Of course, you do not have to define a projection when creating a new coverage. For this example, simply click on the Next button.

You are now prompted for the ArcInfo feature type (point, line, poly, or tic) that will be the main feature type for this coverage. Select Polygon. You can also specify whether the coverage should be stored as single or double precision (discussed in detail elsewhere), as indicated in the following illustration.

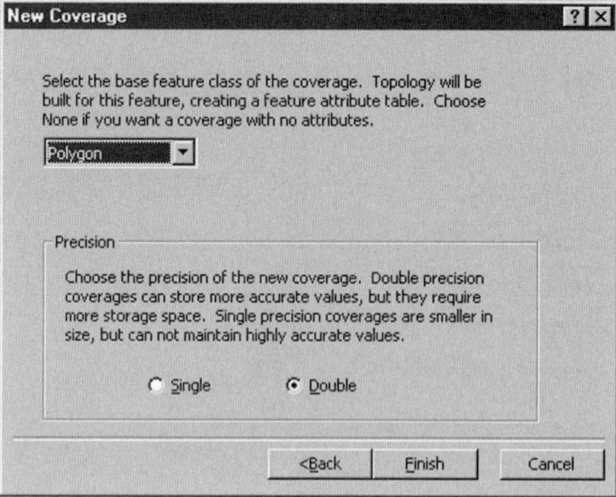

Feature type and precision selection.

You should now see that the *Test* coverage has been added to the coverages folder, as shown in the following illustration. It is an empty coverage, ready for data addition via such processes as digitization.

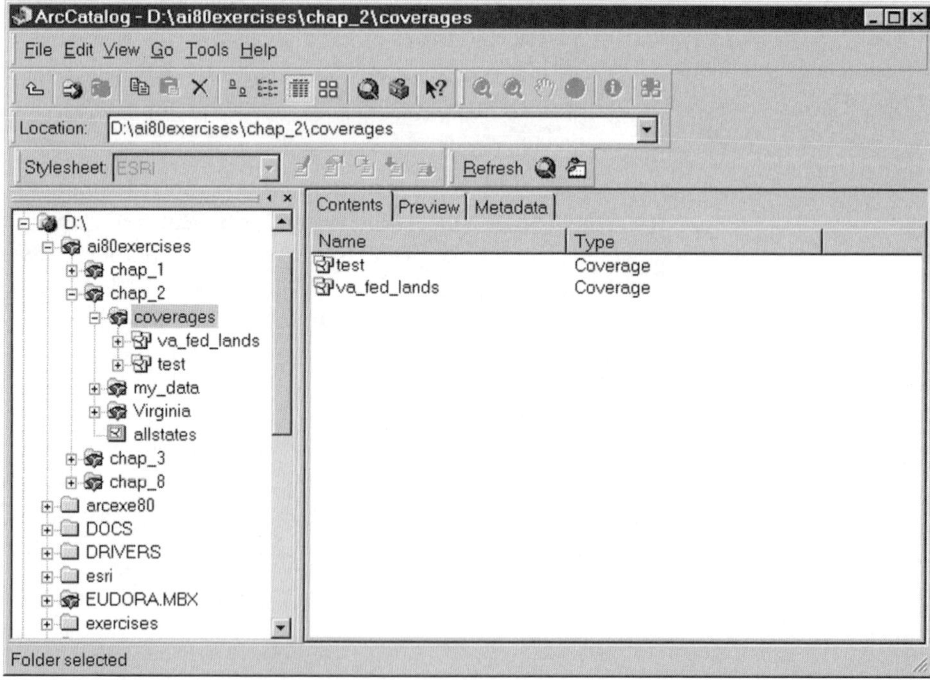

Test *coverage added to coverage folder.*

Customizing the Catalog

Once you are familiar with the look of ArcCatalog, consider adding some tools to make it even more useful. You will increase your efficiency, and limit the information you deal with to that which is truly useful. The exercise that follows shows you how to customize toolbars to suit your needs.

Exercise 2-2: Customizing a Toolbar

ArcCatalog's toolbars are well equipped with frequently used tools (commands). However, suppose you perform a lot of work with ArcView shapefiles, and you are an experienced Visual Basic programmer. ArcCatalog's flexibility lets you create and modify your own toolbars. Exercise 2-2, which follows, takes you step by step through customizing a toolbar for working with shapefiles, including accessing the Visual Basic editor.

Step 1: Assigning a Name to a New Toolbar

Click on the Tools ➥ Customize menu, shown in the following illustration. Here you will see the default menu bars ArcCatalog uses. On the right side are several buttons, starting with New.

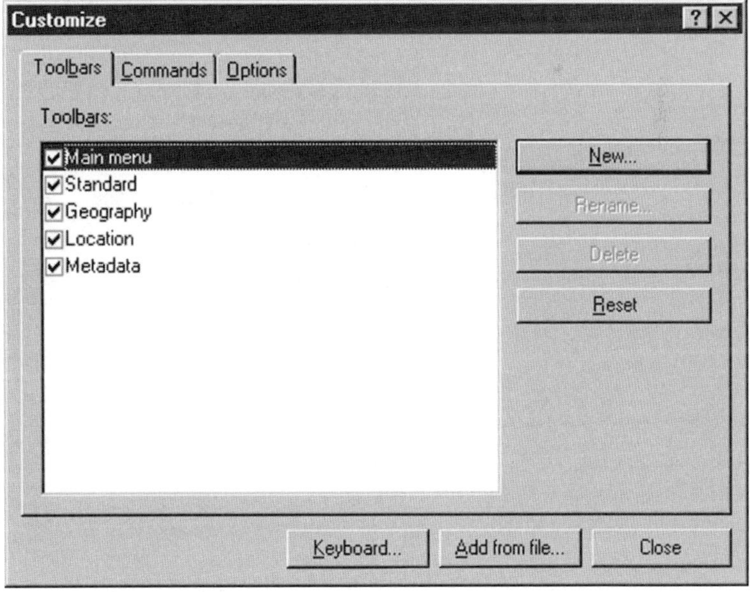

Tools ➥ Customize menu.

Click on New to assign a name to your custom toolbar. The New Toolbar window appears. Type *My Tools* into the Toolbar Name text box, as shown in the folowing illustration.

New Toolbar screen.

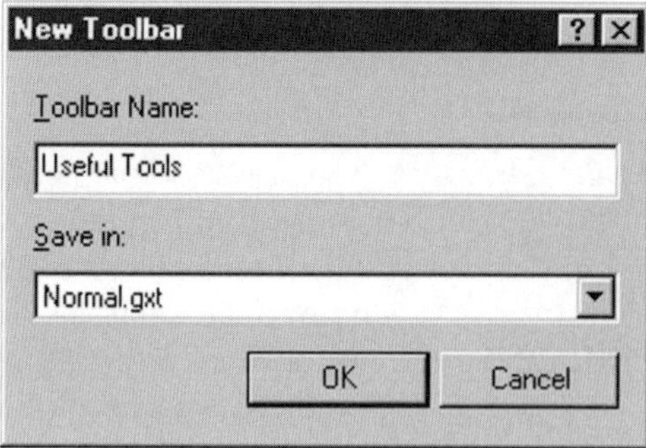

Now click on OK. A blank toolbar appears. This is the new Useful Tools toolbar.

Step 2: Assigning Tools to a Menu Bar

Now select the Commands tab in the Customize window. Two scrollable windows named Categories and Commands appear, as shown in the following illustration.

Categories and Commands windows.

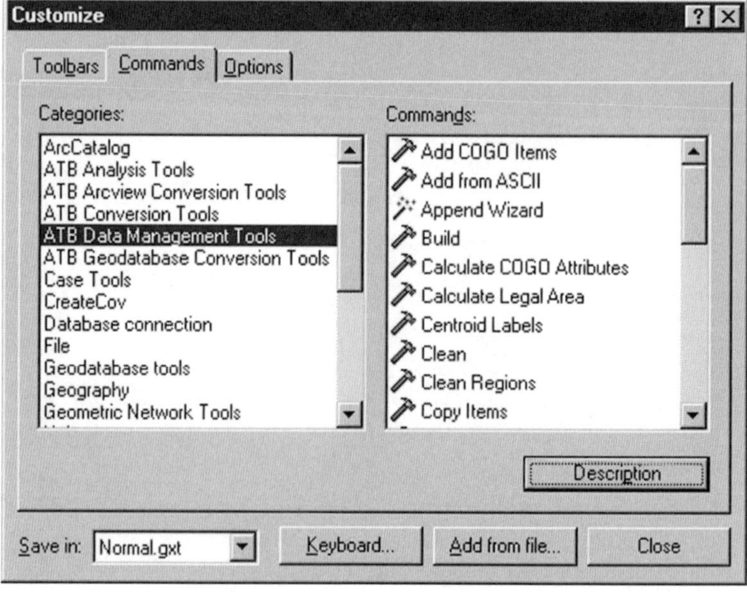

Take a moment to explore the category listings. As you select a category, lists of commands relating to it automatically appear in the Commands window. Each of these commands is available for inclusion in your own custom toolbars.

Add some commands specific to working with ArcView shapefiles to your My Tools toolbar. Select the ATB (ArcToolbox) Conversion category; then select the "Shapefile to coverage" command. Now hold down the left mouse button and drag the "Shapefile to coverage" command until the mouse is on top of the blank toolbar you just created (My Tools). Release the mouse button and the tool will be added to the toolbar.

Now add another tool. Go to the Tools category and select the Visual Basic editor. Drag this command to the new toolbar. Your main ArcCatalog window should now look like that shown in the following illustration.

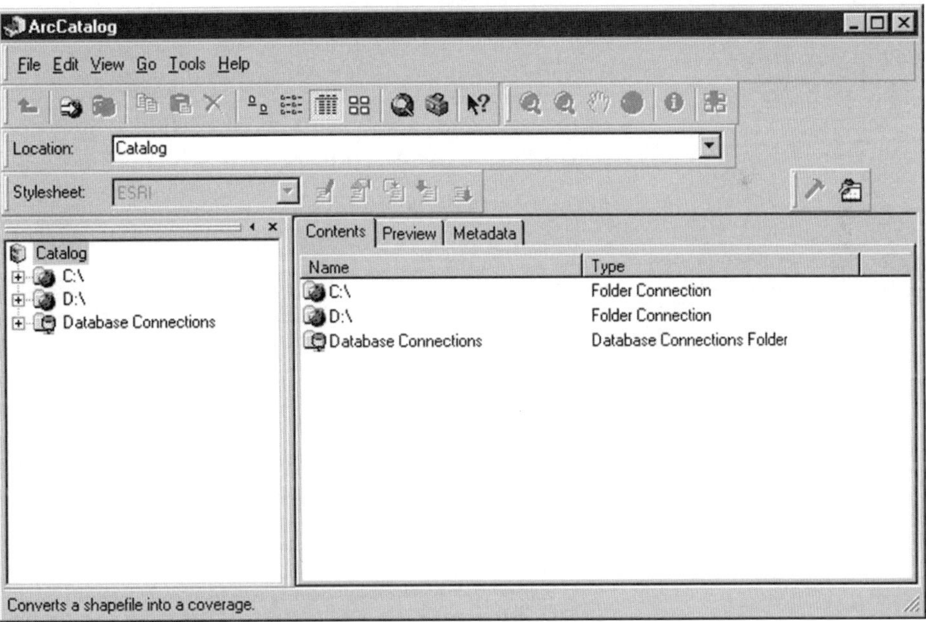

Revised main ArcCatalog window.

Practice adding other tools, such as the ArcMap start-up icon. You can add and delete tools as needed. You can also move the location of the My Tools menu bar by grabbing its toolbar handle (the line on its left side): hold down the left mouse button and drag the bar to its new location. All of the menu bars can be rearranged in this way.

Customizing and arranging the toolbars helps you get to what you need quickly. Individuals in your organization can customize the ArcCatalog menu bars to best fit the particular work they perform. This increases everyone's efficiency by saving command and data search and access time.

Menu Commands

The following sections discuss the various menus and the options within them. The File, Edit, View, Go, and Tools menus are explored.

File Menu

File menu.

New: Select this option to create new folders, ArcInfo workspaces, ArcInfo coverages, and personal geodatabases.

Connect Folder: Establishes a connection to another hard drive on your PC, or to a shared drive somewhere on your network.

Disconnect Folder: Similar to the previous option, this selection disconnects you from a hard drive or shared drive.

Delete: Use this to delete coverages, workspaces, and data files.

Rename: Allows you to rename folders, coverages, and files.

Properties: If you have a hard drive selected in the Tree pane, selecting Properties will reveal details about the hard drive, such as available disk space. Highlight a coverage and this choice will provide information about the coverage's features, projection, and so on.

Exit: Ends your ArcCatalog session.

Edit Menu

Edit menu.

Copy: Copies the highlighted selection to the Clipboard.

Paste: Takes the content of the Clipboard and pastes it in a location you specify.

View Menu

View menu.

Toolbars: Use this option to change the toolbars appearing in the ArcCatalog window. You may turn on or off the following menu bars: Main, Standard, Geography, Location, and Metadata. All commands accessed through these menus are also available from the pull-down menus.

Status Bar: Turns on or off the message that displays in the bottom of the ArcCatalog window each time you move the cursor over a toolbar button. This is strictly a matter of personal preference. Some like to see a message that describes the use of each button, whereas others find it annoying.

Catalog Tree Bar: Toggles on or off the Tree panel. Turn it off if you want to work closely with an item present in the View panel.

Refresh: Select this option to refresh the screen.

Go menu.

Go Menu

Up one level: Use this option to view the content of the next folder up.

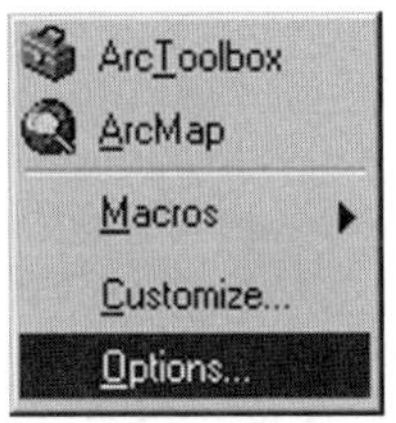

Tools menu.

Tools Menu

ArcToolbox: Starts ArcToolbox.

ArcMap: Starts ArcMap.

Macros: Select this option to run or record a macro. A macro is a "shortcut" command that executes a collection of commands you may use frequently. You can also gain access to the Visual Basic editor included with Version 8.0 of ArcInfo. Programs in Visual Basic can call and execute ArcInfo commands. Visual Basic can be a useful tool for tying the functionality of ArcInfo into other programs/processes your organization uses. Programming in Visual Basic, however, is outside the scope of this book.

✓ **TIP:** *Create macros to efficiently execute series commands you use frequently.*

Customize: Allows you to create new menu toolbars and to assign commonly used commands to keystrokes. You can increase your efficiency by taking advantage of these customizations.

Options: Lets you control the data types that appear in ArcCatalog. This selection also contains control options for how to handle imagery data. Use this to control what data types appear in the Tree Panel window. This can quicken your search for a particular data layer (a coverage, for example) by only showing items of that data type. If you regularly work with only one or two types, using this selection can save you time in having to search through all of the data types your organization uses.

Help Menu

Help Menu.

The Help menu provides access to ArcInfo's on-line help. Use it to find detailed information about ArcCatalog. The menu contains the following options.

- *ArcCatalog Help*: Select this option to start the on-line help system.
- *What's This?*: Select this option and click on a feature of the ArcCatalog window. A description of the feature will be displayed.
- *About ArcCatalog*: This option displays the version number of ArcCatalog.

Summary

ArcCatalog provides you with an intuitive interface for organizing and managing your GIS data. This chapter discussed the types of data ArcCatalog can access, its menu interface, and how to customize a toolbar. Exercise 2-1 showed you how to use ArcCatalog to explore various data types, and exercise 2-2 showed you how to customize a toolbar. Once familiar with ArcCatalog, you can streamline your data organization and management operations.

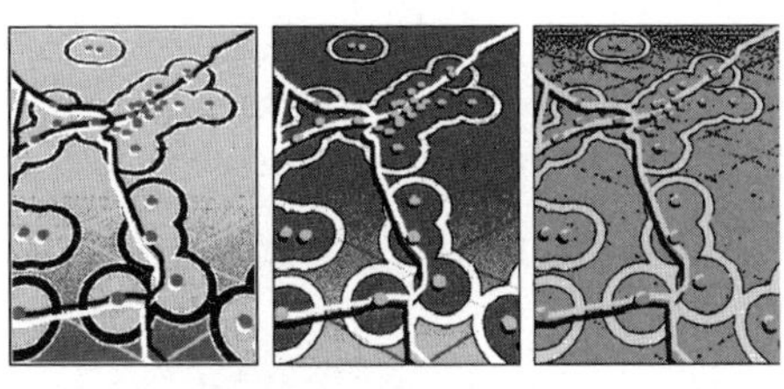

Chapter 3

ArcMap

INTERACTIVE MAPPING

ArcMap is ArcInfo's new data visualization and on-screen map composition system. It combines a graphical user interface with many built-in tools to help you display and query data, and to create maps quickly and easily. ArcMap also incorporates extensive query capabilities to help you answer questions about where something is, how much of something is at a location, and so on. Like the other components of Desktop ArcInfo, ArcMap is fully customizable.

ArcMap will especially please users familiar with ArcInfo Workstation. ArcInfo veterans will be surprised at how quickly they can now create maps. It replaces map composition processes in ArcPlot with a dynamic and easy-to-use interface. In addition, ArcView users will see many similarities between ArcMap and the ArcView layout. For many users, ArcMap will be the most-used component of Arc Desktop.

This chapter assumes that you have read Chapter 2, on ArcCatalog. You should be comfortable using ArcCatalog to access data and navigate your directory structure. If you need to learn to make a map quickly, go to the PowerStart exercise in Chapter 1. The following aspects of ArcMap are presented in this chapter.

- Getting started with ArcMap
- Exercise 3-1: Displaying and Querying Features Using ArcMap
- Displaying multiple data layers

- Using ArcMap templates
- Exercise 3-2: Creating a Map Composition
- Basics of mapmaking
- ArcMap menu commands

This chapter contains two exercises. Exercise 3-1 shows you how to display and query a data layer, how to create a "mouse-over," and how to create a simple report of your query. Exercise 3-2 takes you through creating a map composition using an ArcMap template and multiple data layers. In it you will display a coverage and a shapefile, and you will hyperlink your map composition to an image.

Getting Started with ArcMap

This section introduces you to the start-up options of ArcMap, and takes you through the screen layout. Follow along on your own computer. To start ArcMap, click on Start ➥ Programs ➥ ArcInfo ➥ ArcMap.

Upon starting ArcMap, a dialog box opens that requires you to select one of the following: open an existing map, open the last map worked on, create a new map, or create a new map from a map template. The dialog box is shown in the following illustration.

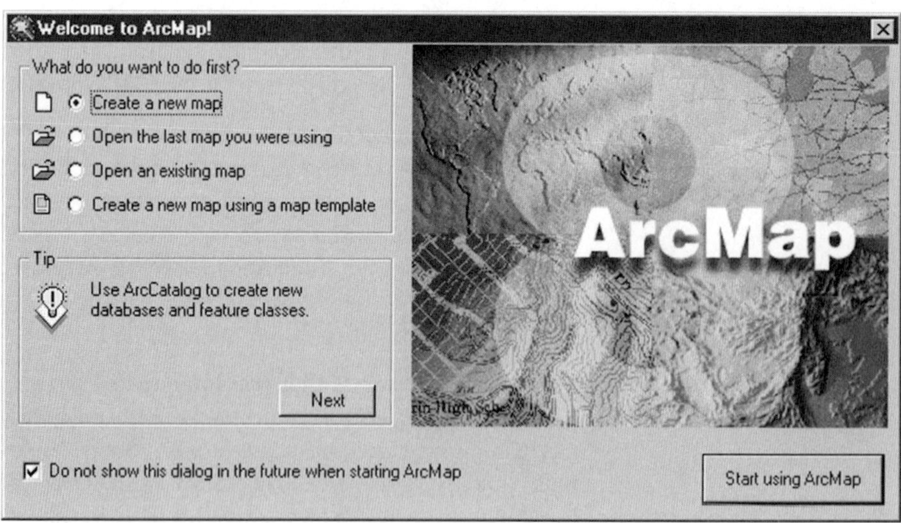

ArcMap start-up dialog.

To create a new map, select the "Create a new map" option. To open an existing map, select "Open an existing map." Select "Open the last map you were using" to work on the last map you had open. Select "Create a new map using a map template" to create a map using one of ArcMap's supplied map layouts. These layouts provide map compasses and predefined areas on the map for titles and legends. Layouts are discussed later in this chapter.

ArcMap provides a different tip in the dialog box each time ArcMap is started. These tips are helpful reminders about using ArcMap. To see additional tips, click on the Next button in the tip section of the dialog box.

✓ **TIP:** *If you become tired of seeing the dialog box each time ArcMap starts, select the "Do not show this dialog in the future when starting ArcMap" box. Starting ArcMap again takes you directly to ArcMap, skipping the dialog box display.*

Once you have made a selection to create a new map or open an existing one, click on the "Start using ArcMap" button in the lower left corner of the window to proceed. Depending on your choice, either a new blank map or a list of existing maps will be presented. For this section, click on "Create a new map," and then on "Start using ArcMap."

The Screen Layout

The opening screen in ArcMap reveals two panes split vertically. The left pane is called the table of contents and the right pane is the map canvas. The main screen is shown in the following illustration.

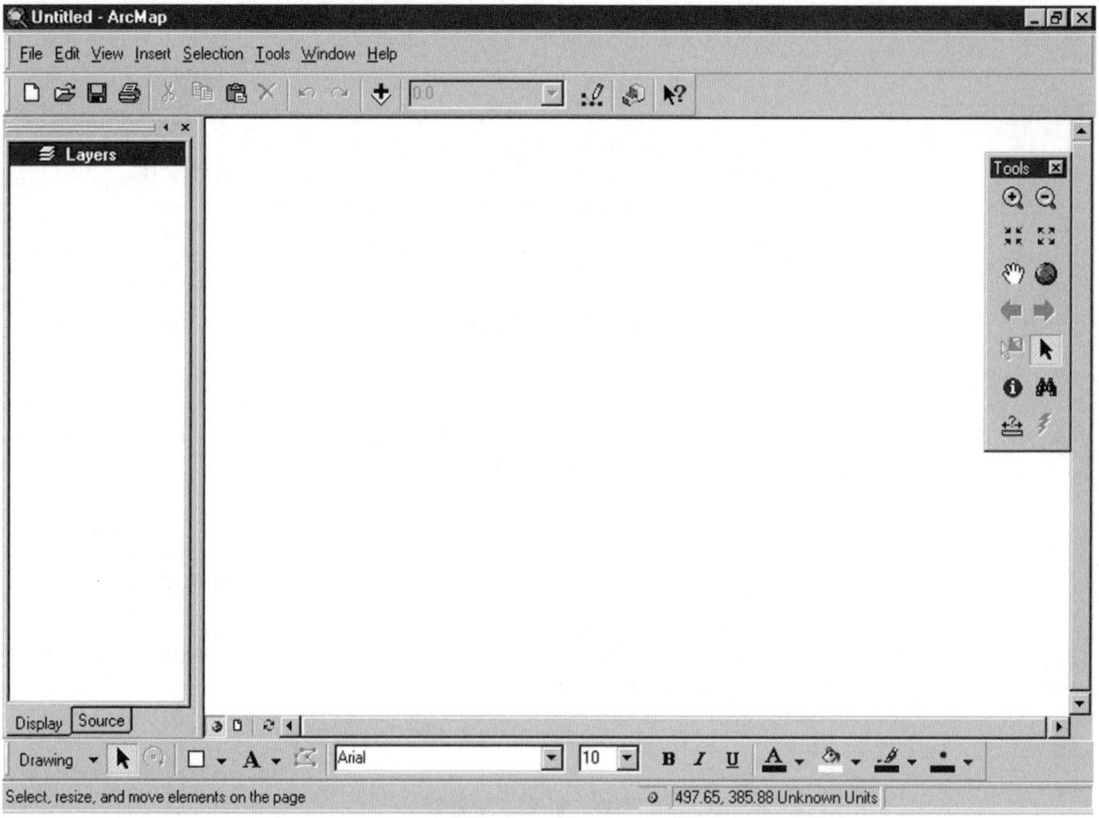

ArcMap main screen.

You will see many tool icons above and below the screen. By default, toolbars exist in three places in ArcMap. First, a floating menu called Tools appears in the map canvas window. This toolbar contains display commands that enable you to zoom in or out, pan around the canvas, and click on items to see descriptive information about them, as well as a command for measuring distances on the map. The Tools menu icons are discussed in the material that follows.

Zoom In/Zoom Out tools.

The Zoom In and Zoom Out tool buttons, shown at left, are used to zoom in or out of a map canvas. To zoom in, click on the button on the left (the magnifying class with the plus sign in it). Once it is selected, click once in the map canvas or use the mouse to drag a box covering the area you would like to see close up. To zoom out, select the right-hand tool (the magnifying glass with a minus sign in it). Click on a location in the canvas to zoom out with the selected spot as the center of the new view.

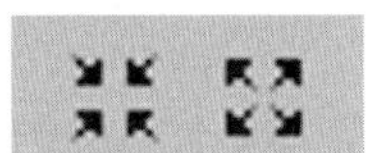

More Zoom In/Zoom Out tools.

The More Zoom In and Zoom Out tool buttons, shown in the illustration at left, zoom in or out based on the center point of the current map extent. Select the left-hand button to zoom in and the right-hand button to zoom out.

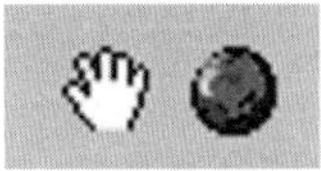

Pan and Resize tools.

The Pan and Resize tools, shown in the llustration at left, are used to pan around the map canvas and to resize the map extent to its original size. Use the tool on the left (the small hand) to pan around the canvas, and the tool on the right (a globe) to reset the view to its original size.

Switch to next or last map extent.

The two tools shown in the illustration at left change the map canvas to the previous or next map extent. Select the left-hand tool (the left arrow) to reset the map extent to its last extent (before you either zoomed in or out or panned about the view). Select the right-hand tool (the right arrow) to return to the current extent.

The Selection tools, shown in the illustration at left, are used to select spatial features and graphic elements such as text and freehand drawings in the map view. Use the tool on the left to select spatial features from coverages or shapefiles. Use the tool on the right to select graphics elements such as text or freehand drawings.

Selection tools.

Use the Identification tools, shown in the illustration at left, to select spatial features and to see descriptive information about them. The tool on the left, the Identify tool, lets you select a feature using the cursor and then displays information about that feature. The tool on the right, the Find tool, allows you to type in information. This information is then searched for throughout all spatial databases in your table of contents. Any feature containing the typed text will be selected and highlighted on the screen.

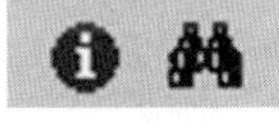

Identification tools.

Measure and hyperlink tools.

The left-hand tool in the illustration at left is used to measure distances in the map canvas. Select it, and then choose two points in the map canvas window. The distance in map units between points will be displayed. The right-hand tool, Hyperlink, is used to select and link features to elements such as web pages, digital photographs, word processing documents, and so on.

Along the top of the screen is a toolbar containing icons for basic commands such as opening files, saving files, and printing. Other icons add data to your map composition (the bold "+") or remove it (the "X"), or edit it (the pencil). You can query map composition features using the arrow-and-question-mark icon, or even launch ArcCatalog by clicking on the "filing cabinet" icon.

The toolbar along the bottom of the screen (the Drawing toolbar, shown in the following illustration) contains drawing tools that allow you to draw freehand in the map canvas pane; control the size and color of text, lines, and shapes; and add text to the map. The icons in this toolbar are discussed in the material that follows.

Drawing toolbar.

This toolbar contains all the tools for adding freehand graphics and text. The term *graphics element* is used to describe freehand text and graphics; that is, items that are added interactively and that are not part of a spatial data set. For example, you might use them to create map titles and annotation.

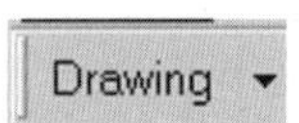

Alignment tool.

Use the Alignment tool, shown in the illustration at left, to group a number of graphics elements so that an operation such as move, change size, or delete occurs on all elements in the group. This tool also performs tasks such as nudging a graphics element to either side, or up or down.

Selection tool.

Use the Selection tool, shown at left, to select graphics elements, and to move or resize them.

Rotation tool.

Use the Rotation tool, shown in the illustration at left, to select and rotate graphics. Once a graphic element is selected using this tool, you can rotate it either clockwise or counterclockwise by moving the mouse in either direction.

Shape tool.

Select the Shape tool, shown in the illustration at left, to draw a shape on the map canvas. Clicking on the down arrow next to this button reveals a number of predefined shapes: rectangles, squares, circles, ellipses, and freehand lines.

Text tool.

The Text tool, shown in the illustration at left, is used to add text to a map composition. Select it, position the mouse where you intend to add text, and type the text. Click on the down arrow to the right of this tool button to add text that follows a line, or to have text that appears as a callout box (the text appears to the side with a line between it and a point on the canvas that you indicate). Callout boxes are helpful for calling attention to a feature that, perhaps, you cannot add text to because it is surrounded by too many features.

Edit tool.

Use the Edit tool, shown in the illustration at left, to edit freehand lines you have added. You can change the shape of a line by moving or deleting vertices with this tool.

Font window.

The Font window, shown in the illustration at left, lists the fonts available for use on your computer. Any text you add will display in the currently selected font.

"Font size" window.

Use the "Font size" window, shown in the illustration at left, to change the text size. Any text added will be in the current font size.

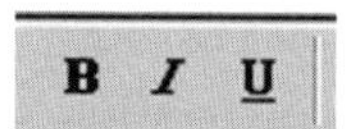

Font characteristics tools.

The "font characteristics" tools, shown at left, represent choices of bold, italic, and underlined text. Select the B to make text bold, the I to make text italic, and the U to make it underlined. You can also select any combination of these.

Font color tool.

The "Font color" tool, shown at left, controls the font color. Select the button to view a list of available colors. Text added will appear in the currently selected font color.

Fill color tool.

The "Fill color" tool, shown at left, specifies the color used to fill shapes created using the Shape tool, previously described.

Line color tool.

Use the "Line color" tool, shown in the illustration at left, to specify the line color and line size used when adding shapes or freehand lines.

A series of pull-down text menus is also available along the top of the ArcMap screen. These menus provide another means of accessing all of the commands available in the toolbars. Each is described in the reference section later in this chapter.

Two blank areas also appear on the main ArcMap screen. The map canvas, the right-hand pane, graphically displays all map information as you create it. Here you see what your map looks like as you create it. Note that as you work on a map, the map may be saved or printed at any time.

The left-hand pane, the table of contents, lists all layers displayed in the map canvas. A layer is simply a data set, such as an ArcInfo coverage, a shapefile, or a CAD file. Layers may be listed but not necessarily displayed; that is, you may turn layers on or off for drawing at will. Layers may even be grouped into data frames. By grouping them, you may turn many layers on or off for drawing.

For example, you may have a map of highways, secondary roads, and side streets in a county. You can create a data frame of secondary roads and side streets, which you can display when you want to see detailed street information or turn off when you want to see only major roads and highways. Data frames thus allow you to group many individual layers together to control their display.

Customizing ArcMap

Once you have worked awhile with ArcMap, you will find its user interface straightforward and fairly clear. You can also customize the interface to meet your needs. Customization of the interface works the same way as that performed in ArcCatalog. (See Chapter 2, including exercise 2-1, on customizing ArcMap's look and feel.) One of the great benefits of Arc Desktop is that the customization procedure is the same in ArcMap and ArcCatalog. Rather than spending your time learning to work with the software, you learn how to make the software work for you.

You are now familiar with the basics of ArcMap's main screen layout, the functions of its toolbars and icons, and customizing the user interface. Next, in exercise 3-1, you will use ArcMap to display and query a street data layer to create a report listing roads within 500 meters of a tanker truck accident.

Exercise 3-1: Identifying Features Using ArcMap

ArcMap does more than just display spatial data. It incorporates some very good commands that allow you to query spatial data for information. In this exercise you are an emergency response operator who has just been informed that a tanker truck carrying inflammable liquids has overturned on Hessian Road in Charlottesville, Virginia (an entirely fictional scenario). Assume that standard procedures for this type of accident require that residents within 500 meters of the road be alerted by police for possible evacuation.

You will use ArcMap to first find Hessian Road and to then find the streets that fall within 500 meters of it. You will create an on-screen map. You will also add a "mouse-over." Before beginning, be sure to copy the data needed for the Chapter 3 exercises from the companion CD-ROM to your hard drive.

✓ **TIP:** *Remember that all exercise data should be copied from the companion CD-ROM to your hard drive. All data for Chapter 3 will be under a folder named* D:\ai80exercises\chap_3. *If you loaded the exercise data to another hard drive, supply the appropriate drive letter, instead of D.*

Step 1: Drawing City Streets

Start ArcMap. Select "Create a new map" from the start-up window. The main ArcMap screen appears.

Now, add data (in this case, city streets) to the map composition by using the mouse to select the Add Data button, shown in the following illustration.

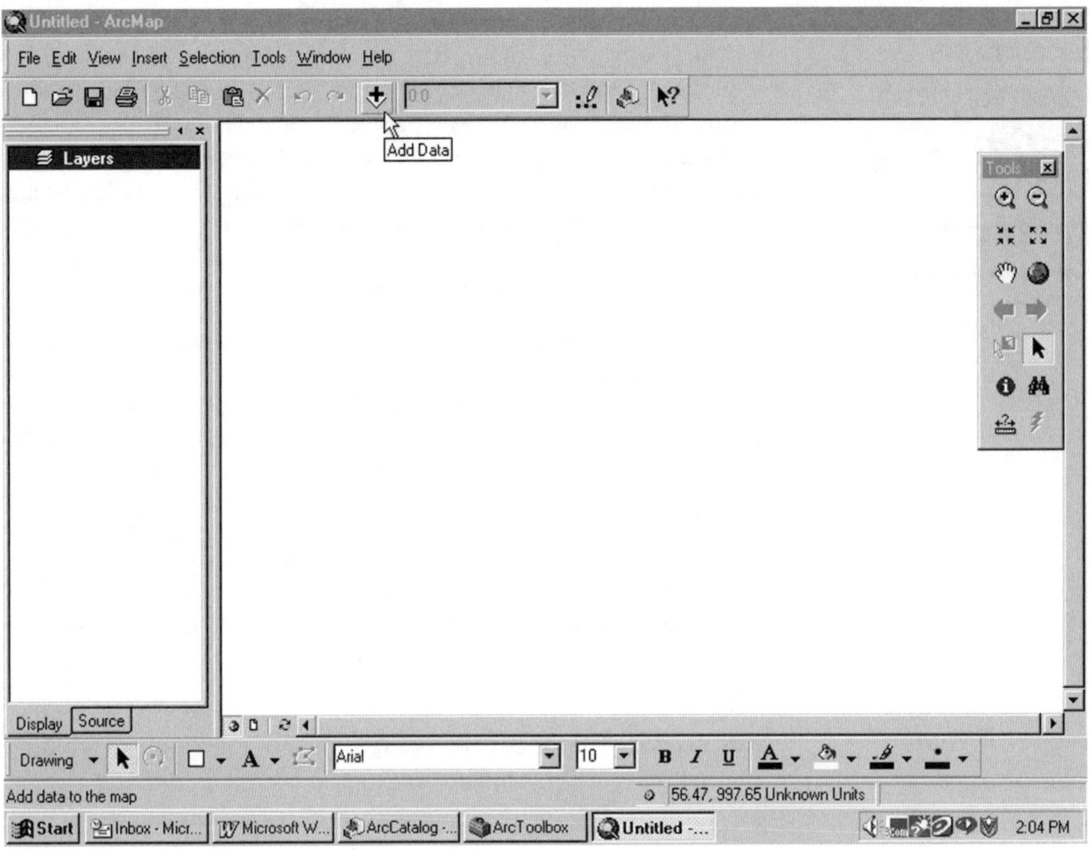

Add Data button in ArcMap.

The Add Data window appears. Navigate to your workspace (assumed to be *D:\ai80exercises\chap_3*) and you will see three coverages listed, as shown in the following illustration. Select the one called "street" and click on the ADD button.

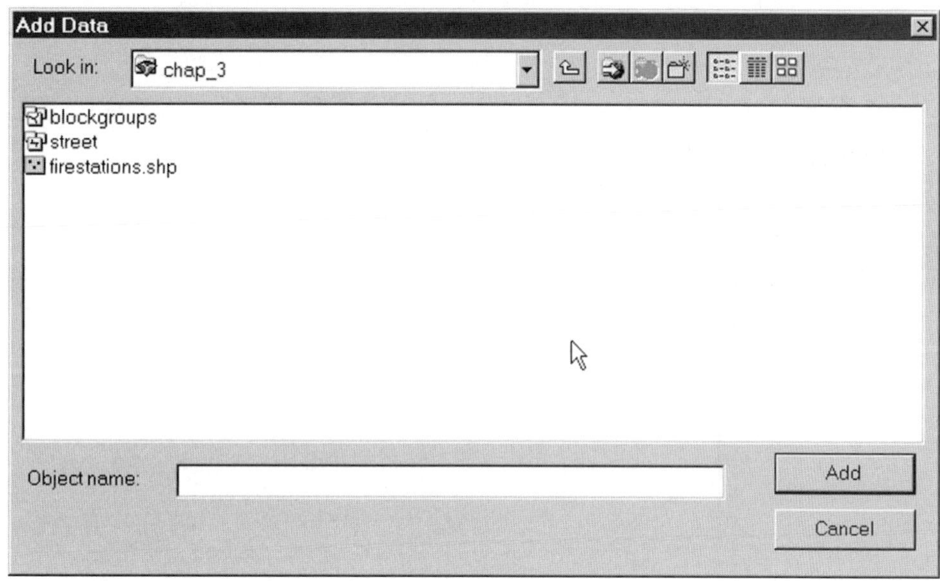

Open file window in ArcMap.

All roads in the city are now displayed in the map canvas pane. Your screen should look similar to that shown in the following illustration.

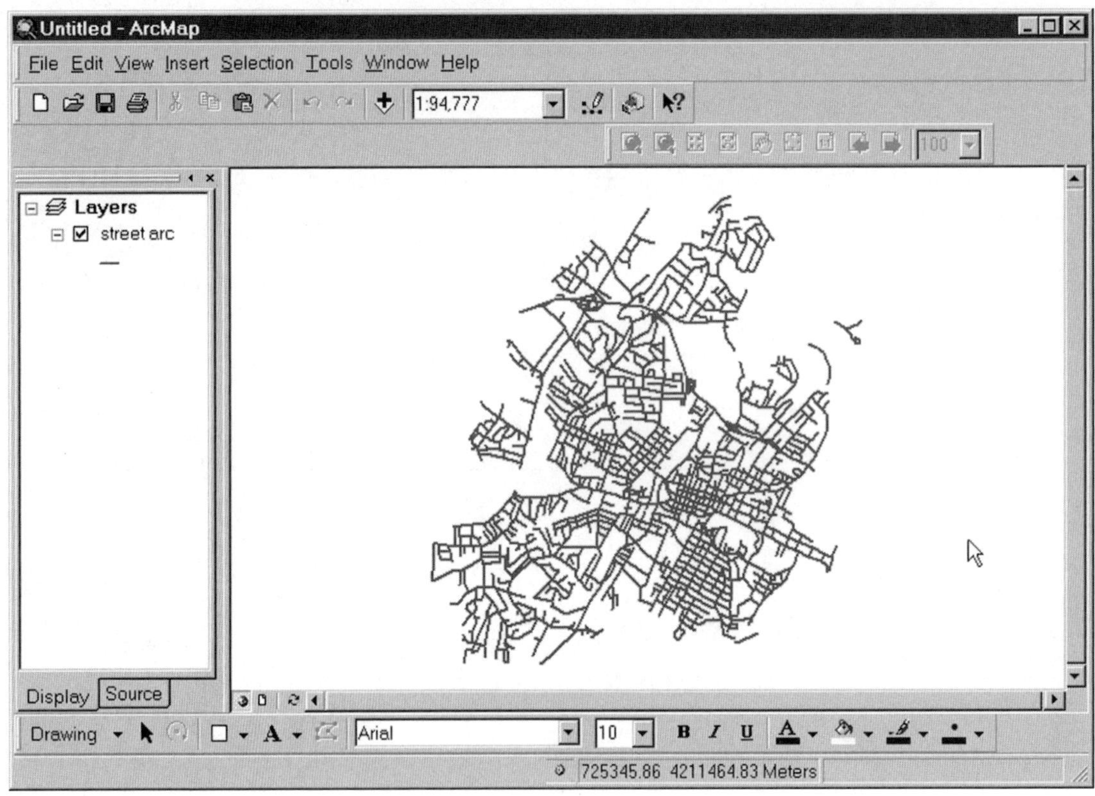

Roads displayed in the map canvas pane.

✓ **TIP:** *ArcMap assigns random colors to data layers. You may see red features the first time you add a layer, and blue features the second time you add it. Colors are easy to change, which you will learn how to do in exercise 3-2.*

Step 2: Finding Hessian Road

To find Hessian Road, use ArcMap's selection commands to query your coverage. Go to the Selection ➥ Select By Attribute menu. The Select By Attribute pop-up menu appears, shown in the following illustration.

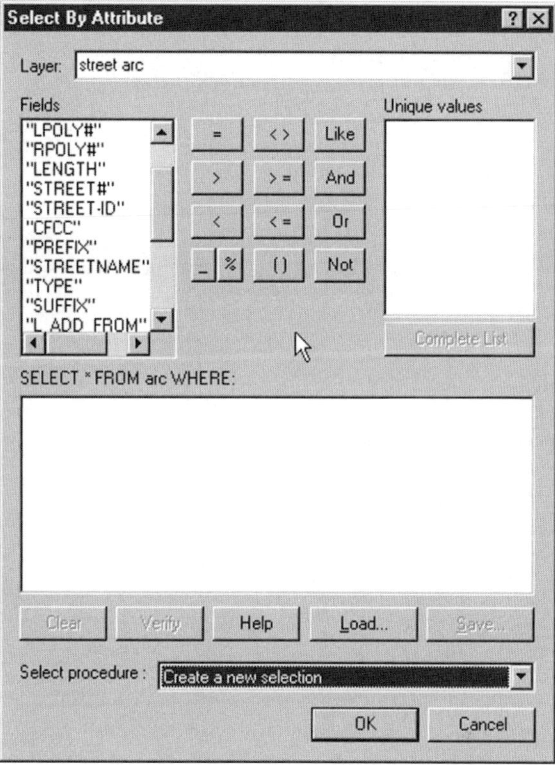

Select By Attribute menu.

The menu consists of several parts. The Layer box lists the data layer's name and feature type. The left-hand window lists all fields for the chosen feature type (in this example, it lists the .AAT items in the street coverage). Buttons representing logical operators such as "+", "-", and "=" appear to the right of this window. Next, the "Unique values" window displays all unique values for the field highlighted in the Fields window. Below these three sections is a window that displays the selection criteria you specify.

To find Hessian Road, scroll in the Fields window until you find the STREETNAME field. Double click on it, and STREETFIELD will be displayed in the selection criteria window beneath. Click on the "=" button. Now, scroll in the "Unique values" window until you find "Hessian" and select it. The Select By Attribute menu should now look like that shown in the following illustration.

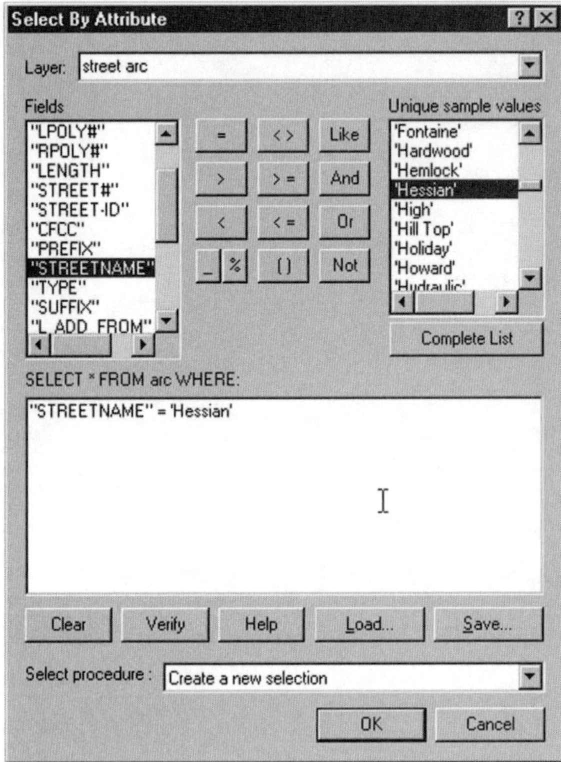

Selecting Hessian Road using the Select By Attribute menu.

Click on the OK button. ArcMap now highlights Hessian Road on the screen. Your screen should resemble that shown in the following illustration.

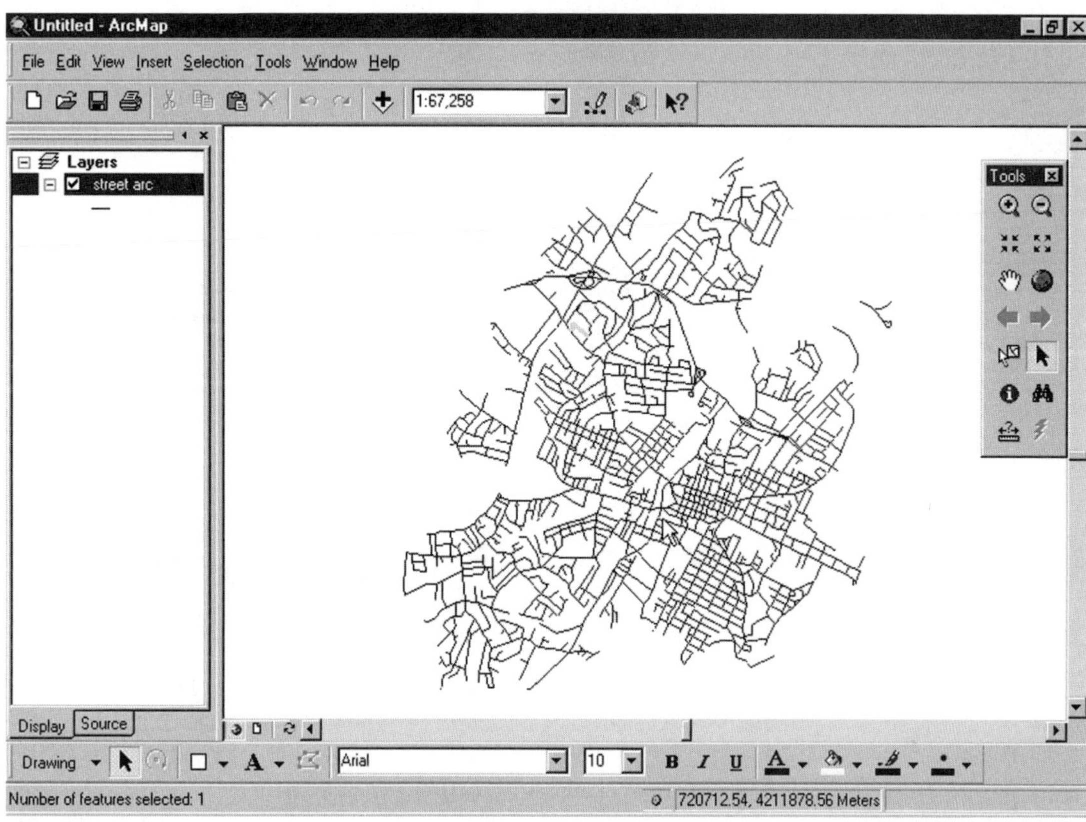

Hessian Road highlighted in ArcMap.

> **NOTE:** *The symbol used for Hessian Road has been changed to make it appear more prominent for use in this chapter. You are more likely to see the selected feature (Hessian Road) appear as a different color but using the same line weight as the rest of the roads.*

Step 3: Finding Roads Within 500 Meters of Hessian Road

Now that the road where the tanker accident occurred has been found and is still highlighted (selected), you must find all roads that fall within 500 meters of it. You will use a procedure similar to the Select By Attribute procedure. This time, use the Selection ➡ Select By Location command from the pull-down menu. The Select By Location menu is shown in the following illustration.

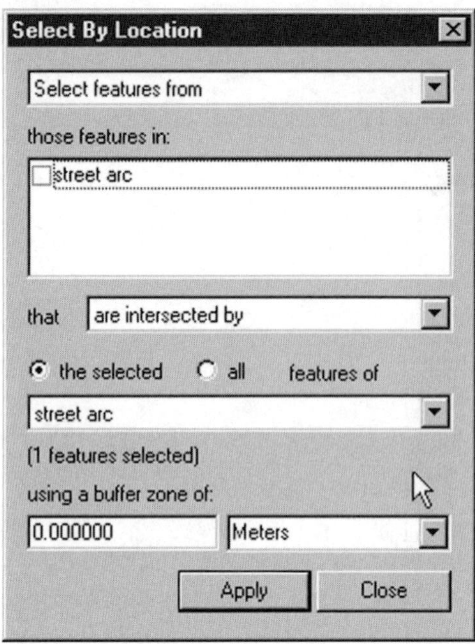

Select By Location menu.

This menu selects features based on geographic location. To find every road within 500 meters of Hessian Road, check the box next to "street arc" to tell ArcMap to search arc features in this coverage. Select "Are within a distance of" in the criteria box. This tells ArcMap to search for features within a specific distance of Hessian Road. Finally, type *500* in the "Using a buffer zone of" box and select "meters" as the unit. Remember, Hessian Road is still the highlighted feature in your map composition. The Select By Location menu should now look as it does in the following illustration.

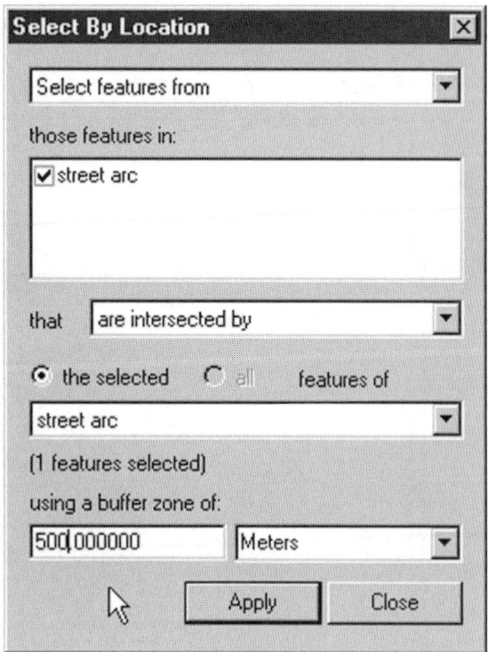

Select By Location menu with criteria for finding roads within 500 meters of Hessian Road.

Click on the Apply button. ArcMap makes the selection and highlights the features meeting the selection criteria. Click on the Close button to end the application. The map canvas should now display all roads that fall within 500 meters of Hessian Road, as shown in the following illustration.

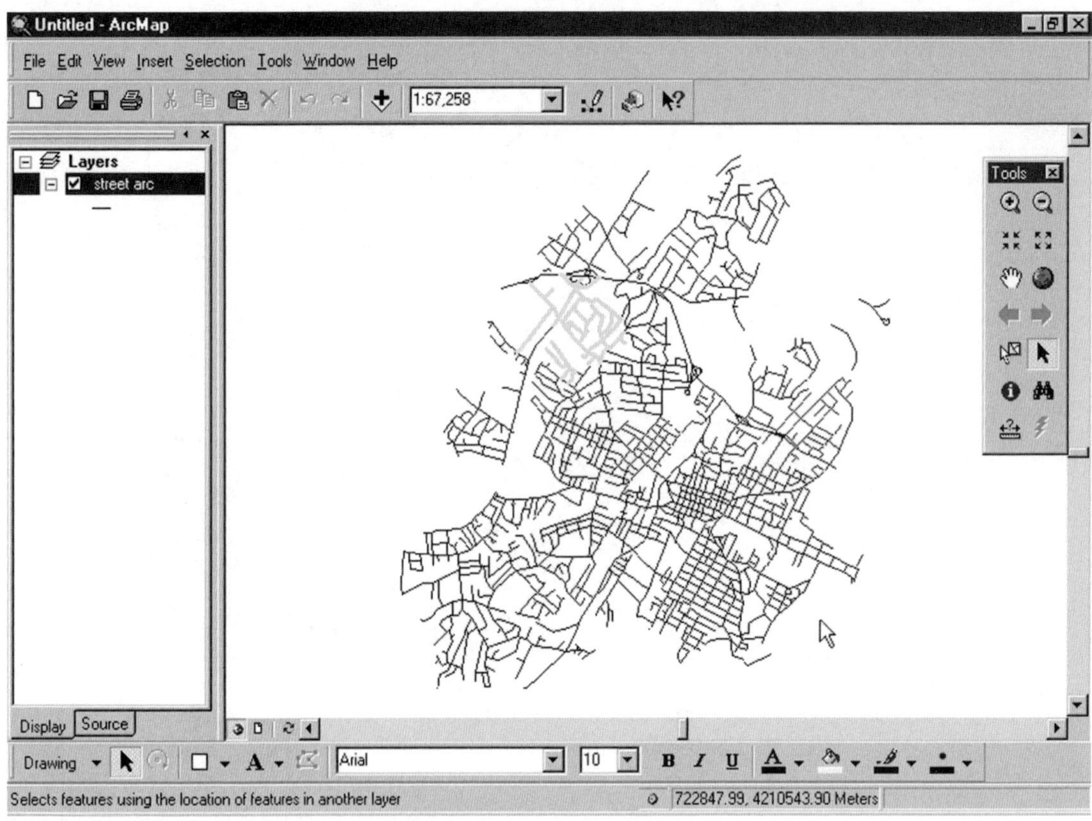

Roads within 500 meters of Hessian Road.

Step 4: Adding a "Mouse-over" and a Hyperlink

When creating an on-screen map composition such as this, it can be helpful to include extra information for viewers. Two ways to do this are mouse-overs and hyperlinks. ArcMap includes a new capability for displaying text information about features simply by positioning the mouse on a feature. For this exercise, it will be useful to have nearby fire stations shown on the map composition. Adding a mouse-over with firehouse names and fire chief names provides a quick means of contacting these entities in case of emergency.

Using the plus (+) button, add a shapefile theme named "firehouses." This step shows how to make the station name and station chief name appear when moving the mouse over a station, something called a "mouse-over."

Double click with the left mouse button on the *firestations* theme in the Table of Contents. The Layer Properties window appears. Select the Display tab, and the screen shown in the following illustration will appear.

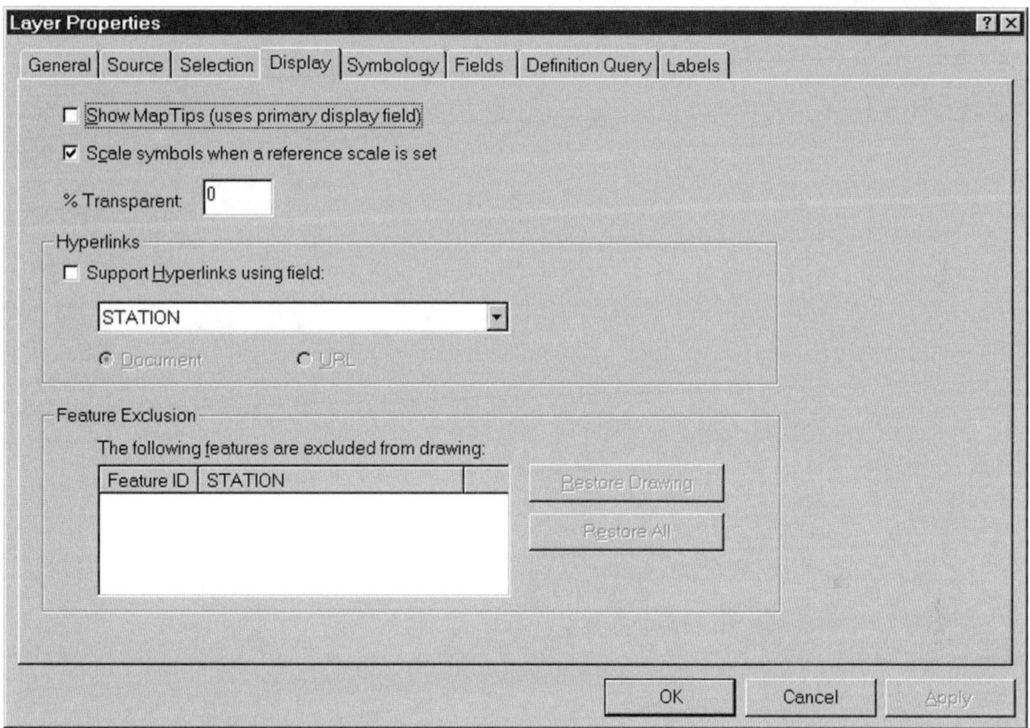

Display properties in Layer Properties window.

Click on the Show Map Tips option. This enables the "mouse-over" capabilities of ArcMap.

Next, specify the field that will display when the mouse moves over a fire station. Select the Fields tab at the top of the Layer Properties window. A listing of available fields appears in the window, as shown in the following illustration.

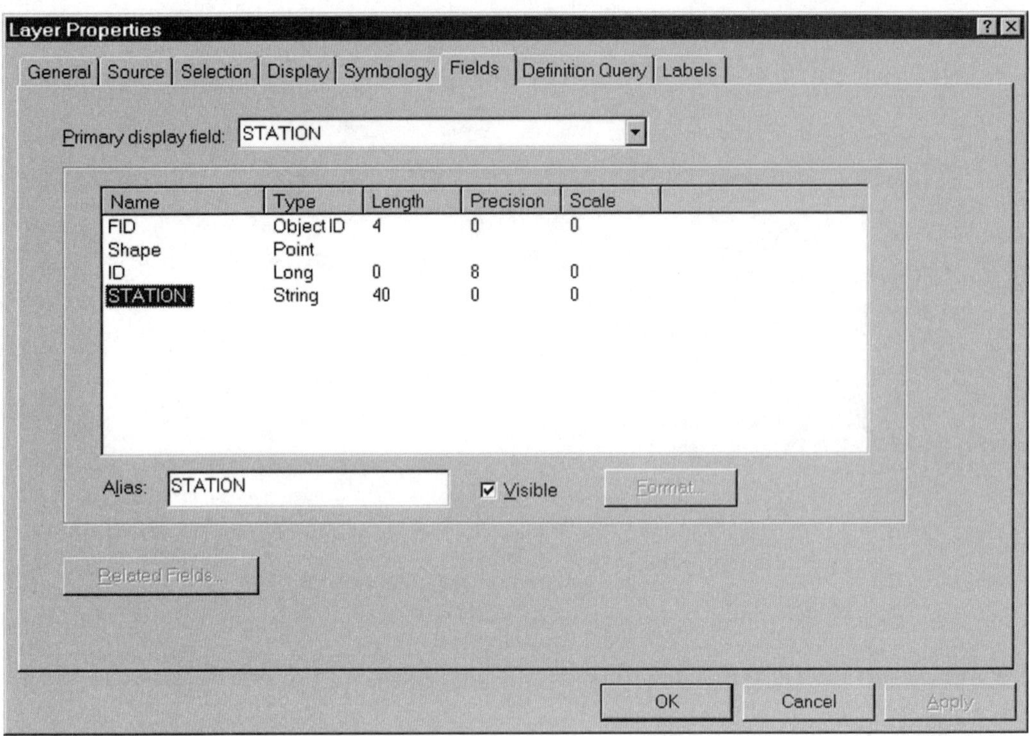

Field properties in Layer Properties window.

Select the STATION field, which contains firehouse and fire chief names. Click on the OK button.

Now, move the mouse until it rests on the firehouse symbol in the map canvas. The text in the STATION field will now display, as shown in the following illustration.

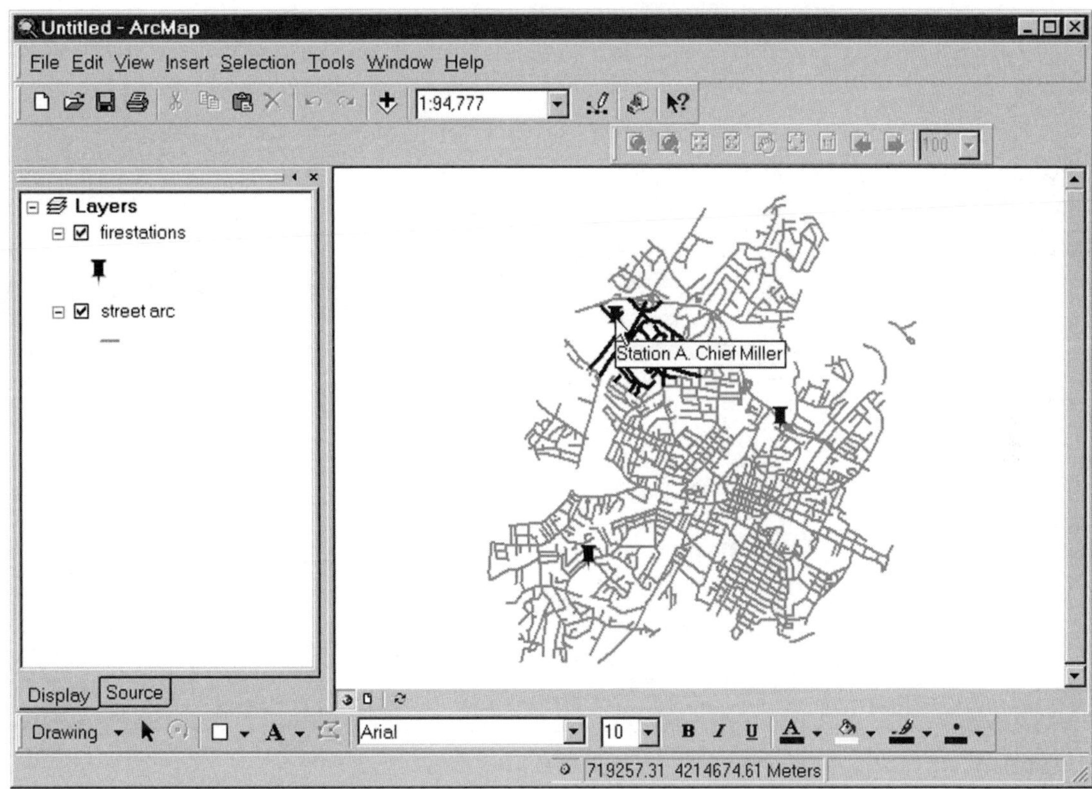

Showing a label describing a firehouse.

Step 5: Creating a List of Streets Within 500 Meters of Hessian Road

You must now notify the police with a list of which roads are within 500 meters of the Hessian Road tractor-trailer accident. Use ArcMap's report facility to generate a quick report of street names.

Select the Tools ➡ Make Report menu selection from the pull-down menus. The Report Properties window is shown in the following illustration.

Report Properties window.

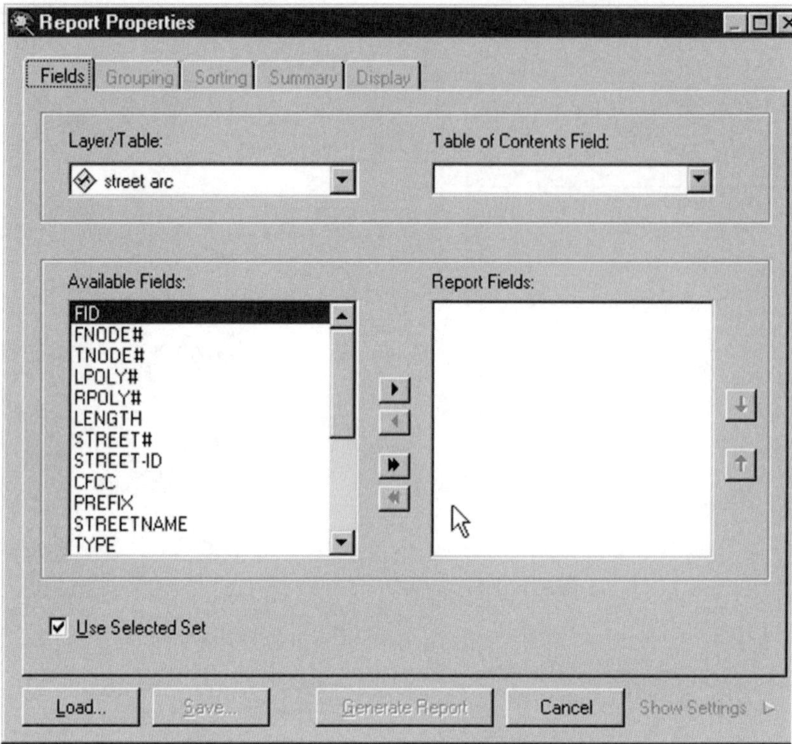

Select street names by double clicking on STREETNAME in the scrollable Available Fields window. You can add other fields to your report. For example, add address ranges (for the left side of the road) by double clicking on L_ADD_FRM and L_ADD_TO. Adding the left-hand "address from" (L_ADD_FROM) and left-hand "address to" (L_ADD_TO) fields tells police which street blocks they must visit to inform residents that they may have to evacuate. Those fields will now be transferred to the Report Fields window. Now click on the Generate Report button along the bottom of the Report menu. The resulting report is shown in the following illustration.

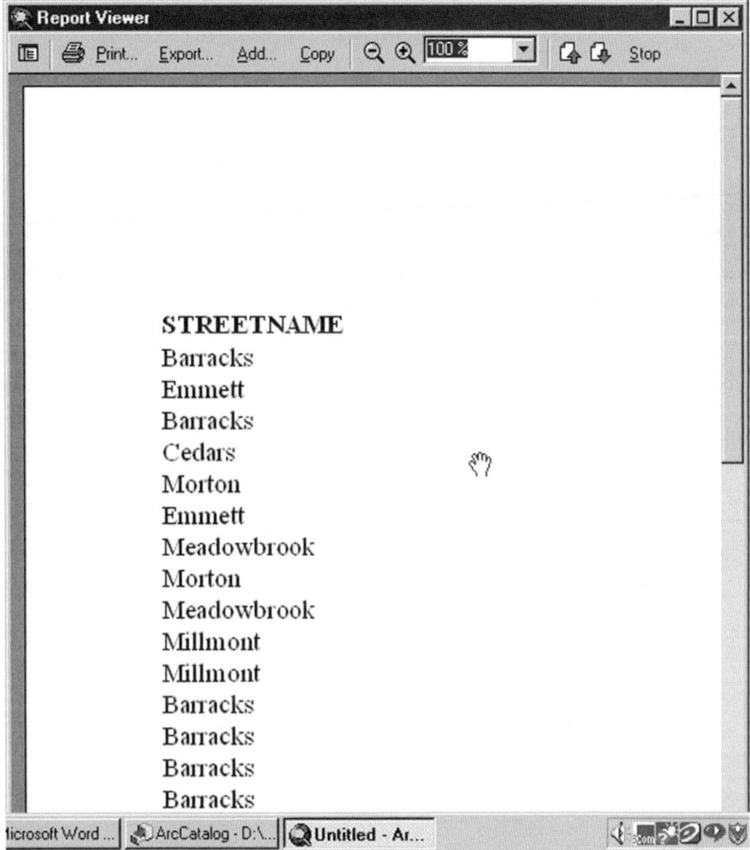

Results of report.

The previous illustration shows street names of roads within 500 meters of Hessian Road. Look carefully at the report. Some names are listed more than once; for instance, Barracks Road is listed quite a few times. Why? Because each arc segment of Barracks Road gets listed separately, and there are many arcs making up the whole length of Barracks Road.

This exercise showed you how to display and query a data layer (the street coverage) using ArcMap. You also learned how to perform a "mouse-over" using the *firestations* theme, and how to generate a simple report using ArcMap's report tool.

The sections that follow discuss displaying more than one data layer, and using ArcMap's existing map layouts, called templates, to easily create professional-looking maps.

Displaying Multiple Data Layers

Up to now, you have used only one data layer in ArcMap: the ArcInfo line coverage named *street.* ArcMap can display many data layers (such as coverages, shapefiles, and CAD files) in the same map composition. This section shows you how to work with two coverages in an ArcMap session. The principles covered here are true regardless of how many data layers you use. Follow along on your computer.

First, examine the screen shown in the following illustration. Notice that the table of contents lists two coverages: *street* and *blockgroups.* As you know, *street* is a line coverage containing the city streets of Charlottesville, Virginia, and *blockgroups* is a polygon coverage containing the 1990 census block delineations of Charlottesville. Although both layers are checked to make them active, the map canvas screen is showing just the block groups. Where are the roads?

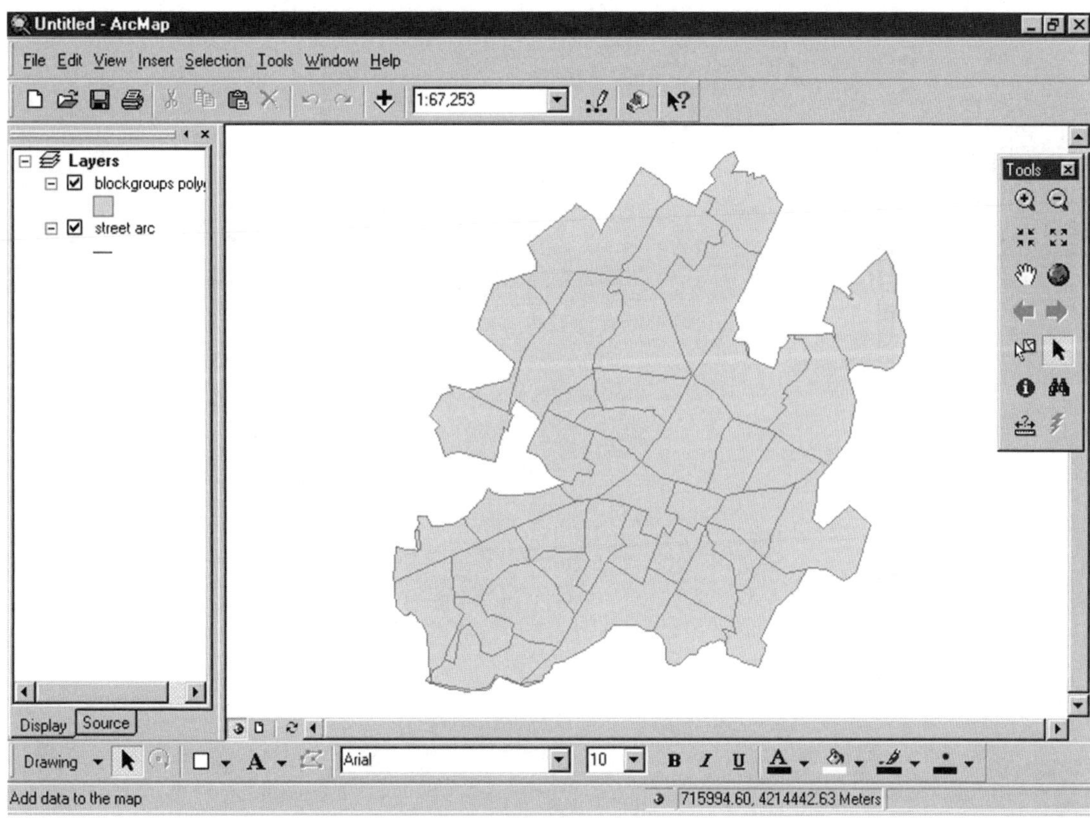

ArcMap with two coverages but showing only one.

The key to understanding how ArcMap displays data is in realizing that the map canvas screen "stacks" the data layers in the same order they are listed in the table of contents pane. In other words, *blockgroups*, at the top of the list, is drawn on top of street. Thus, the *street* coverage is not visible on screen. An easy way to remember this is "top of contents equals top of map."

How do you make the roads and the block groups both appear? The easiest way is to change their order in the table of contents. Click once on the *street* coverage and, while holding down the left mouse button, drag it to the top of the list in the table of contents. ArcMap instantly refreshes the screen, displaying both data layers, as shown in the following illustration.

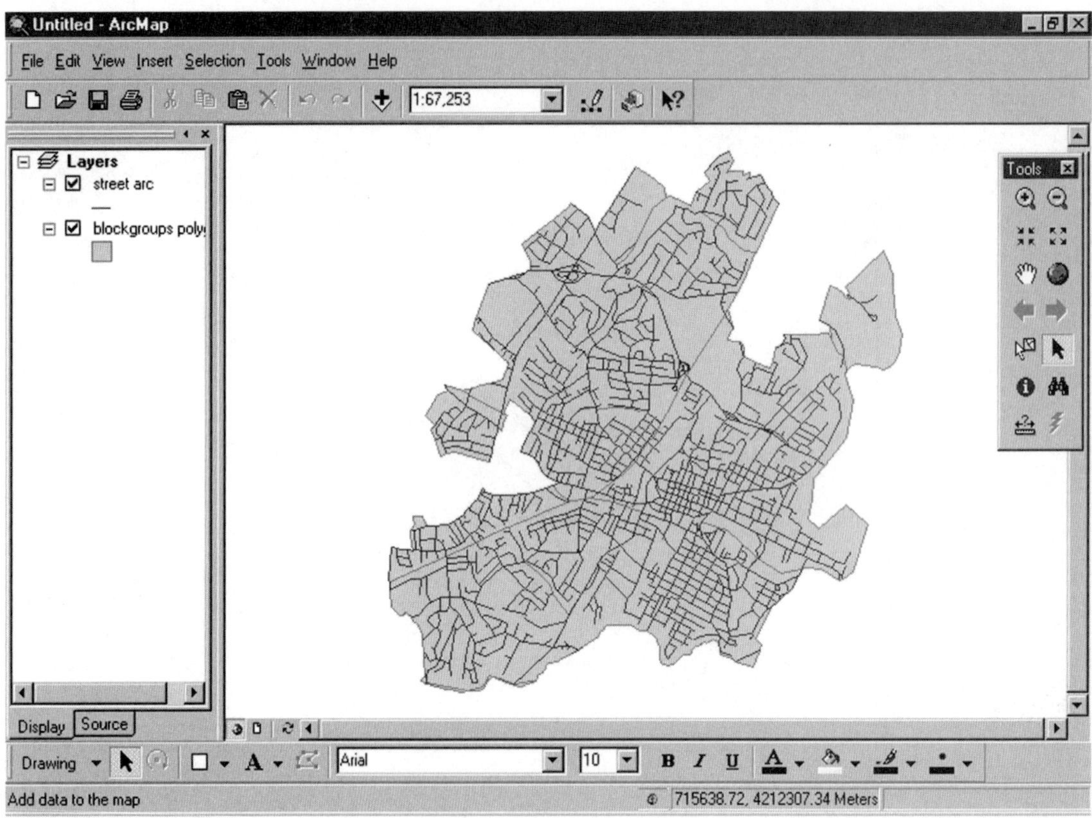

ArcMap displaying two coverages at once.

A second way to make both appear at the same time is to remove the polygon shading from the *blockgroups* coverage. This way, only the blockgroup boundaries and the streets are displayed. To turn off the polygon shading, double click on the blockgroup legend in the table of contents. The Symbol Selector window appears. In the Fill Color scrollable window, located on the right side, select No Color. Click on the OK button to apply the new shading scheme. The new screen is shown in the following illustration.

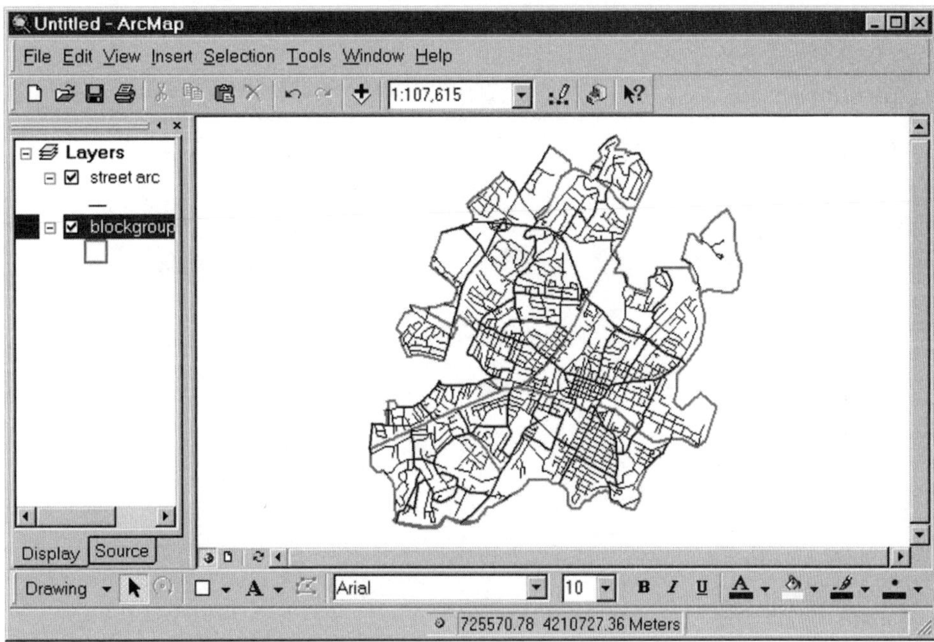

Turning off polygon shading for blockgroups.

There is a third way to make the roads in the street coverage visible, but without changing the drawing order or turning off polygon shading. ArcMap allows you to set the transparency of a feature so that data layers beneath it become visible. To change *blockgroups'* transparency, use the View ➥ Toolbars ➥ Effects pull-down menu. The Effects window appears, shown in the following illustration.

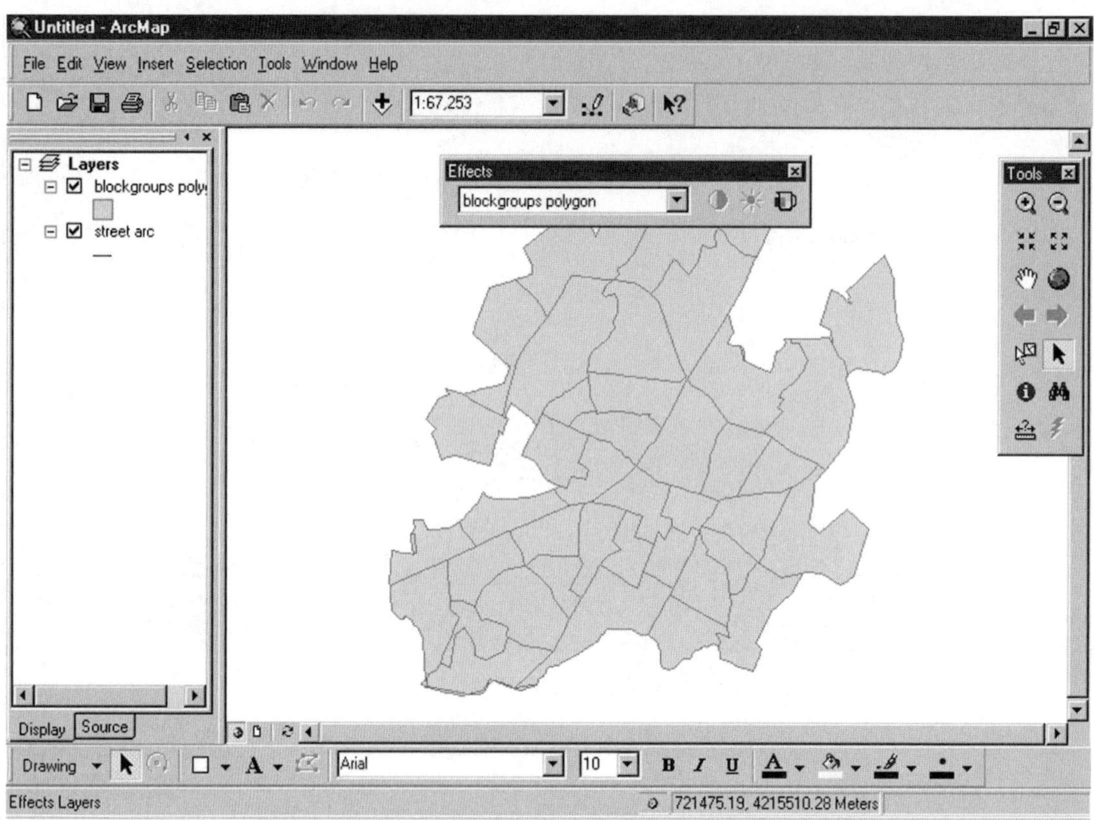

Effects window.

There are three options available, as well as a scrollable window to let you select the feature whose transparency you want to affect. Click on the pitcher icon (the right-most button on the Effects window), and a slider bar appears. By default, the transparency is set to 0, meaning that it is solid. Use the slider to change the value to 50%, and the *blockgroup* coverage becomes transparent enough to see through, as shown in the following illustration.

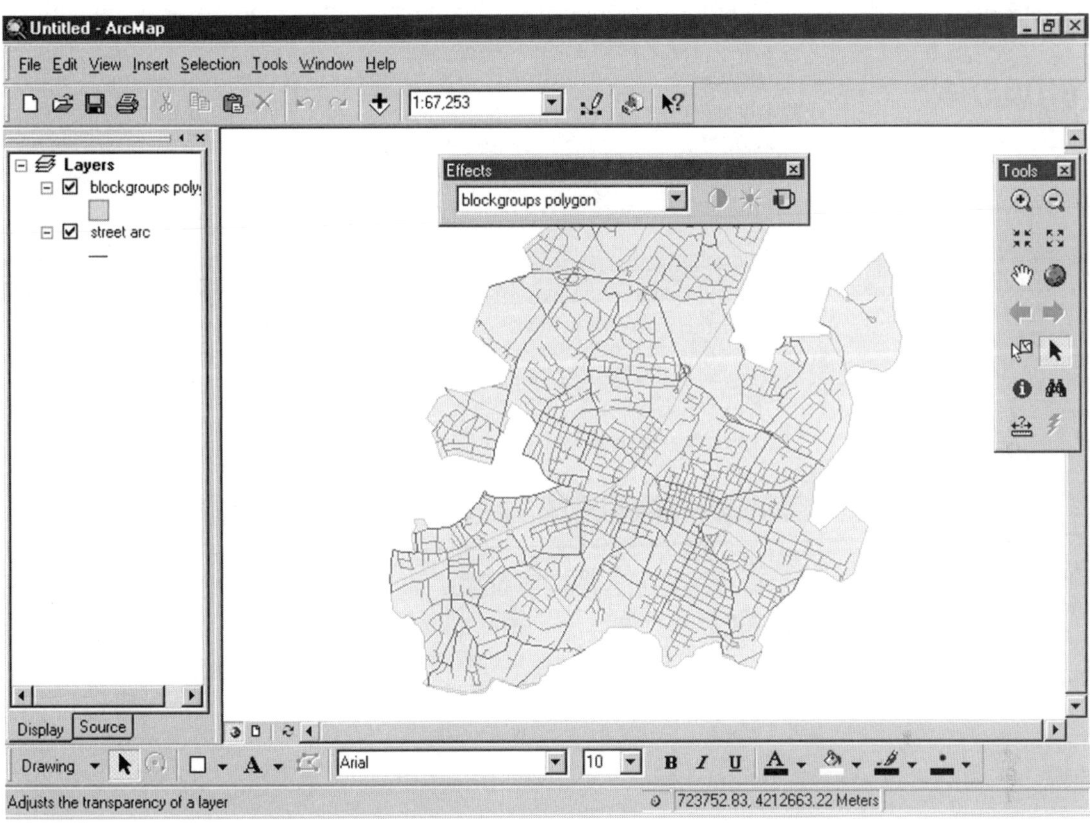

Transparency set to 50% for blockgroups *coverage.*

The other two buttons control saturation and hue of the color used for displaying the top data layer's features. Experiment with changing the saturation and hue of *blockgroups.*

✓ **TIP:** *If you have more than one spatial data set in your map and one or all of the data sets are not displayed, check the projections of all data sets. They must all be the same in order to be shown together. If any are not, use the Projection Wizard in ArcTools to make all data sets use the same projection.*

This section discussed displaying two or more data layers. You can use the order in which the layers are listed to control their display in the map canvas pane, and you can choose to show polygon data layers as either outlines or solid areas. You can also control the transparency of the top layer. The next section discusses ArcMap's map layout templates.

ArcMap Templates

Making maps has never been easier in ArcInfo. ArcMap includes a number of preexisting map layouts to help you quickly compose professional-looking maps. You are not confined to using the templates, however. ArcMap allows you to create your own maps from scratch. The templates are great to use for maps you need to produce quickly. This section examines ArcMap's templates and discusses how to use them to create new maps.

Templates in ArcMap are predefined map layouts that have places reserved for North arrows, legends, titles, and the graphic content of a map. ArcMap contains nine time-saving templates.

To view the template list, use the File ➥ New pull-down menu. A "New" pop-up window appears, which lists available templates. This list is shown in the following illustration.

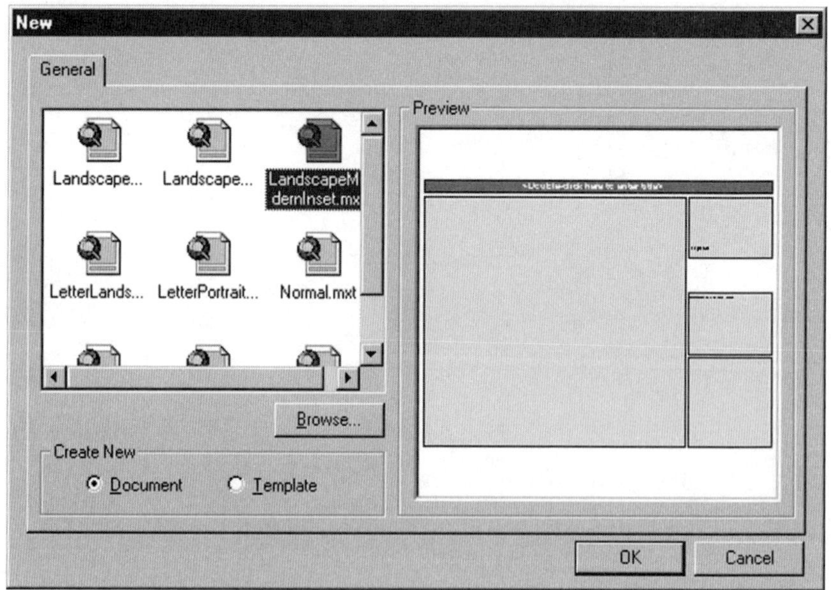

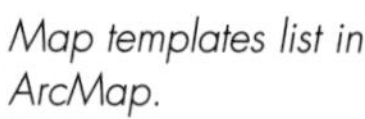
Map templates list in ArcMap.

To preview a template, use the mouse to click once on a template. The template will display in the window's preview pane. Notice that one of the templates is named *normal.mxt.*, which is blank. Select this template if you intend to create your own map layout. Selecting any of the other templates generates a basic map containing a North arrow, legend area, and so on, as shown in the Preview box of the previous illustration. In exercise 3-2 you will create a map using an ArcMap template.

Exercise 3-2: Creating a Map

Continuing from exercise 3-1, say that you are an emergency response officer who has been asked to now make a map showing the streets within 500 meters of Hessian Road, nearby fire stations, and a topographic image of the area where the tractor-trailer carrying inflammable materials overturned. This exercise shows you how to do so using an ArcMap template. It also introduces two additional aspects of ArcMap: changing a feature's color and labeling features. This exercise assumes you have completed exercise 3-1, and that you have loaded the data from the companion CD-ROM to your workspace (assumed to be *D:\ai80exercises\chap_3*).

Step 1: Selecting a Template

Start ArcMap by clicking on Start ➥ Programs ➥ ArcInfo ➥ ArcMap. Select the "new map" option. Next, click on File ➥ New to display the template list. Select the template named *LetterLandscape.mxt* and click on the OK button. The following illustration shows the new ArcMap screen.

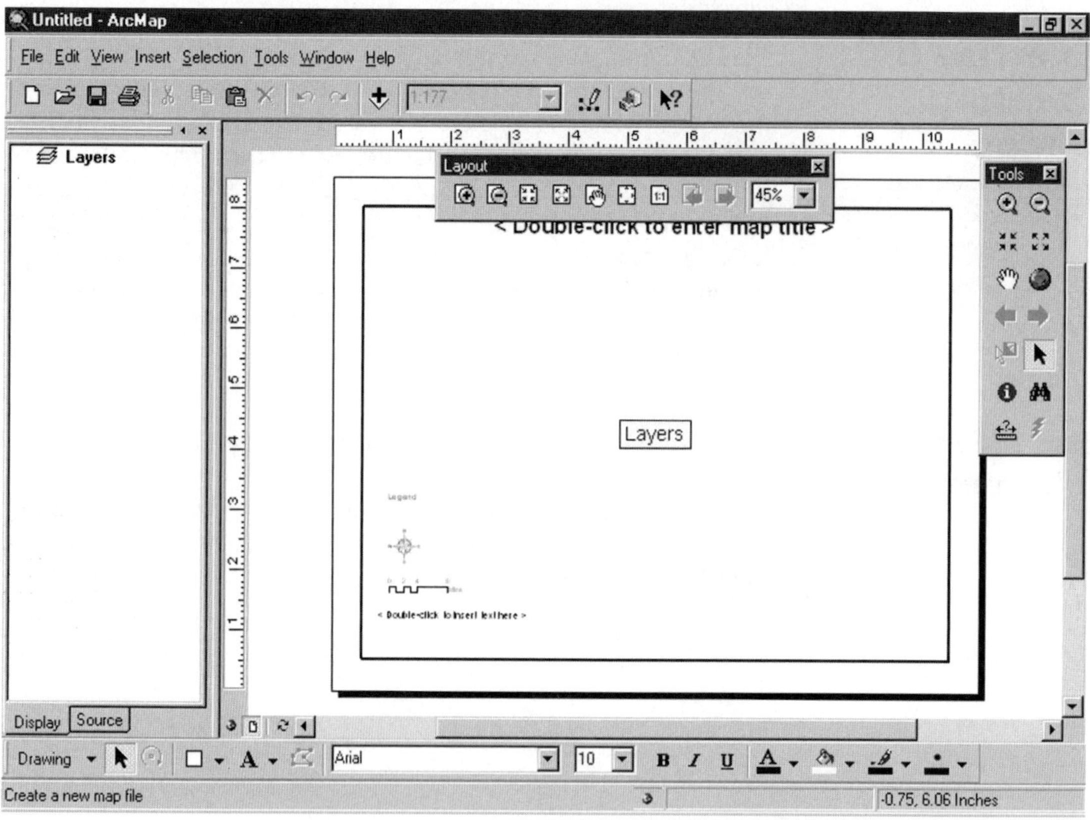

LetterLandscape.mxt *template in ArcMap.*

The template creates a basic map layout. Take a moment to examine the screen. A pop-up toolbar named Layout appears. It contains tools for zooming in and out and for navigating in the layout. The zoom tools work like the zoom tools in the Tools menu previously described. There is an additional tool you can use to type in a number specifying the amount you will zoom in or out. This can be helpful when you need to zoom in or out by just a small amount.

✓ **TIP:** *Zoom in to add fine-print text that may not be visible when you are zoomed out.*

Step 2: Adding Data Layers to a Map

Notice that the center of the map contains the word *Layers* in a yellow box. This represents the space where coverages, shapefiles, or any of the other data types will appear. Now, add the Charlottesville city streets to the map and once again select for all roads within 500 meters of Hessian Road (you are repeating steps 1 through 3 of exercise 3-1). Next, add the fire station shapefile data layer named *firestation*. Upon completion, your screen should look like that shown in the following illustration.

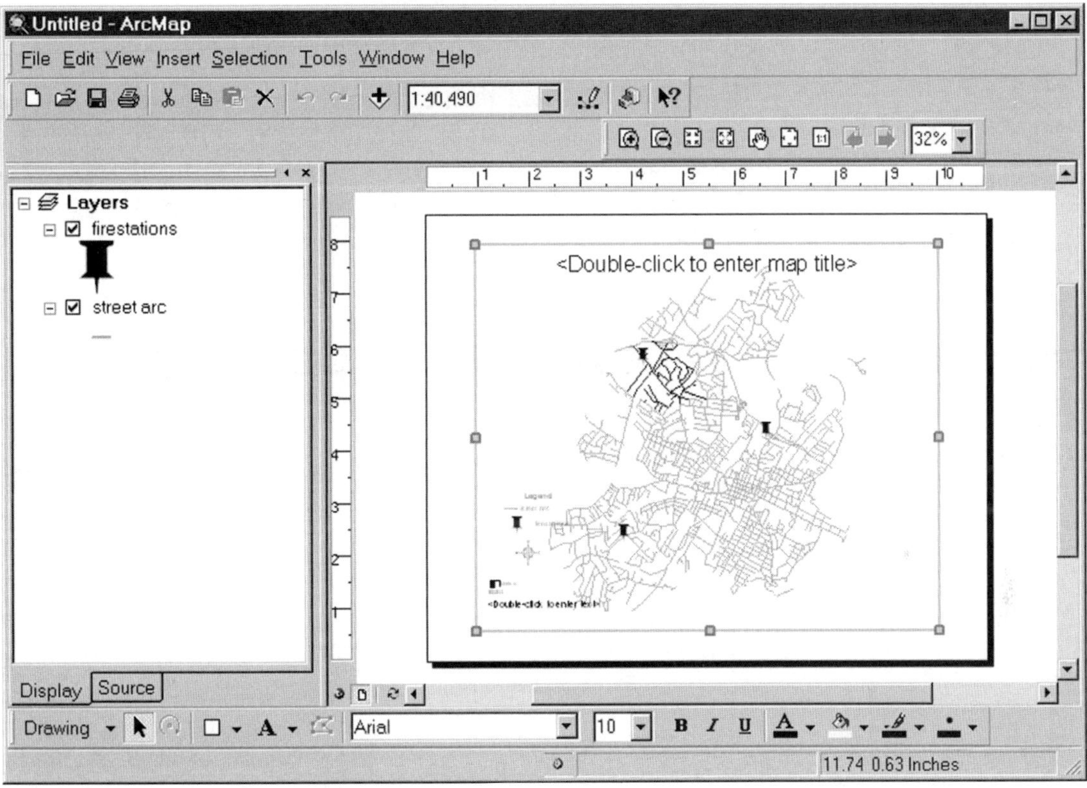

Streets within 500 meters of Hessian Road.

The map now shows all streets in the city, including those potentially affected by the tractor-trailer accident. Make this map more usable to police (who need to notify residents about the accident) by zooming in on just the selected features. To do this, click on View ➥ Zoom Data ➥ Zoom to Selected Features from the pull-down menu. Notice that the map canvas pane now displays just the area containing the roads within 500 meters of Hessian Road.

Step 3: Resizing the Map Area

After zooming in, you will notice that the title area along the top of the map has been overwritten by the *street* data layer. Use the Tools menu to resize the area in which the map graphics are drawn. To do this, select the solid black arrow (the fifth button down in the right-hand column) from the Tools menu. Now click once in the area where streets are displayed. A box, with blue squares at each corner as well as at the midpoint of each side, now surrounds the streets.

Use this box to resize the area in which the streets are displayed. Because you need to shrink the road area vertically, move the mouse so that it is on top of the blue square in the middle of the top line of the box. The mouse now looks like a two-headed arrow. Hold down the left mouse key and drag the top of the box down until it is just below the title box. The new map layout should look like that shown in the following illustration.

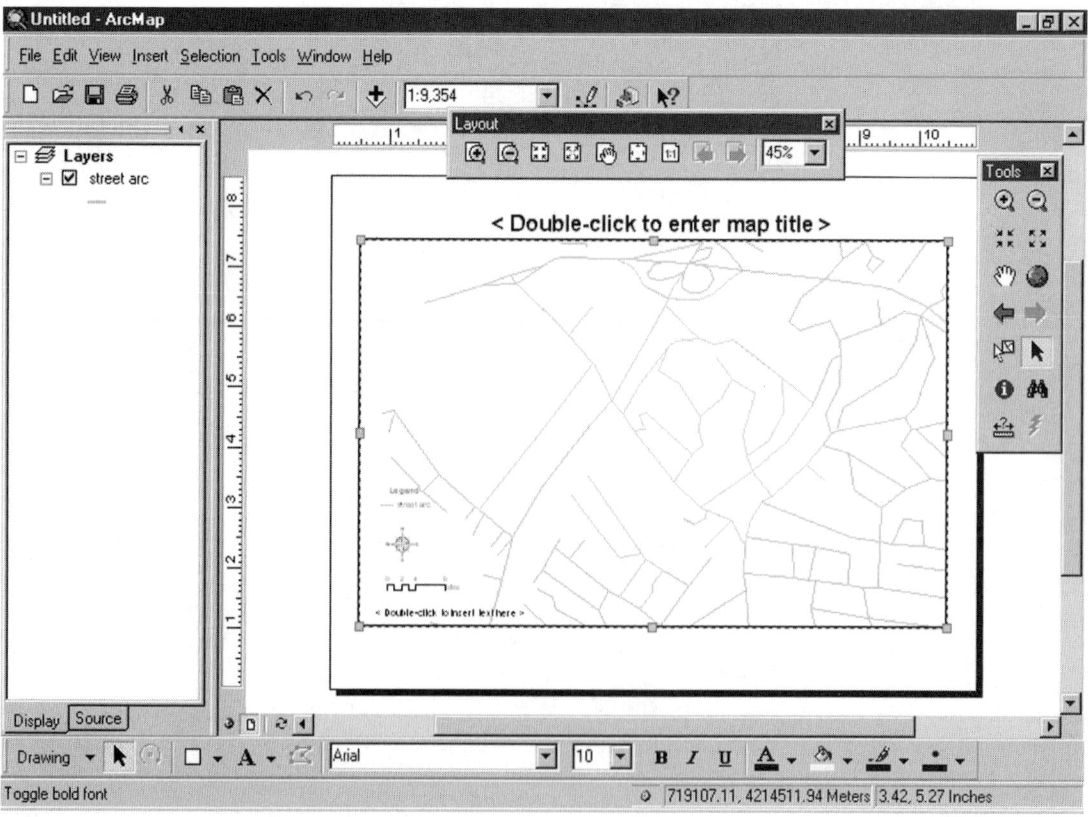

Resizing the map graphics area.

Step 4: Changing Symbols, Labeling Streets, and Adding a Title

Now make the map easier to read by making the streets line symbol a thicker line. To do this, double click on the line symbol beneath the *street* coverage listed in the table of contents pane (the left-hand window). The Symbol Selector menu appears, which is shown in the following illustration.

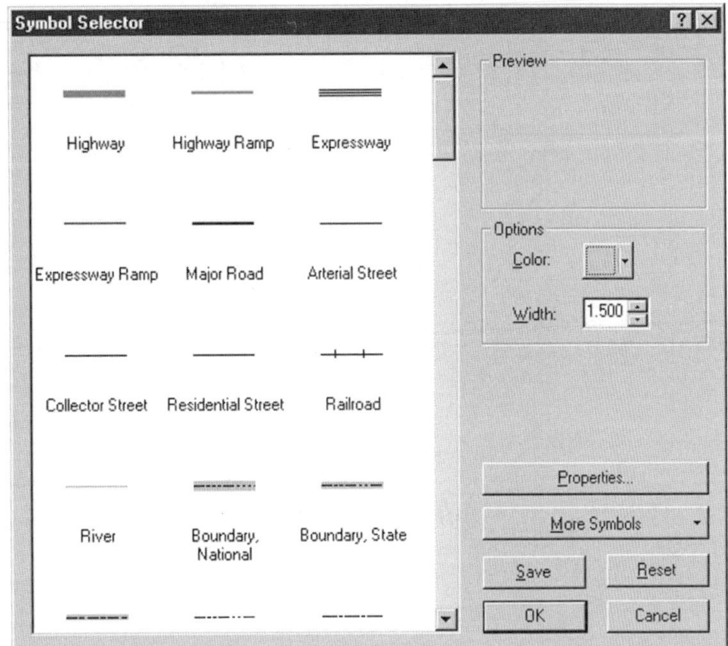

Symbol Selector menu.

The Symbol Selector menu provides many line pattern options, ranging from those traditionally used to depict transportation features such as roads and railroads to political boundary lines used to separate counties, states, and countries. For this exercise, select the symbol named Major Road and specify a width for this symbol of 1.5. Changing the width to 1.5 makes the roads appear as a thicker line. The map now appears as shown in the following illustration.

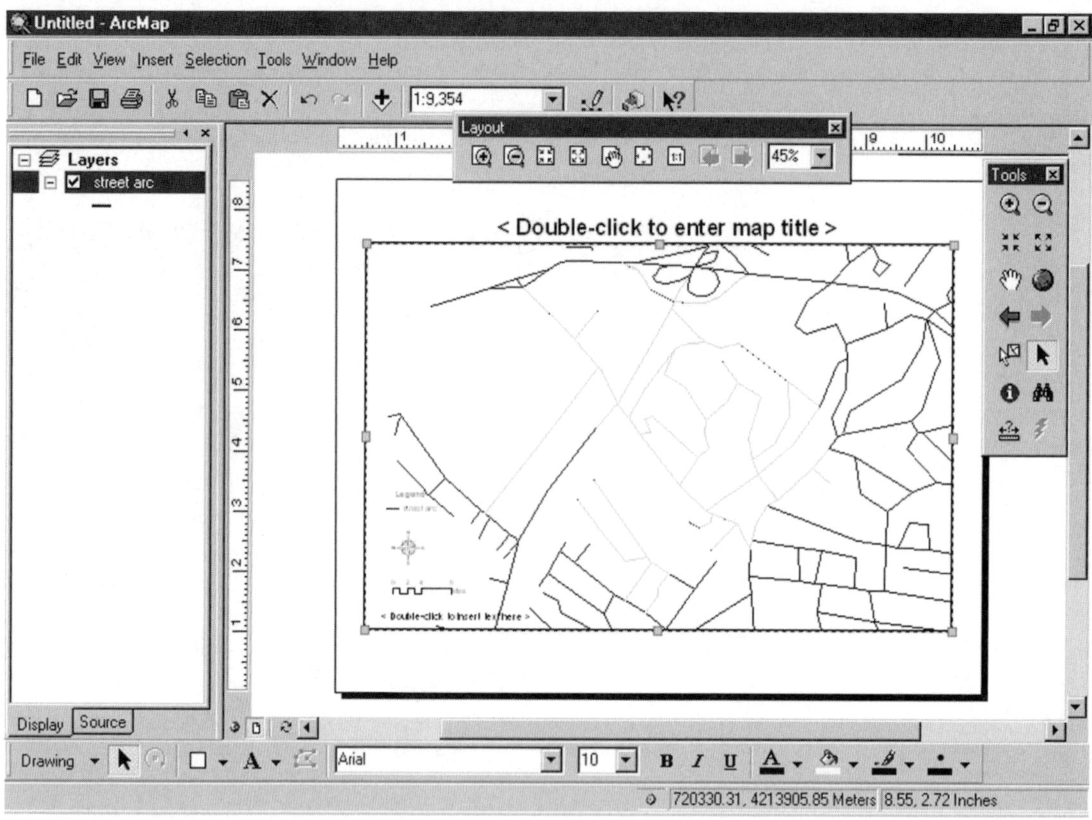

Roads changed to Major Road symbol and line width changed to 1.5.

Now you will label the streets. To do this, double click on the *streets* coverage in the table of contents. The Layer Properties window appears, which is shown in the following illustration.

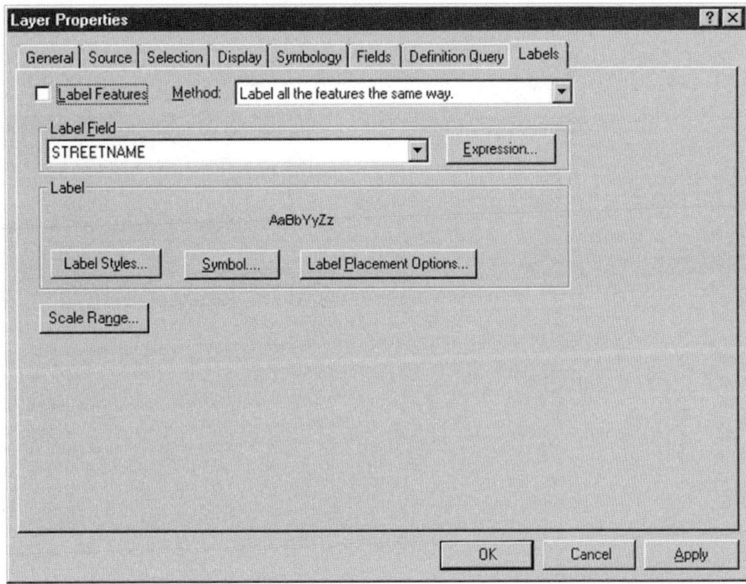

Layer Properties window for the streets coverage.

To label features, simply click on the Label Features check box. How does ArcMap know to label the streets using the names stored in the STREETMAP field? ArcMap defaults to the field used to make a selected set of data. In other words, when you selected Hessian Road, you selected it using an item named STREETNAME. ArcMap remembers that and uses it as the default field for labeling data. To label the streets using a different field, simply click on the scrollable window where STREETNAME appears and select a different field. After clicking on the Label Features check box, the map composition should look like that shown in the following illustration.

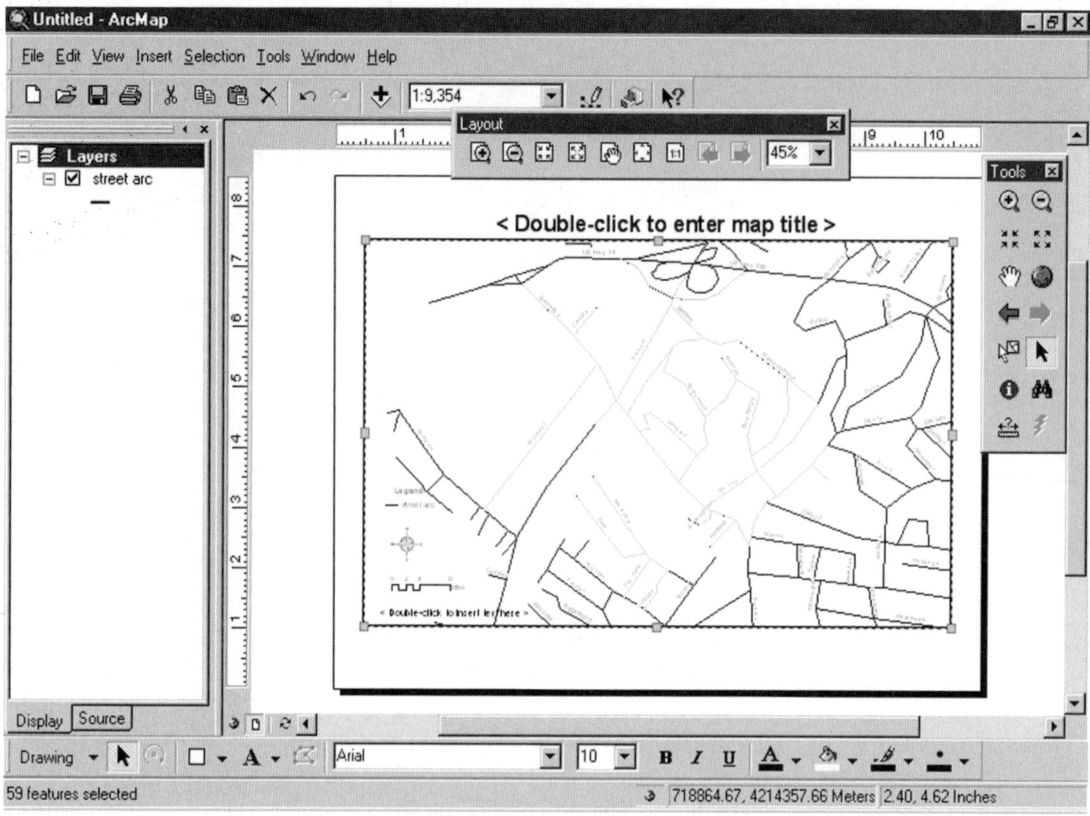

Revised map composition.

To add a title to the map, double click on the "<Double-click to add map title>" line at the top of the map. The Text Properties box appears, which is shown in the following illustration.

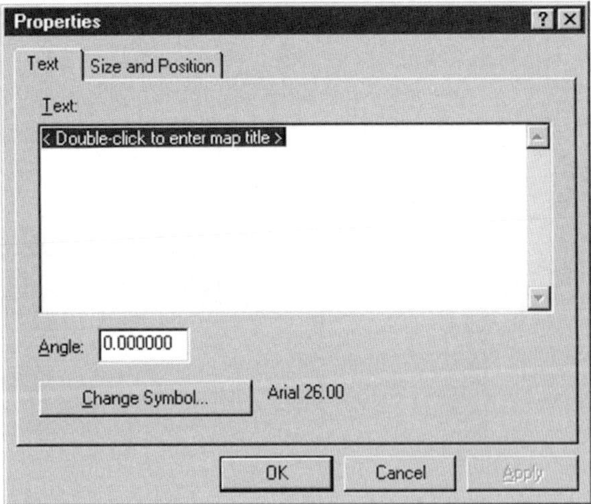

Text Properties box.

This box is where you type the title. By default, the title will appear in Arial font at 26-pt size. To change the font or size, click on the Symbol button, which gives you access to all fonts stored on your computer. There is also a box named Angle, which has a default value of 0.00. If you change this value, the text will appear on a diagonal, depending on the angle you supply. For now, just type *Streets Within 500 Meters of Hessian Road* and the map should look like that shown in the following illustration.

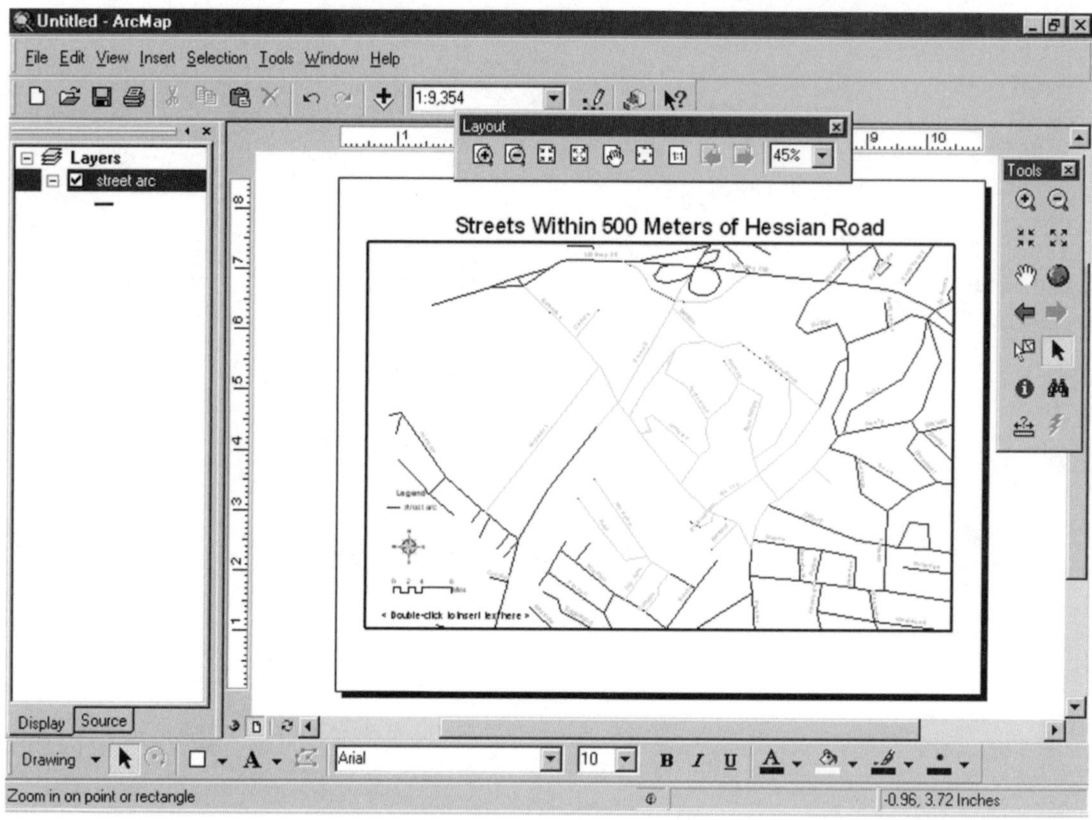

Adding the title.

Step 5: Saving a Map

Save your new creation by simply accessing the File ➥ Save pull-down menu. You will be prompted to enter a file name. Type *hessian road* and click on the OK button. The map is now saved. You may send it to a printer, e-mail it to police stations (provided they have ArcMap, they can view it on screen), or even post it to a web site for access by the police. You can also work on it again.

✓ **TIP:** *When you save a map, ArcMap saves the path names to the directory locations in which your data layers reside. If you later copy or move your map to a new folder, ArcMap could lose its knowledge of where the data are. Make copying map compositions easier by keeping copies of all data layers in the same folder as the map. This way, ArcMap always knows where the data are.*

Step 6: Creating a Hyperlink

You can use hyperlinks in ArcMap to link a feature in a coverage or shapefile to an object. By clicking on a feature, the linked object appears on the screen. Objects can be aerial photography, digital photos, maps, or even word processing documents.

This step shows you how to create a link between a fire station and a 7.5' topographic sheet of the area in which the fire station resides. The topographic sheet is a digital version of the standard U.S. Geological Survey topographic maps. It is stored as a TIF graphic, which is a common graphics file format usable by ArcInfo.

To add a hyperlink to a feature, first click on the Identify button in the Tools toolbar. Then click on the one fire station present on the map canvas. This will display the Identify Results window, shown in the following illustration.

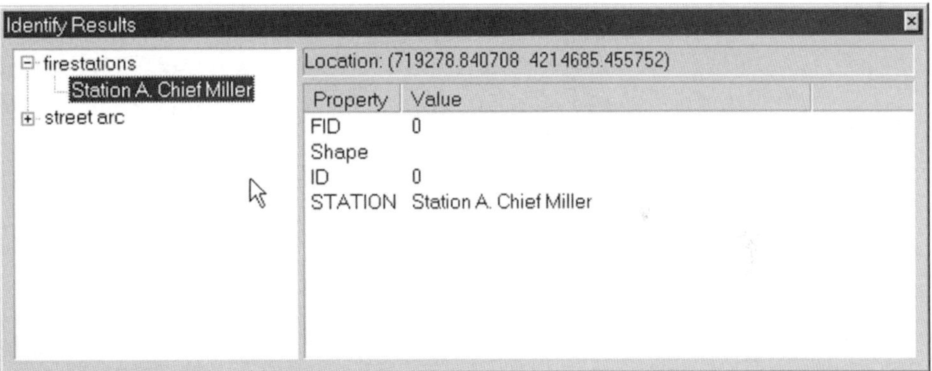

Identify Results window.

The Identify Results window displays data about the selected fire station. Now right click on the fire station data in this window. Select "Add hyperlink" in the new menu that appears. This will display the Add Hyperlink menu, show in the following illustration.

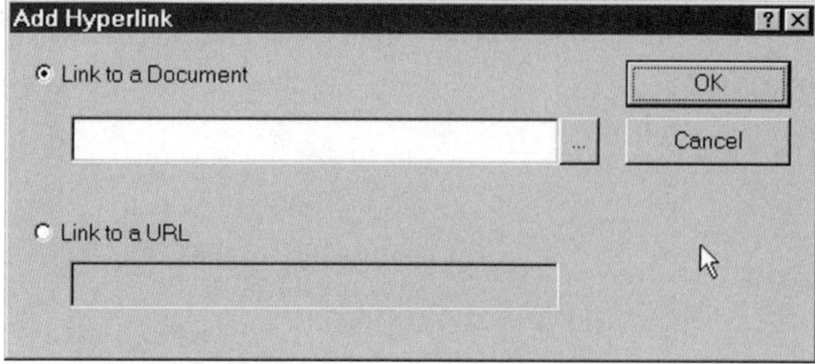

Add Hyperlink menu.

The Add Hyperlink menu is used to create a link to either a document (an image, a word processing document, and so on) or web site. For this exercise, create a link to the topographic map named *O38078A4.tif*, located in *D:\ai80exercises\chap_9*. In the Link to a Document box, type the name of the topographic map.

To use the hyperlink feature, click on the Hyperlink tool (a lightning bolt) in the Tools toolbar. Then click on the firehouse you just added a hyperlink to. The topographic map will now display as shown in the following illustration.

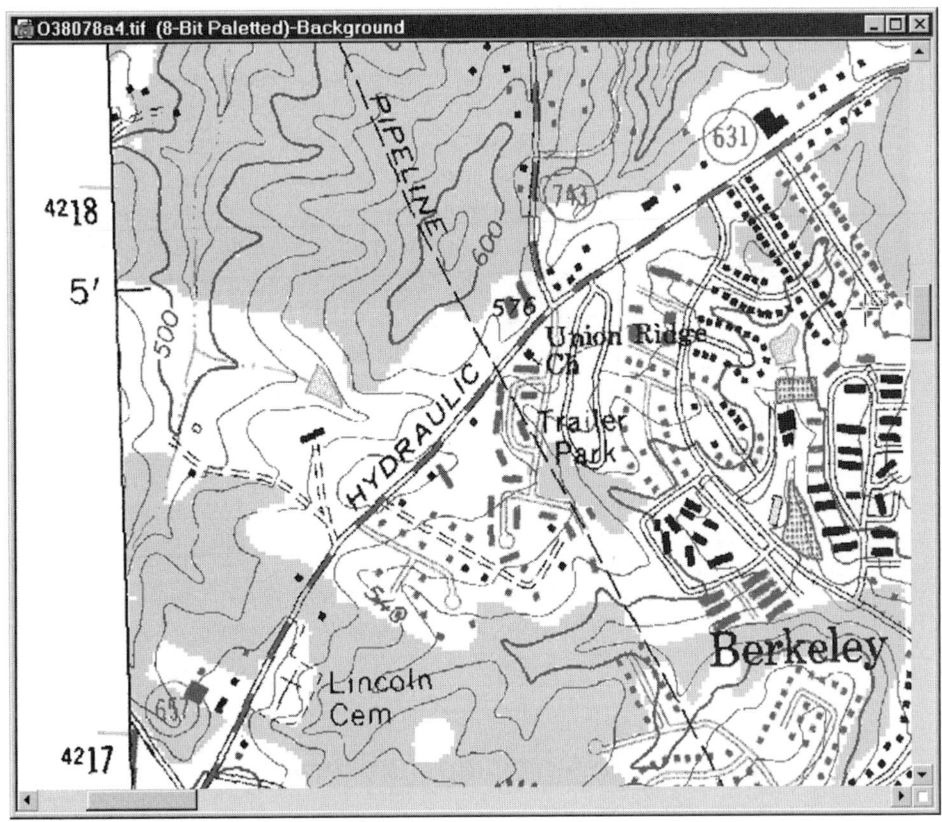

Topographic sheet accessed as a hyperlink.

Notice that the hyperlink, in this case, opened a paint program, not ArcMap, to display the map. This is because any document you create a link to will open a software program your computer thinks created the document. In other words, if you link to a Microsoft Word document, the document will display using Word. If you link to a web site, an Internet browser will open to display the site. In this case, the map is stored as a TIF file, a format commonly used by graphics paint programs. The map will display using one of these programs. At this point, you may print the topographic map from the paint program for distribution to the emergency response team.

Cartographic Principles

This section introduces the ArcMap user to some basic cartographic principles. Well thought out, effective, and interesting map elements such as line weights, colors, and symbol selection will help your map convey its information quickly and effectively. Keep in mind that the art and science of cartography is its own field of study, and this section covers only the basics. The cartographic process using ArcMap can be thought of as involving three primary phases: conceptualizing the map, classifying data, and designing the map.

Although maps contain base data (used to locate features on the earth), those used in decision-making processes generally contain thematic data, which shows the distribution of given items across an area. Most of the maps you will make with ArcMap fall into the latter category. How they are designed can affect the decisions that result from their use.

Conceptualizing a Map

When you conceptualize you form a general notion or idea, which is what you are doing in the initial phase of cartographic design. To conceptualize your map, first determine its purpose and audience, and then develop its content and features.

Determining the Purpose of a Map

The first step in designing any map is to determine its purpose: what will it show, and where? For example, say you need to make a map of rainfall totals across a sub-Saharan Africa study area. In this case, the "what" and "where" are easily determined.

- *What* becomes the total amount of rainfall in a given unit (say millimeters) within a given amount of time.
- *Where* could be the seven countries of Mauritania, Mali, Burkina, Sudan, Niger, Ethiopia, and Chad.

When developing the purpose, it is useful to learn the habit of completing the sentence "The purpose of this map is to show ________." The filled-in blank can even become the map's title. For this example, the sentence might be completed as follows.

```
The purpose of this map is to show rainfall totals in millimeters across the study area
between June 1 and July 10, 1999.
```

Some purposes can seem deceptively well defined, yet need a more thorough breakdown. For example, examine the following statement: "The purpose of this map is to show famine risk areas in the study-area countries at war." This can seem well thought out, but may actually need more work. What, for example, is meant by "famine risk"? Do you already have a coverage named *famine risk* or do you need to assemble one? Can you defend your choice of variables that combine to determine this risk (in other words, are there standard procedures for combining variables such as rainfall and crop production to get "famine risk" areas)?

How are you defining war? As conflict formally endorsed by the countries' governing bodies or as all skirmishes? Be aware that some definitions can be politically touchy. Not every purpose needs to be analyzed this thoroughly, but you as the map compiler have final responsibility for all decisions. Included in the worksheet is a "purpose" statement, and room left for definitions if they are needed.

Audience

The second phase of the ArcMap cartographic process, which in many ways is related to the first and subsequent steps, is to determine who will be using your map. Ask yourself who the audience is and what they know. Do they know, for example, the jargon? If not, you may have to change the titles and labels, or include explanations of them on the map. What do you anticipate their level of understanding to be?

For example, if you show rainfall totals and crop production, can your users infer famine risk areas, or do those regions need to be shown and labeled as such? Is your audience familiar with the area shown? If not, perhaps an inset map should be included. What natural and artificial (man-made) features should be included to alert your user to the map's location (latitude/longitude marks, country capitals, oceans, country boundaries, and so on)? Such decisions are discussed in the sections that follow. Be aware of the limitations, if any, your audience may place on your map design.

Decisions made with respect to your audience also affect the layout and production of your map. How big will the entire page be; for example, 8.5-inch by 11-inch, or wall size? Size also helps determine the level of detail you can show. If your map is to be included in a report, it may be limited to 8.5-inch by 11-inch or 11-inch by 17-inch. These relatively small sizes mean the map must contain less complex data. Where will the map be used? If it will be used outdoors to determine distances and areas, it needs to be produced with fade-proof inks on a scale-stable, weather-resistant base.

Selecting Features for Inclusion

Map features can be roughly classified as either physical or man-made. Generally, physical features refer to mountains, rivers, forests, deserts, canyons, and so on, and man-made features include such things as cities, railways, and roads. The two classes can overlap, however, as in the case of a lake enlarged to create a reservoir. Decide which features will enhance your map's purpose. For the rainfall totals map, physical features such as mountains and deserts are important, as they can affect or be affected by weather patterns.

Man-made features such as railways and highways would not add much to the understanding of your map's purpose. Capital cities, however, could provide a useful orientation for the reader. Man-made features such as political boundaries often overlap with physical features. Questions arise such as what boundaries should be included: international ones only, or both international and regional? If both, what symbols will be used to differentiate them?

Classifying Data

The subject matter of your map is simply what it is trying to show. It must vary across an area or your map will have little of interest to convey. These variations must be classified logically, or your map will have no inherent meaning. In the rainfall totals example, the totals might be classified by every 10 millimeters, such as in the following.

```
0-10 = Color A on map

11-20 = Color B on map

21-30 = Color C on map
```

Notice that the numbering for the categories does not overlap (that is, they are not 0–10, 10–20, and so on), but are mutually exclusive. Because they are all categories of the same variable, colors A, B, and C could be different shades of the same color family (blues and purples, for instance). Intuitively darker colors imply more: the darker the sky, the more clouds are in it. Thus, the highest rainfall total could have the darkest color shade and the lowest total the lightest shade.

Different types of data require different methods of classification and are best displayed using various symbols. The sections that follow delineate various options involved in two methods of classifying data: by dimension and using measurement data.

By Dimension

0 dimension: Point data, such as cities on a small-scale map. Although some data are truly point data, other data are collected at points and extrapolated to areas (such as rainfall totals).

1 dimension: Line data, which have length but no width, such as country boundaries. Roads and other transportation routes are classified as line data because they are negligibly thin (they have no significant area).

2 dimensions: Areas having length and width. These areas, such as famine risk zones and land-use blocks, clearly take up space on the map.

3 dimensions: Areas having length, width, and height. Maps of ocean depths or mountain heights, or crop yields over an area, are good examples.

Measurement Data

Nominal data: Categories that are either equal to or different from each other. There is no relationship between categories.

Ordinal data: Has an implied order. That is, one can tell if something is more or less than another, but not by how much.

Interval data: Can show if the categories are equal, not equal, greater or less than, or positive or negative relative to each other. Mathematical functions cannot be performed on interval data, however.

Ratio data: Just like interval data but incorporates a zero. Examples would be temperature, latitude, and longitude.

Designing a Map

It is during the layout process that the cartographer's creativity can come into play. Now that the purpose, audience, and content have been determined, you are ready to create a graphic image that is both efficient and pleasing to the eye. Although one cannot quantify the creative process, there are several standard considerations to take into account. These are discussed in the sections that follow.

Map Page Size

How big will the map page be? 8.5-inch by 11-inch or wall size? Will it be horizontal or vertical? Generally speaking, if the area you are mapping has a greater east-west length than north-south height, the map page should be horizontal.

California, for instance, looks better on a vertical page with its great north-south length. With a little creativity, however, California could be put on a horizontal page by balancing the other map elements around the state outline.

Standard Inclusions

There are a few standard elements that most cartographers agree should be included on every map. All maps need a title (stating the purpose of the map), a scale (showing the map's relationship to reality), and a key or legend (that the map viewer can use to interpret the symbols and colors).

The name of the cartographer or the organization responsible for the map must be included to give the map validity, and a statement as to the source of the data not only gives credit where credit is due but adds further credibility to the map. Finally, a compass rose or North arrow should be included to help orient the map user (many organizations have a custom compass rose), and a map page border adds a polished touch to the finished graphic product.

Putting It All Together

You may want to sketch the basic layout of your map page first, and place your map graphics, title, North arrow, and so on. Balance the items around the true center of the page and be aware that the overall shape of an item can "point" in a direction, either leading the viewer's eye to another map item or off the page.

You want your map to hold the viewer's attention. Therefore, take care to have your items "point" to other items. Attempt to balance "white space." That is, the portion of the map page not covered by map items should be evenly distributed across the page. Do not crowd your map elements. Either make them smaller or increase the size of the page if necessary.

Choosing Line Weights, Colors, and Symbols

This section is a brief summary of traditional cartographic selections of line weights, colors, and symbols. Generally, the more important the line feature is to achieving the purpose of your map, the darker and thicker is its line weight.

Traditionally, the map page border gets the heaviest line, with international boundaries, national boundaries, and city outlines having progressively lighter weights. For instance, the map page border would be in a heavy, black line, international boundaries in a thinner black line, and a country's internal boundaries shown with an even thinner black line, a black dashed line, or perhaps a thinner green line. Dashed lines are generally used to indicate a greater degree of cross-transit. Therefore, national boundary lines are often shown as dashed, whereas international boundaries are solid.

Hydrologic features are generally shown in blue. However, if you are creating a black-and-white map, be sure that rivers, streams, and similar elements can be easily distinguished from transportation lines. Intermittent streams are generally shown as a dashed line. Lines for streams should be thinner than those indicating rivers. Transportation lines are often shown in a different color and/or with a different line symbol. Railways generally have a hatched line, whereas highways can have a double or a thicker line (indicating more lanes). Roads are then shown as thinner, single lines.

If distinguishing between types of roads is an important part of your map, refer to a road atlas or a topographic map for ideas on the use of road symbols. Note that the foregoing recommendations are generalities. If, for example, your map is intended to show railways of a country, the line symbol for railways would be darker and thicker than other symbology, giving that feature prominence over other linear features such as highways. Table 3-1, which follows, is a chart suggesting line weight choices that may be helpful to you in the map designing process.

Table 3-1: Features and Line Symbols

Feature/Line	Thickness	Color	Symbol
Map page border	Thickest	Black	Solid
International boundaries	Thick	Black	Solid or dashed
Internal boundaries	Medium	Black	Solid or contrasting dashed
City limits	Thinnest	Black	Solid or contrasting dashed
Rivers	Thickest	Blue, black	Solid
Year-round streams	Medium	Blue, black	Solid or dashed
Intermittent streams	Thinnest	Blue, black	Dashed
Railways	Dependent on importance of feature	Contrasting	Hatched
Highways	Thickest	Black	Solid or double
Roads	Thinnest	Black	Solid, contrasting dashed, or other symbol

Designing Black-and-White Maps

Although black-and-white maps may seem limiting, they can actually be used to convey a lot of information quickly and effectively. Two main thoughts must be kept in mind: replace colors with patterns and eliminate extraneous features.

The human eye can only distinguish about eight shades of gray. Adding black and white gives you a possible total of ten shades. If you are filling in areas, however, it becomes difficult to distinguish between gray shades that are close on the shade scale. It would be much better to choose three or four shades of gray and several patterns.

Say you have three shades of gray, black and white, and four patterns. In this case, if you were designing a map with one variable (say, crop type), you could have up to nine categories of crop type represented that did not repeat either a shade (including white and black) or pattern. This approach works best with nominal data or data with two dimensions. You can also show two variables quite effectively with a black-and-white map, as long as there are a small number of categories for each variable.

For instance, if you are showing soil types and crop yield, you can have one pattern per soil type and one shade (the darkest shade indicating the most, the lightest the least) per crop yield category. Again, the number of categories for each variable must be small; this obviously would not work well with 20 soil types and six categories of crop yields. Notice that one variable (soil type) is nominal or two-dimensional and the other (crop yield) is ordinal (or interval) or three-dimensional. It would be quite difficult to have two ordinal or interval variables, because both the shades and the patterns would have to have a darker to lighter scale.

Finally, be sure to eliminate features that are not necessary in achieving your map's purpose. Although it is nice to include features such as rivers, railways, and so on, they can clutter a black-and-white map, especially if the map has prominent linear features. If you include lakes or oceans, it is a good idea to shade them in order to distinguish them from white space, but be sure not to use that shade or a very similar shade for one of your categories. In addition, use white as a category shade only if the entire map will be shaded, so as not to confuse white space with a variable value.

The foregoing suggestions should enable you to create effective black-and-white maps that are easily and inexpensively reproduced. If you are making a map that needs immediate impact and must grab the reader's attention, however, color representation of variables is the most efficient and effective method.

Color Maps

When it comes to cartography, the value of color cannot be underestimated. Since the earliest days of mapmaking, color has been used to differentiate features, provoke reactions, add interest to catch the map reader's eye, and to take mapmaking beyond its role of graphic representation and into the realm of art. Another reason is that if the map is going to be used in a decision-making situation, color will get its point across very quickly, and has the added benefit of holding the reader's eye longer. One problem that has remained with cartographers through the centuries, however, is reproducing color maps. Even with today's technology, color map production can be costly and time consuming.

Because color maps cost more to reproduce than black-and-white maps, it pays to use color wisely. One main reason color should be chosen over black and white is that the map displays several variables. For example, you may have eight variables, each with four categories. Assigning each variable a color and each category a shade (perhaps of the same color) allows you to clearly display at least 32 combinations.

Although most of us do not realize it, color can speak to us louder than words. Someone driving a red car might be saying something about his or her personality: he or she likes attention, is a risk taker, or so on. Mourners wear black. We use phrases such as "in the pink" and "singing the blues." Colors definitely provoke an emotional response, and when it comes to making a map, colors can highlight certain variables over others, help support a point of view, and generally add interest and hold the reader's attention.

This section provided you with a glimpse into the discipline of cartography. Representing where things are on the earth is one on the fundamental means of communication. Although this can now be done easily and faster than ever, the advantages technology offers cannot replace the human thought behind a well-made map.

ArcMap Menu Commands

As you have seen, ArcMap is reasonably easy to use and is an excellent means of creating maps quickly. This section provides a reference to the menu commands available in ArcMap. Each pull-down text menu is shown, and its command functions described. You may want to follow along on your own computer.

File Menu

The File menu, shown in the following illustration, gives you access to some of the basic functions of ArcMap, such as saving and opening existing map files. It also contains features for exporting a map to a file format suitable for posting on a web page or saving to the Windows Clipboard for insertion in a program such as Microsoft Word. The options within this menu are described in the material that follows.

File menu.

New: Select this option to start a new map composition. Choose from an existing ESRI-provided map template or create your own map from scratch.

Open: Opens an existing map composition file.

Save: Saves a map composition file.

Save As: Saves a map composition file under a different name.

Add Data: Adds new spatial data to the map composition. This functions the same as clicking on the Add Data button on the toolbar.

Page Setup: Select this command to select a printer for printing. You can change paper orientation from landscape to portrait, and you can specify the size of your map as it will appear on the page. Some plotters can print on paper 40 inches wide, for instance, so use this to specify large dimensions for your paper map.

Print Preview: Displays on the computer screen a version of the map as it will appear when printed.

Print: Sends the map to the printer or plotter.

Map Properties: Select this command to enter some descriptive information about your map, such as who created it, when it was produced, what its purpose is, and so on. This information is helpful when you examine maps created some time ago or by someone else.

Export Map: Exports the map to a number of different graphics formats, including JPG, TIFF, BMP, EPS, and PDF. Once the map is saved as a graphics file, it can be placed on a web page or used by other programs such as Microsoft Word.

Copy Map to Clipboard: Makes a copy of the map and places it on the Windows Clipboard. You may then place the map in another program such as Microsoft Word.

Edit Menu

The Edit menu, shown in the following illustration, provides commands used while creating a map composition. These commands allow you to "undo" an action such as resizing text or moving a

Edit menu.

North arrow to a new location. You can perform typical tasks such as copying, cutting, and deleting items in the map composition.

Undo: "Undoes" the last command issued. It is helpful when you either make a mistake or move something such as a title to a new location, and then decide that it looked better in its original place.

Redo: "Undoes" your "undo." If you use "undo" to move a text back to its original place and then decide it really looked best in its new location, use "redo" to move the text back to its newer position.

Cut: Removes the selected graphics element.

Copy: Copies the selected graphics element to the Clipboard. Use Paste to place this copy on the map.

Paste: Places the previously copied or cut element on the map.

Paste Special: Lists all graphics elements that have been cut or copied. Choose which one you want to paste on the map.

Delete: Deletes selected graphics elements from the map.

Find: Use this command to search for a user-supplied string of text through all databases associated with the data layers in the table of contents.

Select All Elements: Selects all elements in the map. This is useful when you want to delete everything or copy or move everything to a new location.

View menu.

View Menu

The View menu, shown at left, contains commands for controlling how data are displayed on the screen, such as zoom in and out commands, as well as how the ArcMap interface is configured, such as which toolbars are shown. The options within this menu are described in the material that follows.

Data View: Displays the spatial data used to create your map. No other map information (such as legend, title, or North arrow) is shown. This view is handy when you are querying data, but do not need to actually create a paper map.

Layout View: Displays the data used for your map, as well as all other map elements, such as legend, scale bar, and title. Use this view when creating a map to be printed.

Zoom Data: Use this command to zoom in and out of the data you are displaying. It works when you are in either Data View or Layout View.

Zoom Layout: This command is also used to zoom in and out of a map, but works only when you are in Layout View. Use it to quickly zoom to a view of the entire map, or to just portions of the map. It is similar to Zoom Data, except that this command only allows you to zoom to either the entire page or to the following percentages of the page: 25%, 50%, 75%, or 200%.

Bookmarks: Bookmarks may be set so that you can always return to a specified area on the map at any time. For instance, you may want to zoom in on a portion of the map again and again. Bookmarking it allows you to come back to that view easily.

Toolbars: Use this command to turn ArcMap toolbars on or off.

Table of Contents: This command turns on or off the table of contents pane. Once you have added all data layers to your map, turning off the table of contents pane gives you a larger screen with which to work.

Status Bar: Turns the status bar on or off.

Overflow Labels: This command is useful when you are labeling features in a map (for example, labeling streets with their names), and some labels overlap. ArcMap takes all labels (street names) that overlap and stores them as annotation. Stored as annotation, you may then manually place the labels so that they do not overlap others.

Scrollbars: Hides or shows the scrollbars along the sides of the map canvas pane. If you zoom in closely on a map, portions of the map may appear off the screen. Turn on the scrollbars so that you can use them to pan around the map, or turn them off to have a slightly larger screen to work with.

Rulers: Turns rulers on or off.

Guides: Guides are useful for snapping freehand graphics elements to a grid. When snapping guides are turned on, things you freehand draw are snapped to the nearest grid point.

Grid: Turns on a grid to help you line up features. It is useful, for example, when you need to left justify lines of text. Turning on the grid will show you where to begin each line of text.

Insert menu.

Insert Menu

The Insert menu, shown at left, is used to insert common map elements such as titles, scale bars, and text in the map. It can also be used to add a data frame to the table of contents.

These commands add North arrows, scale bars, and scale text. Each command opens a new window that lists many styles of arrows, fonts, and so on. The Insert menu contains the following options.

Data Frame: Adds another data frame to the table of contents. Group data layers into data frames to easily control which features are shown and which are not.

Title: Select this command to add a title to the map.

Text: Use this command to add text anywhere on the map.

Legend: Adds a legend to the map.

North Arrow: Choose this command to add a North arrow to the map.

Scale Bar: Adds a scale bar to the map.

Scale Text: Adds a scale as text instead of as a scale bar. An example would be "1 inch equals 24,000 feet."

Picture: Use this command to insert a graphics file, in a common format such as JPG or TIF, to a map.

Selection Menu

The Selection menu, shown in the following illustration, contains commands that allow you to select features from the data layers displayed on your map. You can, for example, select all major roads from a coverage of roads. You can interactively use the mouse to select individual features on the map to see how they are labeled. Both ArcInfo coverages and ArcView shapefiles can be used. The options within this menu are described in the material that follows.

Selection menu.

Select By Attributes: Displays the feature attribute table associated with your coverage. You may select features from the coverage based on a criterion you specify, such as STREETNAMES = "Hessian."

Select By Location: Selects features in a coverage based on their location to an already selected set of features.

Select By Graphics: Use this command to select features using the mouse. You may select features one at a time or hold down the mouse key to create a box that selects all features within the box.

Zoom To Selected Features: Select this option to change the map extent so that it encompasses the currently selected features.

Statistics: This command is used to create summary statistics on the currently selected set of features.

Set Selectable Layers: Displays a list of features currently in your map composition. You can specify which are available for selection and which are not.

Clear Selected Features: Clears the selected features and makes all features available for new selection commands.

Interactive Selection Method: This command is used to modify the way features are selected. By default, each selection command creates a new set of selected features. Use this command to specify that items will be selected from the currently selected set or that new features will be added to the currently selected set.

Options: This command is used to change the color of selected features (blue, by default) and to specify the selection search distance when using the mouse. The default search distance is 3 pixels. You can change this so that it is more precise (less than 3 pixels) or less precise (more than 3 pixels).

Tools Menu

The Tools Menu, shown in the following illustration, contains a collection of commands used to create charts and graphs, perform simple editing of ArcInfo coverages, write macros using Visual Basic, and customize the ArcMap toolbars. The options within this menu are described in the material that follows.

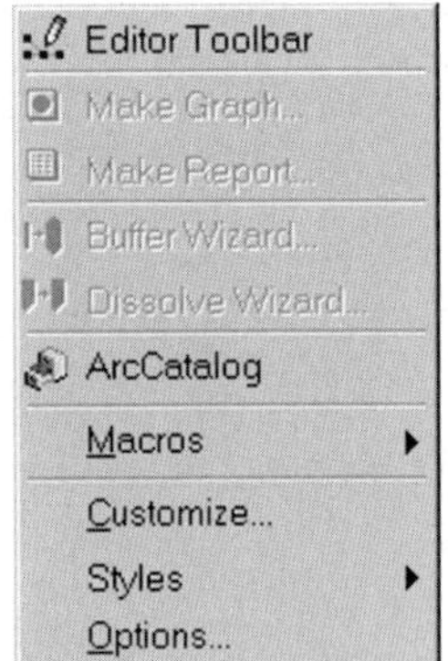

Tools menu.

Editor Toolbar: Choose this selection to display the Editor toolbar. The tools here are used to perform basic editing tasks on coverages, such as adding and deleting arcs, points, and polygons. They are a subset of the ArcEdit commands, which are explained in detail in Chapter 6.

Make Graph: Use this choice to make a simple graph from the currently selected features. You are prompted to choose an attribute from a coverage or shapefile (depending on which you have selected) for the X and Y axes. A graph is then constructed in a new window and may be cut and pasted into a word processing document or into the map canvas.

Make Report: This selection starts the Report Generator, which is described in exercise 3-2.

Buffer Wizard: This command runs the Buffer Wizard, which you can use to create a buffer around the currently selected set of features. The Buffer Wizard is described in the reference section of Chapter 4.

Dissolve Wizard: This wizard aggregates features that have the same value in an attribute. For instance, you can group all polygons in a zoning coverage that have a value of "R-1" in an attribute called ZONING. The Dissolve Wizard is described in the reference section of Chapter 4.

ArcCatalog: Select this command to start ArcCatalog

Macros: Use this selection to record a macro using Visual Basic or to bring up the Visual Basic editor. See ArcMap's on-line help for a discussion of using Visual Basic.

Customize: Select this option to customize the toolbars in ArcMap. Chapter 2 contains a discussion of how to customize the user interface in Arc Desktop.

Styles: Use this selection to access the Style manager. The Style Manager contains a number of ESRI-supplied symbol sets for particular map themes, such as conservation, crime reporting, utilities work, and so on. Selecting a style here makes symbols commonly used in such map themes available for use in ArcMap. Adding a North arrow, for instance, will show North arrows appropriate to conservation mapping.

Options: This selection controls how information is displayed on the screen. Use it to control the following features: the image formats that can be displayed, whether to resize data in Arc Canvas if you increase or decrease the window size, and whether to display the "Getting Started" window each time ArcMap starts.

Window Menu

The Window menu, shown at left, contains three tools used for viewing features in the ArcMap canvas. The first two tools open new viewer windows that aid you in examining your data. The last tool is used to switch between different windows. The options within this menu are described in the material that follows.

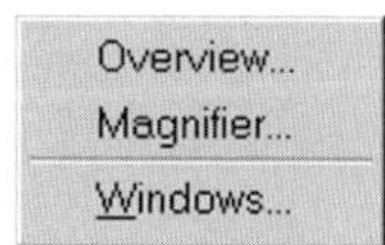

Window menu.

Overview: Opens a viewer window to let you work on your map.

Magnifier: Opens a map viewer tied to your crosshairs, so that you get a magnified view of the map canvas in a separate window as you move the cursor about the canvas. It only works when you are viewing the map composition in Data View (View ➥ Data View).

Windows: Displays the currently open windows. Select one to bring it to the foreground.

Summary

ArcMap is a powerful new addition to ArcInfo, combining a Windows NT graphic user interface with the map creation capabilities of ArcInfo. This chapter introduced you to ArcMap's capabilities. The first section discussed the main screen layout, including its toolbars and icons.

Exercise 3-1 stepped you through displaying, querying, and generating a simple report using ArcMap. Next, displaying multiple layers and using ArcMap templates were discussed. Exercise 3-2, building on the first exercise, showed you how to use multiple data layers and a map template to create a map composition, and how to incorporate hyperlinks. Finally, the text menu reference section covered ArcMap's available pull-down text menus.

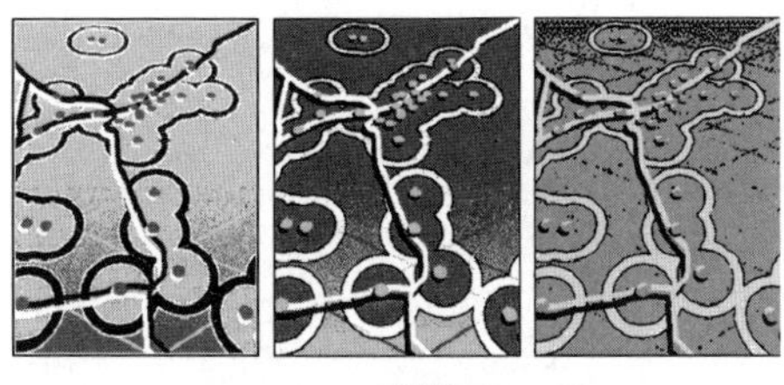

Chapter 4

ArcToolbox

ArcInfo 8.0 has over a thousand commands. How do you find which one to use? And what pieces of information are needed to execute the command once you find it? The new ArcToolbox module provides access to many of the commonly, and not so commonly, used commands in ArcInfo. The toolbox is organized into four sections called tool sets, the commands within each section performing related functions, like the other Arc Desktop modules ArcCatalog and ArcMap. ArcToolbox's easy-to-use pull-down menus are fully customizable, and you may even add your own AMLs, as usable tools.

There are important features tucked into Arc Toolbox, as well, such as the ability to establish connections between coverages and databases (called a "Relate"). Access of the Remote Processing Manager, a new and very powerful feature in ArcInfo 8.0, is also done through the pull-down menus. This chapter includes the following.

- Description of pull-down menus
- How to use the new menu-based commands in ArcToolbox
- An exercise that demonstrates how to use ArcToolbox
- A complete reference section that describes each command and its syntax in ArcToolbox

The main goal of ArcToolbox is to provide an easy-to-use front end to the commands usually found at the Arc prompt in ArcInfo Workstation. Access to these commands is now simpler and more intuitive.

ArcPlot and ArcEdit commands are not available in ArcToolbox, which means that none of the Toolbox commands will draw a map or allow you to perform interactive editing of coverages. These modules' commands are not available because a graphics window is required for using either ArcEdit or ArcPlot.

ArcToolbox does not allow a graphics window to be opened because of an additional new feature, the Remote Processing Manager, which serves as a means of executing Toolbox commands on a remote computer. The remote computer could be more powerful and faster than your own, and not displaying graphics would speed up processing even more.

The Menu System

The ArcToolbox main screen, shown in the following illustration, shows two pull-down menus, Tools contains options for searching for information, establishing relationships between data files, customizing the look of ArcToolbox's window, accessing a remote processor, and so on. Each of these options is discussed in the sections that follow.

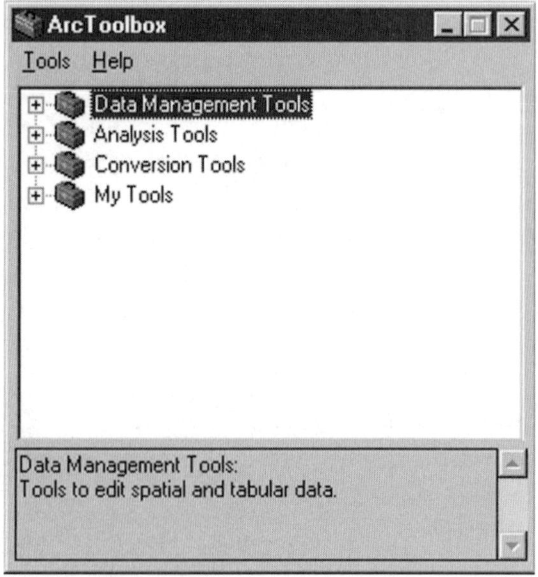

ArcToolbox main screen.

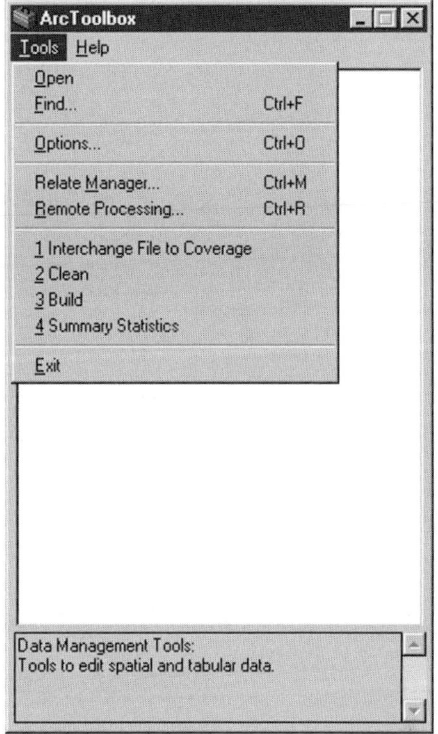

ArcToolbox Tools menu.

Open

This is the simplest command in the menu. It opens the toolbox or tool you have currently highlighted in the ArcToolbox window. You can also open a toolbox or tool by simply double clicking on it with the mouse.

Find

The Find command can be your greatest friend when learning ArcInfo. It searches via name, command, or key word for references to any word you type in. The following are some examples of how the Find command performs these functions. Follow along on your computer.

Example 1: Exploring the Find Tool

Say you want to know which commands address working with topology (topology refers to the storage of spatial relationships in

ArcInfo coverages; see Chapter 1). First select Tools ➡ Find. The Find window, shown in the following illustration, appears. The window contains three tabs: Name, ArcInfo Command, and Keyword. Type *topology* in the Name text box and then click on the Find Now button.

Find window.

You can see that when topology is entered in the Name tab, shown in the first of the following illustrations, the program finds two references. These are instances where the word *topology* exists in the name of an Arc tool. Now double click on the first reference to automatically start that tool. You will see a discussion of the Topology tool set, shown in the second of the following illustrations.

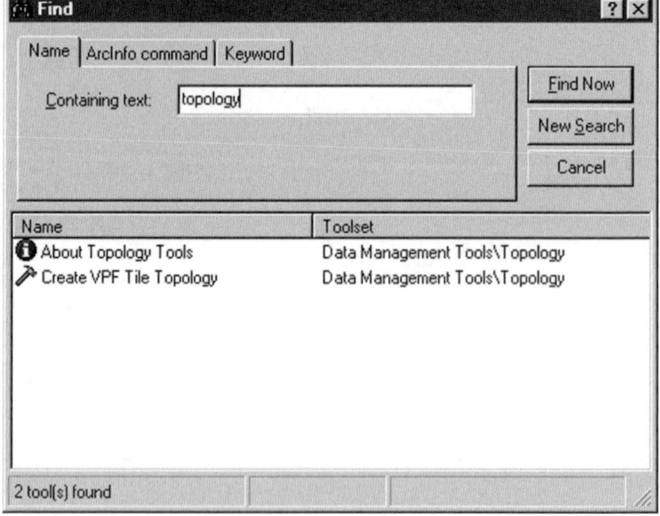

Name tab.

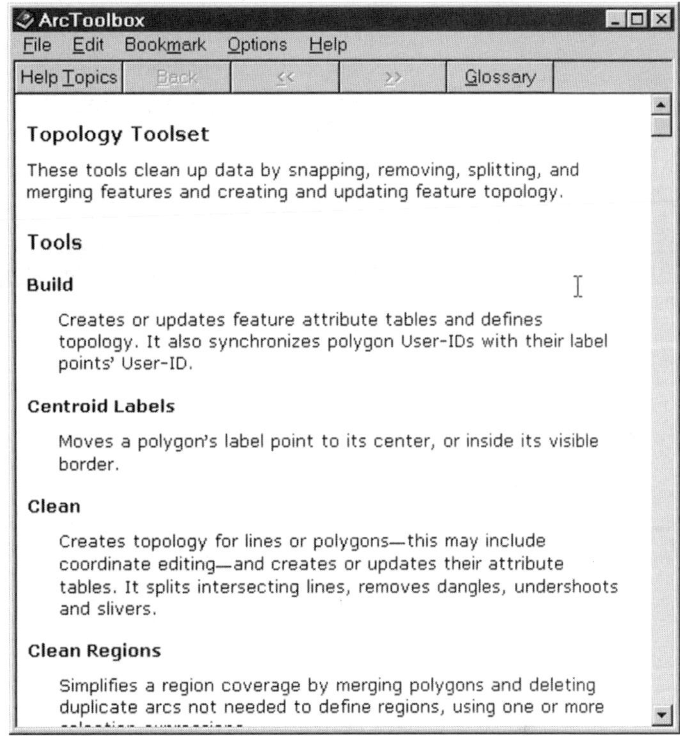

Typology tool set descriptions.

Now dismiss this window and try typing *topology* in the ArcInfo command tab text box, shown in the following illustration, and clicking on Find Now. There are no items listed; therefore, Find could not make a match. What this means is that there are no ArcInfo *commands* called "topology."

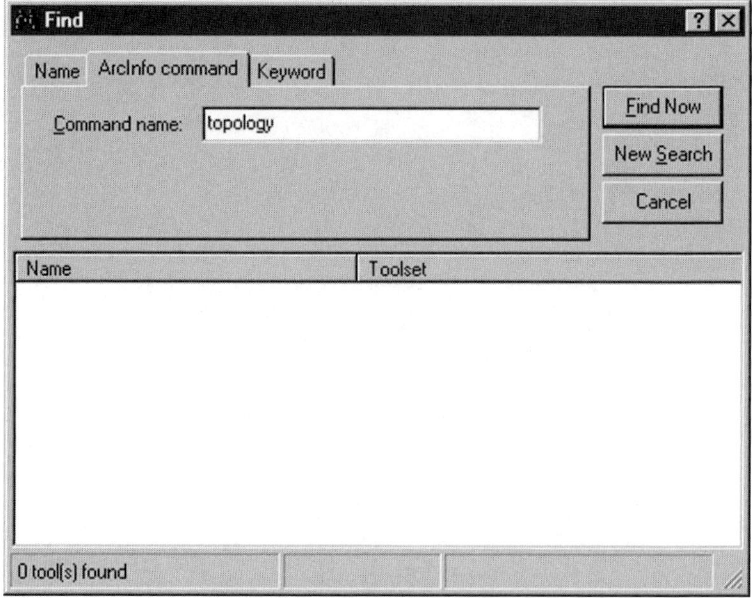

ArcInfo command tab.

Finally, try searching for *topology* using the Keyword tab. Many items are listed, most of which do not contain the word *topology* in their heading. Why did ArcInfo find them? Because either in the tools themselves or in the discussion about how to use the tools, the word *topology* is present. The commands listed may or may not create or update topology, but there is at least a mention of it.

Example 2: Tracking Down a Command

Suppose you used an ArcInfo command a long time ago and remember something about it, but not enough to find it again easily. You think the command was called "update," or perhaps the word *update* was in its description. The following examines how Find deals with such a search.

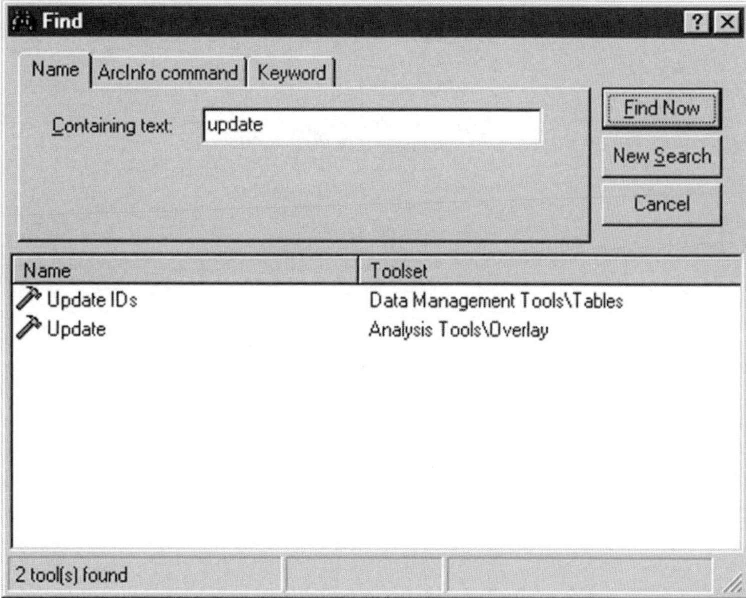

Using Find for help with the Update command.

Select Tools ➥ Find once again, and type *update* in the Names text box. Click on the Find Now button. Find has located two references to "update" using the Name tab. Notice that each instance has listed to its right the tool set where the instance was found. Once you have learned ArcToobox, you will know the Data Management tool set is usually used to update topology. The Analysis tool set contains commands that perform analysis functions on data. Therefore, there is actually more than one meaning to "update" as defined by ArcInfo.

Now, type *update* in the ArcInfo Command tab text box, shown in the following illustration. Entering *update* narrows the search a bit. Remember that you are looking for a command called "update"; ArcInfo finds one instance, and this is the only command that matches what you are looking for. Double clicking on the entry for the instance in the Find window will start its command dialog box.

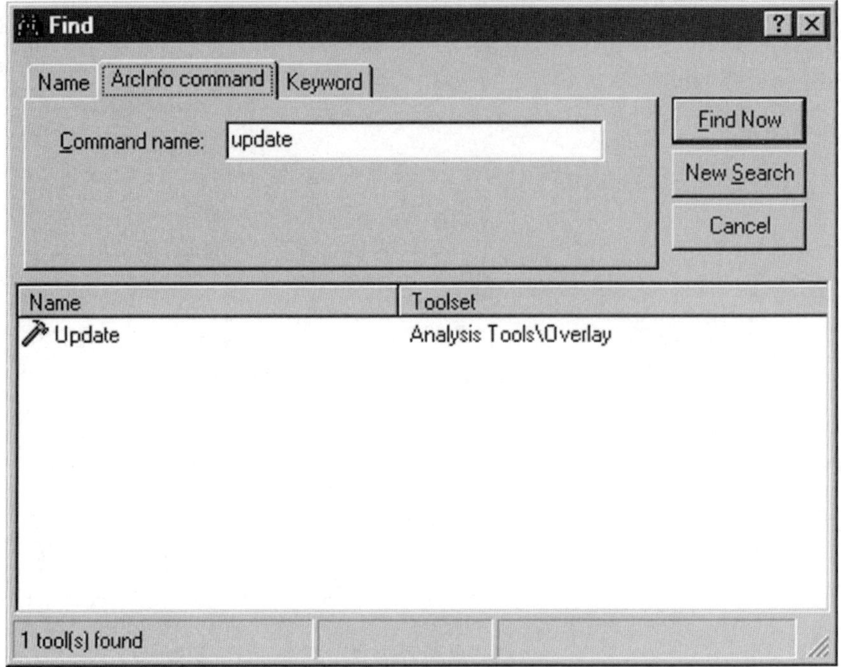

Entering update *in the ArcInfo command tab.*

To finish this search, try entering *update* in the Keyword tab, as shown in the following illustration. There will be two entries listed, one for the Update command and one for Erase. The reason Erase is listed is because there is a reference in the discussion of this command to the Update command. Find has searched for keywords and cares nothing about context!

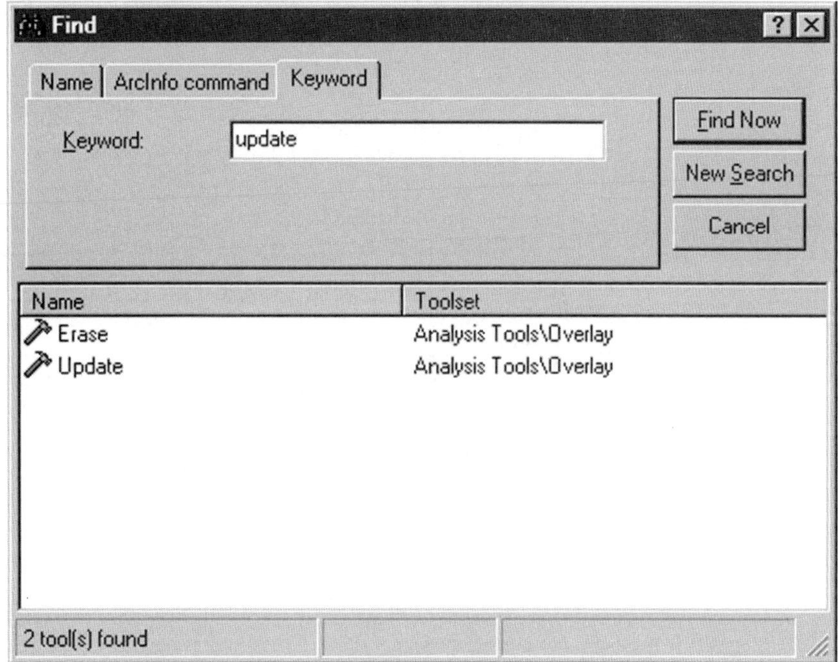

Entering update *in the Keyword tab.*

Example 3: Searching with Key Letters

Occasionally, you may not know how to spell a command or a keyword. Find lets you enter just the first few letters of a word and it finds commands or keywords that begin with those letters. For example, suppose you want to use a command called "clean" but have forgotten how the exact command is worded. Try typing just the first two letters in Find's "ArcInfo command" tab text box, as shown in the following illustration.

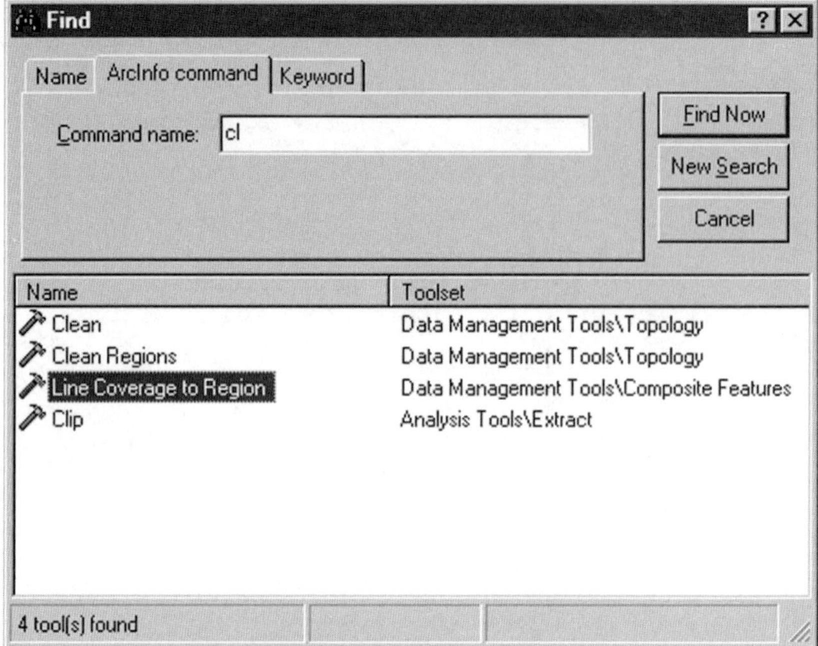

Searching the "ArcInfo command" tab with "cl."

Entering *cl* in the ArcInfo command tab finds a number of commands that begin with "cl," including Clean. Find's ability to search using just a few letters of a command is a great way to find commands or information that would take too long to uncover using the on-line help.

Options

Two very different tasks are brought together under the rubric Options. First, the View tab controls the look of ArcToolbox's main window. Second, the Precision tab controls the precision of any new data sets created during your current ArcInfo session. The Options menu is shown in the following illustration.

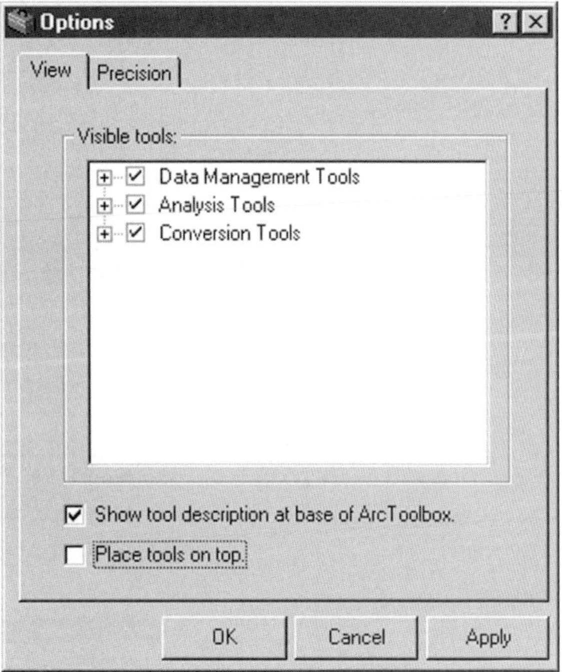

Options menu.

First, examine the ways in which you can customize how ArcToolbox looks. Notice that the three main Toolset categories (Data Management Tools, Analysis Tools, and Conversion Tools) each has a "+" as well as a check box next to it. Click on the "+" and you will see the subcategories of tools held in each Toolset category. By clicking on each successive "+", you get to the individual commands and wizards that constitute ArcToolbox. Click on the minus symbol to remove them from view.

Now, suppose that there are commands you will never use and want to remove them from ArcToolbox. Shown in the first of the following illustrations, simply use the mouse to click on the check box, toggling the check mark off. Next, click on the Apply button. Now that tool set will not appear in the ArcToolbox main window (second of the following illustrations). To turn the tool on again, just click the check box and Apply again. Use this same approach to turn on or off an entire tool set.

Turn off the check box to no longer display Conversion Tools.

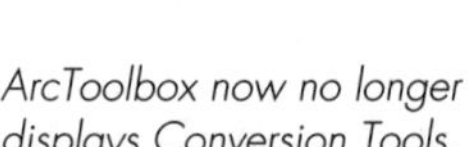

ArcToolbox now no longer displays Conversion Tools.

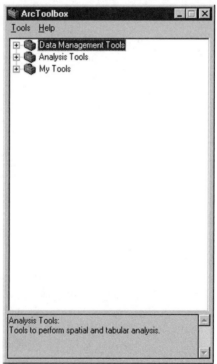

As shown in the previous illustration, there are two additional check boxes located at the bottom of the View window control. The first check box, "Show tool description at base of ArcToolbox," gives you a brief description (when turned on) of the highlighted command. The second check box specifies whether the main ArcToolbox menu should be on top of all other pop-up menus or not.

Data set precision is also controlled within the Options menu, (see following illustration). Data sets can be coverages or grids. There are two methods of setting precision, one for any new data sets created and one for coverages created as a result of an analysis (say, joining two coverages).

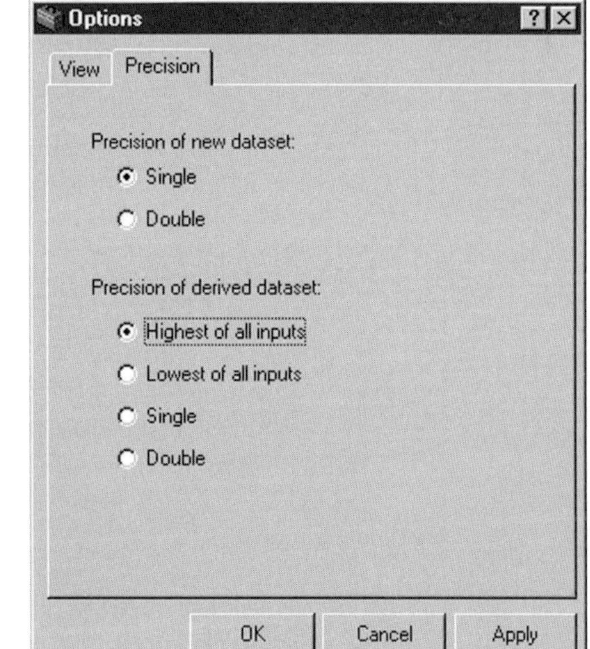

Setting coverage precision.

For new data sets, simply select either single or double precision in the menu window. For data sets derived from tasks such as spatial overlays, you have more choices. You may still choose either single or double precision for your new data sets, or you may choose that new data sets use either the highest or lowest precision of the coverages used to create the new one. That is, you may overlay two coverages, one stored in single precision and one in double precision. If you selected "highest of all inputs" here, the new coverage would be in double precision. If you selected "lowest of all precision," the new coverage would be in single precision.

Precision

What is precision and what is the difference between single and double precision? Precision is how accurately ArcInfo stores spatial coordinates. Single-precision coverages store numbers (up to seven significant digits) used as geographic coordinates. This means that you can have 1-foot accuracy in your data sets if your coverage is up to 1,111,111 feet wide. Although this might seem to cover almost anyone's mapping project, suppose your coverage is stored in, for example, a UTM (Universal Trans Mercator) projection. In this case, your coordinates may be in the millions. In such a case, you will need to use double precision, which uses fifteen significant digits. This will be adequate to accurately map any location on earth.

So why not always use double precision for your data sets? Because along with increased accuracy comes increased storage requirements; that is, your coverages and grids will take up more space on your hard drive in order to store those bigger numbers. Interestingly, ArcInfo does not take any longer to manipulate or perform analysis on double-precision coverages because the software has used double precision internally for years. The following are some rules of thumb as to when to use single precision.

- Data sets do not require high levels of positional accuracy (e.g., for soils, forest, or geologic maps)
- The areal extent of a coverage or grid is fairly small
- Data sets do not have many curves or vertex-rich arcs

The following are examples of situations in which you would use double precision.

- Projection used with data sets requires that coordinates be in very large numbers
- Positional accuracy is very important (e.g., for parcels, roadways, or boundaries)
- The areal extent of your study area is very large

Relate Manager

One of the core functions of any GIS is its ability to work with data from different sources. ArcInfo excels at linking data files from many databases. Use the Relate Manager window to set up links between data in different databases and your data sets. The following describe the components of the Relate Manager window, shown in the following illustration.

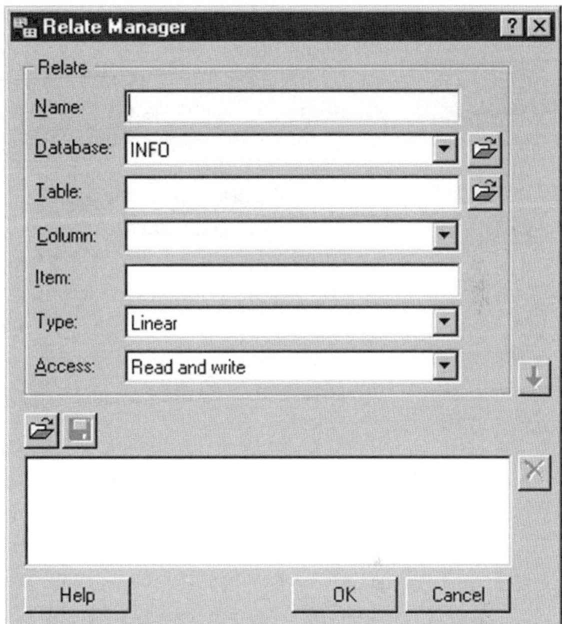

Relate Manager window.

Name: This is a user-defined name that stores the criteria for establishing this link between files. It can be up to eight alphanumeric characters long.

Database: By default, INFO is the selected database manager where ArcInfo will look for a data table. Select the arrow icon to the right of this box to choose another database from a list of third-party databases supported by ArcInfo. Click on the folder icon to navigate to the proper directory.

Table: The name of the table (file) that stores the information to be related.

Column: The destination field (or item) in the related table.

Item: The item in the INFO table on which the relate is created. The two databases must have one item in common, and it must be formatted the same in both.

Type: The type of joining to be made between files. Choices are Linear, Ordered, Link, Table, and First (used when connecting an INFO file to an external database system). For more information on the other choices, see the on-line help.

Access: Denotes the access allowed to the related file. Choices are Read Only, Read and Write, or Auto. Read Only allows the related table to be read but not modified, Read and Write allows data in the related table to be changed, and Auto defaults permissions to be the same as the source INFO table.

The Relate Manager greatly simplifies establishing connections between spatial and tabular data sets. Much of the power of GIS is in its ability to use both types of data together. The Relate Manager provides a means of joining the two types of files to make them work as one.

Remote Processing Manager

The Remote Processing Manager is another new feature in ArcInfo 8.0. It allows you to run your commands on another (presumably more powerful) machine also running ArcInfo. Say you are running ArcInfo at your desk on a mid-level PC. However, say you need to perform a spatial analysis routine that could take hours. Use the Remote Processing Manager, shown in the following illustration, to run this routine on a faster computer.

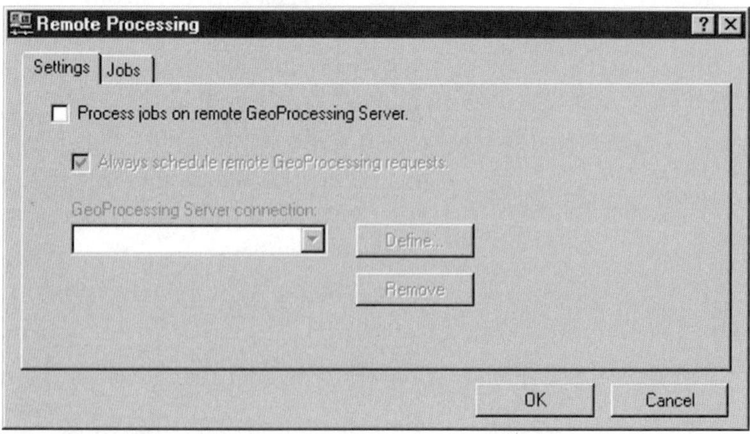

Remote Processing Manager.

Note that once you make a connection to a geoprocessing server (the remote machine that will run your commands), *all* ArcToolbox commands will run there as well. You cannot run some commands locally and some remotely, and the commands you execute *cannot* display graphics. For instance, you cannot run remotely an AML that switches between Arc and ArcPlot. The Remote Processing Manager is unable to handle graphics routines; therefore, the AML will fail. Keep this in mind when deciding if using a remote server for processing makes sense for the task at hand.

Explore this window, by following along on your computer, to learn how to set up a connection to another machine. To enable a routine to run on another computer, you must check the "Process jobs on remote geoprocessing server" box. Then either check on or off the next box, "Always schedule remote geoprocessing requests." When checked, your commands will be scheduled to run at a time when there are fewer users on the remote computer system. If unchecked, your commands will run immediately.

Now, click on the Define button. This displays the Connection Properties menu, shown in the following illustration.

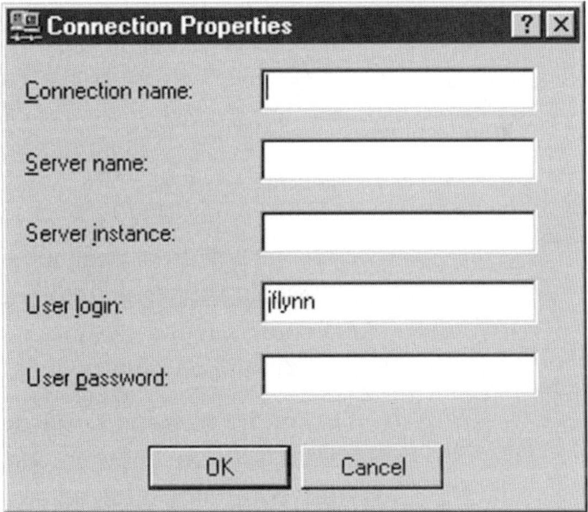

Remote Processing Manager Connection Properties menu.

The Connection Properties menu is where you specify necessary information, such as the name of the geoprocessing server, your account name and password on that computer, and so on. The following describe the components of the Connection Properties menu.

Connection name: A name that identifies this connection. It may be anything you like, but you should make it something that helps you identify it as a connection to a particular server.

Server name: The name of the geoprocessing server to which you want to connect.

Server instance: The instance on the server you want to use. See your system administrator for help with this.

User login: Your user account name on the geoprocessing server.

User password: Your password on the geoprocessing server.

Now, click on the Jobs tab in the Remote Processing window, shown in the following illustration.

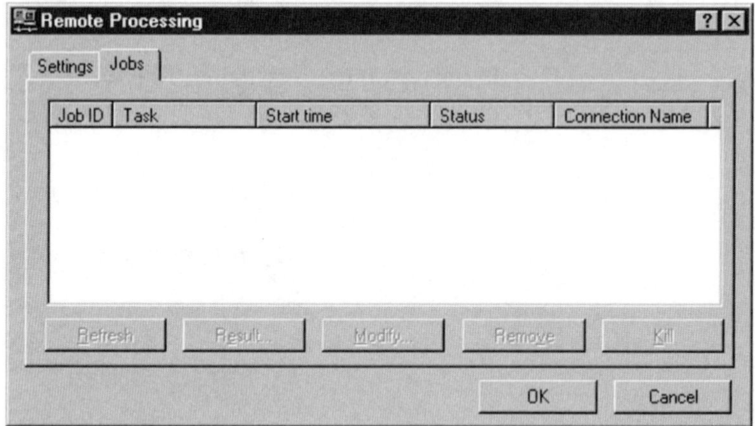

Jobs tab in the Remote Processing window.

Here you will see all jobs you have submitted to be processed remotely, as well as their current status and when they are due to be executed on the geoprocessing server. Here, you can also delete jobs already submitted.

The geoprocessing wizard provides a faster means of executing commands on another computer that might take a long time on your computer. This allows you to run large programs at night or over the weekend.

Using the New Tool Sets

The new tool sets provide menus that guide you through an ArcInfo task. Many of the tools are menus that run traditional ArcInfo commands. Others are wizards that use a sequence of pop-up menus to help you fill in the information needed to execute the command.

✓ **TIP:** *ArcInfo uses wizards for commands that do not lend themselves to single pop-up windows, or for commands that require you to supply a lot of information.*

This section first shows you how the toolboxes are organized, and then takes you through a typical command window. Following this section is exercise 4-1, which demonstrates a spatial analysis process using several ArcToolbox commands and wizards.

The main ArcToolbox menu screen, shown in the following illustration, contains four red toolboxes: Data Management Tools, Analysis Tools, Data Conversion Tools, and My Tools. Just like you may have a toolbox for your automobile, and one for your home, these toolboxes contain commands specific to data management analysis, or conversion. The My Toolbox is customizable; it is where you can store commands, wizards, and Arc Macro Language programs (called AMLs) that you or your organization use frequently.

ArcToolbox main screen.

✓ **TIP:** *You can have several custom toolboxes (for example, one for each project or person) under the My Tools toolbox. These can help organize your GIS operations and save time in hunting down commands and AMLs.*

Within each of the four main toolboxes are other toolboxes, each containing groups of tools that perform related functions. Think of these *tool groups* as being sets of tools that perform related functions, similar to the set of screwdrivers or wrenches you may have in your home toolbox. This is the model used in organizing ArcToolbox.

Open a toolbox by clicking on the "+" symbol next to it. Follow along on your own computer, and open the Analysis Toolbox by clicking on its "+" symbol. You will see this toolbox symbol change from closed to open. This is shown in the following illustration.

The open Analysis toolbox.

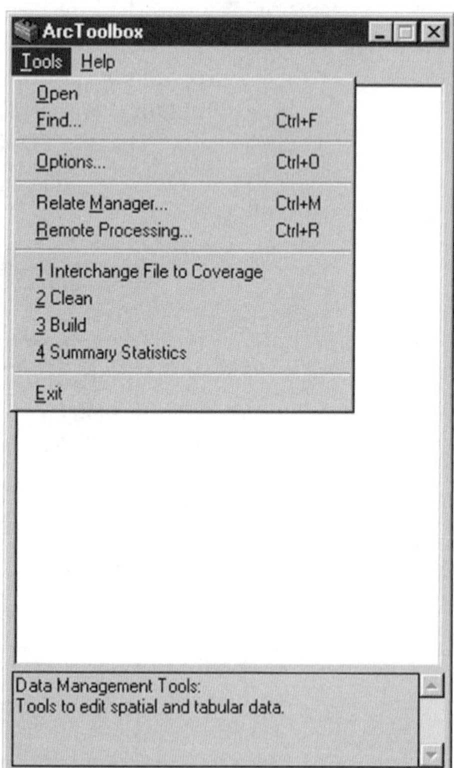

The first listing—the blue circle with the white "i"—is the *information* choice. Double click on this now, and the ArcToolbox help window appears, with the Analysis Tools description showing. Here you will find short descriptions of each toolbox and tool group. You can also perform further searches for information using this window. Take a few moments to explore this help window, and then dismiss it by clicking on the X in the upper right-hand corner.

Now open the first toolbox under Analysis Tools (the Extract toolbox). Notice it has its own information button. It also contains severa tools (symbolized by the small hammers) and one wizard (symbolized by rhe magic wand). Remember that tools have one pop-up window, whereas wizards have a series of pop-up windows. Now double click on the Clip hammer to open its window.

Clip

The Clip command uses the polygon boundary of a clip coverage to extract features from an input coverage. Only those featues in the input coverage that fall within the polygon boundary of the clip coverage are retained. A new coverage is created from the retained features. The command window for Clip, shown in the following illustration, is representative of what you will see in most of the windows. You will use this command in exercise 4-1.

Clip window.

Tool menus typically contain text boxes into which you enter coverage names. You may type in the coverage name, or alternatively click on the yellow folders to the right of the text boxes to see listings of coverages in your current workspace. You can also navigate to other workspaces. This way you can use the mouse to select a coverage rather than having to type its name.

Scrollable windows are another type of input box. The "Clip feature" box has a downward-pointing arrow to its right. This means that you must click on this arrow to view available choices. For Clip, your choices will be coverage feature types such as Polygon or Line. For these input boxes, you must select a feature type with the mouse; you cannot type it in.

There is also a menu button on the lower right of the Clip window named Batch. Many tool menus contain this button. Selecting the

Batch button adds a new part to the Clip window, as shown in the following illustration.

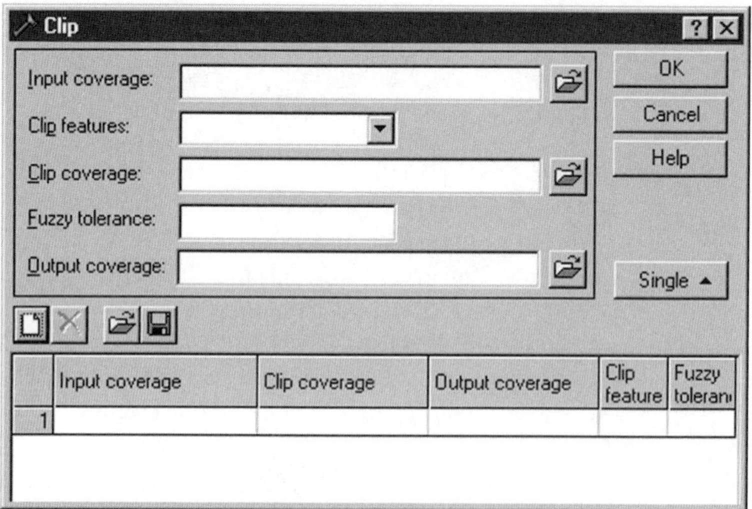

Clip menu with Batch option selected.

Here you see what looks like a spreadsheet with four new buttons above it. The column labels refer to each of the input boxes above the spreadsheet. Clicking on the first button (the white sheet of paper) adds a row, each row being one version of the Clip command. The slash X button deletes a row. The yellow folder button lets you navigate workspaces to add coverage names.

There is one more nice feature about the Batch button. Clicking on the small icon that resembles a diskette lets you save as an AML all information used to execute the Clip commands. You can do this and then incorporate it into a larger AML program. Letting ArcToolbox write the small AML for you saves you time and effort.

This section introduced you to the organization of ArcToolbox's tools and menus. It also showed you the components of a typical command window, using the Clip tool as an example. Next, in exercise 4-1, you will use several tools to perform a spatial analysis.

Exercise 4-1: Exploring the Tool Sets

In this exercise you will use tools and wizards from Analysis and Data Conversion toolboxes to perform a spatial analysis. For this exercise, assume you work for a national travel agency that books tours of scenic areas. One of the criteria you use in locating nice places to visit is to determine if there is much public land there. National forests, national parks, and so on all fit the criterion of public land. You have been asked if Virginia would be a good state to consider for addition to the list of tour destinations. The data sets you have available for your analysis are in shapefile format (a data type used by ESRI's ArcView software). These follow.

- *Fedlands*, a shapefile that contains federally owned land in Virginia, North Carolina, West Virginia, Maryland, and Kentucky
- *Allstates*, a U.S. state boundary shapefile
- *Vacounty*, a county boundary shapefile for Virginia

Before beginning, be sure to copy the data for this exercise from the companion CD-ROM (*D:\ai80exercises\chap_4*). As always, it is best to take a few moments to plan what steps you need to take before getting started. To come up with truly useful results, you need to identify those areas in Virginia that contain significant portions of federal land. After thinking about this, you decide on the following course of action.

1. Import all three shapefiles as ArcInfo coverages.
2. Create a coverage of just the Virginia state boundary. Name it *Virginia.*
3. Use this coverage to clip the *fed_lands* coverage to create a new coverage named *Fed_Virginia.*
4. Overlay *Fed_Virginia* with the Virginia county coverage.
5. Calculate summary statistics to determine which counties in Virginia have the most federally owned land.

Step 1: Importing Shapefiles as ArcInfo Coverages

Because you are starting out with shapefiles, it is important to convert them to ArcInfo coverages. Why? Because although ArcInfo can display shapefiles, you must still use coverages to perform spatial analysis.

First, use ArcToolbox's "Shapefile to Coverage" command in the Data Conversion tool set to convert each of the three shapefiles.

The first shapefile to convert is *Fedlands.* Name the new ArcInfo coverage *Fed_Lands.* In the "Input shapefile" text box, type *FEDLANDS.SHP.* In the "Output coverage" text box, type *FED_LANDS.* Your screen should look like that shown in the following illustration.

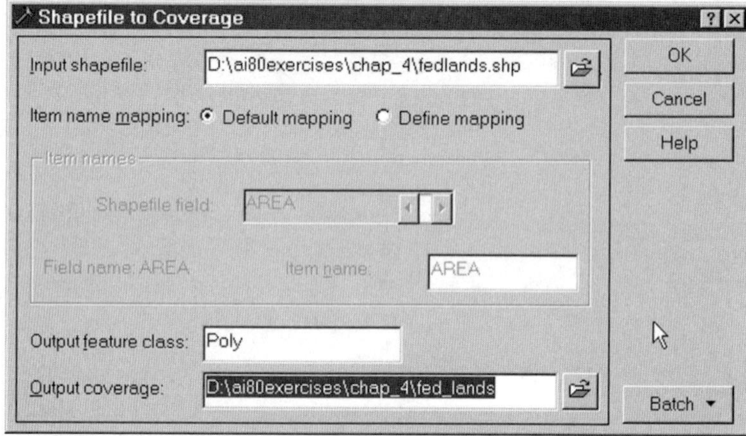

Shapefile to Coverage dialog box.

Now, perform this same operation twice more, once for *Allstates* (call the new coverage *All_States*) and *Vacounty* (call the new coverage *Va_County*). When you are done, you should have the three original shapefiles, plus the three new coverages, in your directory.

Step 2: Creating a Coverage of Virginia's Boundary

Next, you need to create a coverage of just Virginia's boundary. This new coverage will be the "cookie cutter" used to clip just the federal lands within Virginia from the *fed_lands* coverage. Use the Extract Wizard within the Analysis tool set. This wizard will take you step by step through the procedures needed to create a polygon coverage of Virginia.

✓ **TIP:** *The Extract Wizard is a nice front end to the Reselect command used in earlier versions of ArcInfo, as well as in ArcInfo 8.0 Workstation.*

Once the Extract Wizard starts, it prompts you for whether you want to create a coverage from an existing data set of lines or points or one from area features. Because you are looking for Virginia, use the mouse to select "Area features," as in the following illustration.

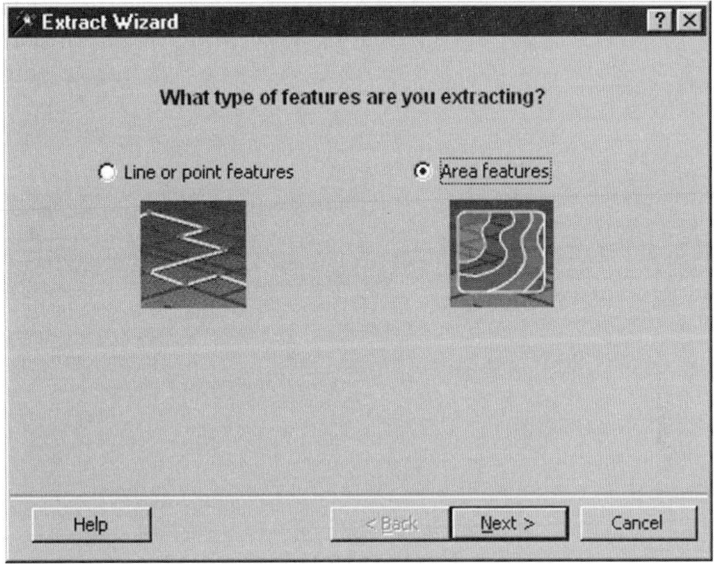

Extract Wizard.

The next screen, shown in the following illustration, can be confusing to the new user. The wizard wants to know if you are selecting data from just one coverage or from regions that may be in one or in many coverages. Here you are working with a single polygon coverage of state boundaries; therefore, use the mouse to select the first option, "As ONE data layer in ONE coverage."

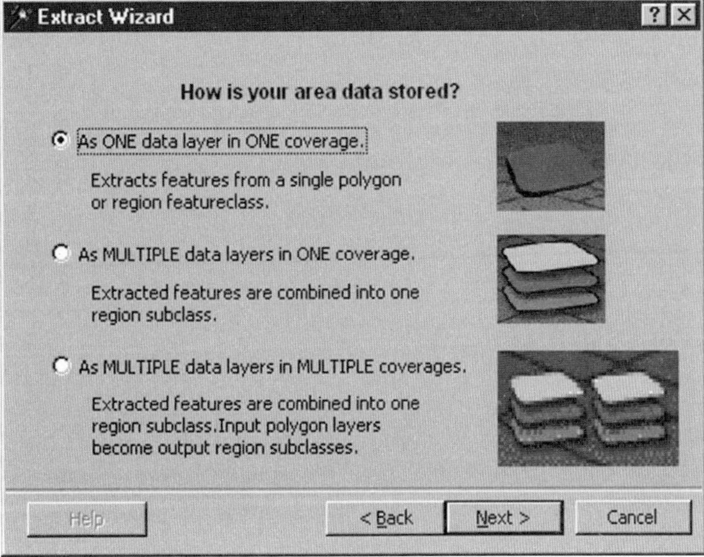

Feature type selection in the Extract Wizard.

Now type the coverage name *all_states* into the "Input coverage" text box. Select "poly" as the feature type to extract. Your screen should look like that shown in the following illustration.

Enter input coverage and feature type in the Extract Wizard.

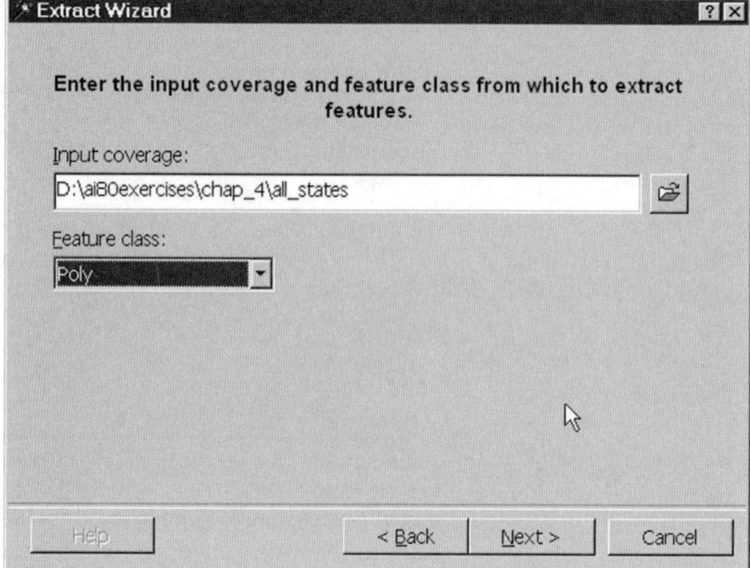

The next screen is the crucial step where you tell ArcInfo to pull out just the state of Virginia. Click on the "Build a query" button to define the criteria needed to find Virginia. The following illustration shows the Build Query screen.

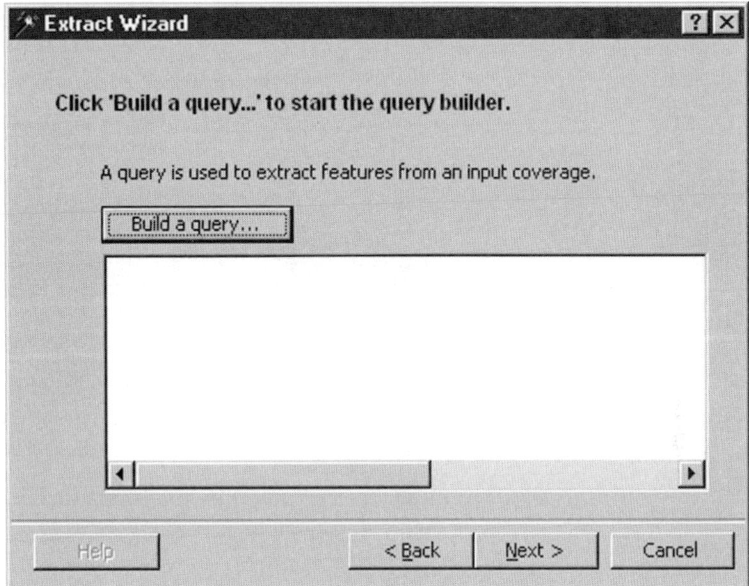

Build Query screen.

ArcInfo starts the Query Builder, which is its tool to let you view the data items and their values in coverages. By specifying a query, ArcInfo will select data from the *All_States* coverage to create a new coverage. A query is little more than a mathematical expression. Continuing with this example, you will see how to build a query.

Use the mouse to select the STATE item in the left-hand window, and you will see that all unique state values held in that item are listed in the right-hand window, as shown in the following illustration.

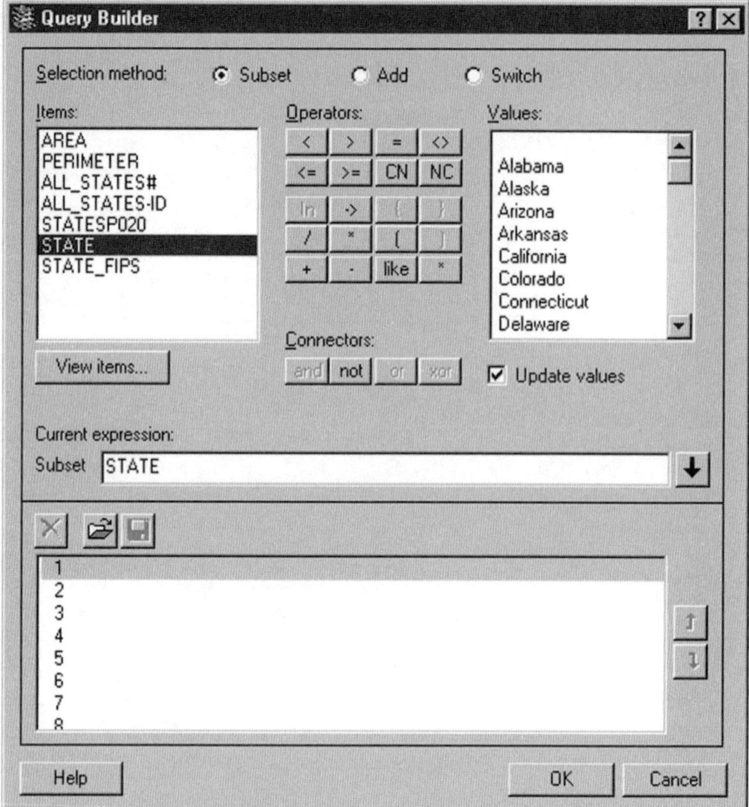

Query Builder showing unique values for STATE item.

Note that the item STATE is now present in the Current Expression box just below. Now, click on the "=" button in the menu's calculator box, and then click on Virginia in the Item Values box. The following illustration shows what the Query Builder screen should look like.

Query Builder with "STATE = Virginia."

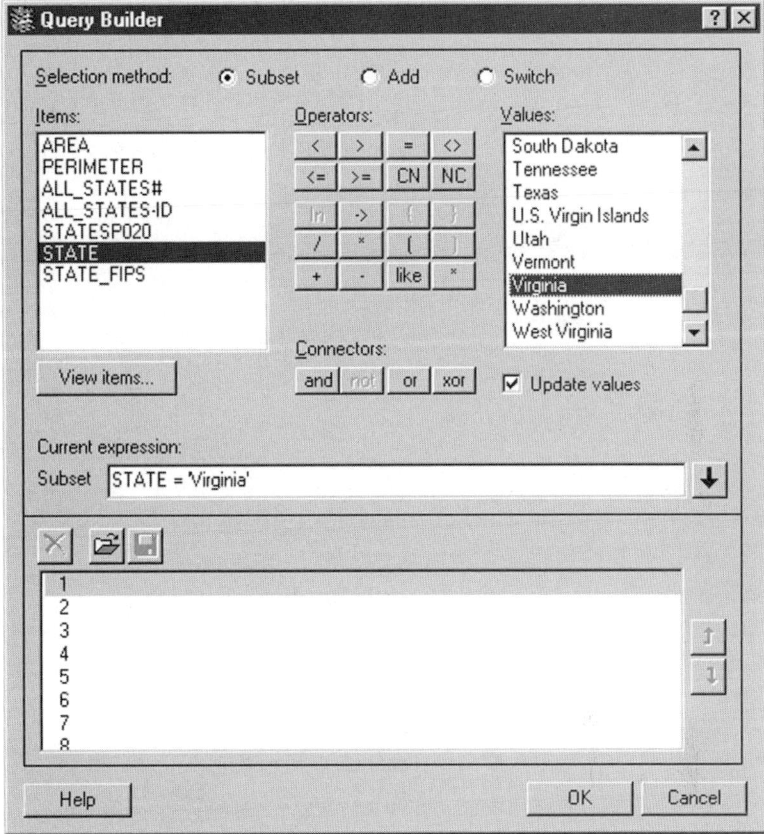

Notice that Virginia is now in single quotes. ArcInfo places single quotes around all non-numeric items. Remember this when you are typing your own expressions! The expression is now complete, so click on the down (↓) arrow button to the right of the Current Expression box. This adds your selection to the Query Builder, as shown in the following illustration. You can see that it is possible to add quite a few queries at one time. Click on the Next button.

Query added to the Query Builder.

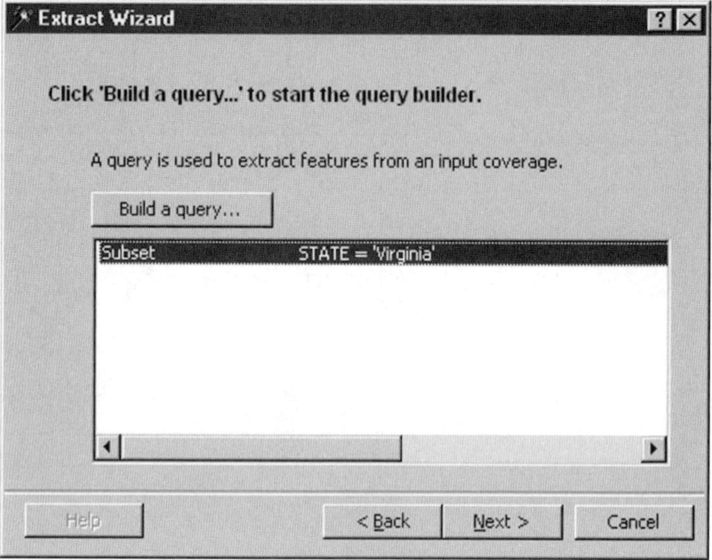

The Extract Wizard has now recognized your selection criterion. Proceed by clicking on the Next button. The wizard prompts you for the name of your new coverage, as shown in the following illustration.

Enter name of output coverage in the Extract Wizard.

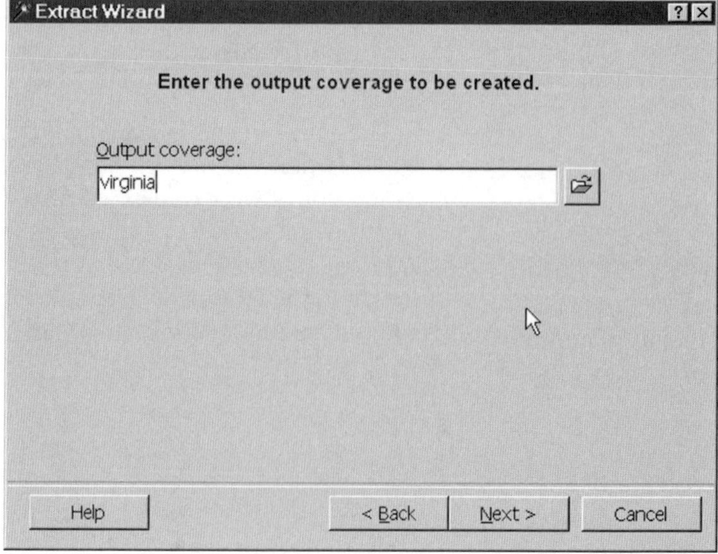

Type *Virginia* (without the quotes) in the "Output coverage" text box as the name for your new coverage.

The Extract Wizard now presents a summary screen, shown in the following illustration, displaying the query you entered and the name of the coverage that will be created. If something is not correct, you can always click on the Back key to return to a previous screen where you can correct it. Note the Save to AML button. This handy feature will save the query you have created so that you can run it again without having to work your way through all screens.

Summary screen of Extract Wizard.

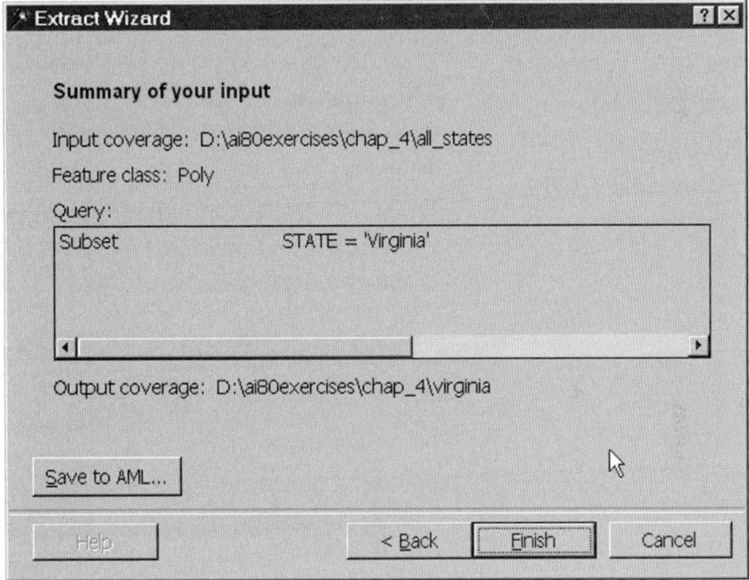

Step 3: Creating a Coverage of Federally Owned Land in Virginia

Now that you have a Virginia state boundaries coverage, use ArcToolbox to select just those federal lands that fall within Virginia's boundaries. Use the Clip tool within the Data Analysis tool set.

Type *fed_lands* in the "Input coverage" text box and *Virginia* in the "Clip coverage" text box. In the "Output coverage " text box, type *fed_Virginia*, the name of the new coverage that will act as a "cookie cutter" and extract any features from the input coverage that fall within the clip coverage. Your screen should look like that shown in the following illustration.

Clip

Input coverage: D:\ai80exercises\chap_4\fed_lands

Clip features: Poly

Clip coverage: D:\ai80exercises\chap_4\virginia

Fuzzy tolerance: 1.99999994947575E-05 Undefined units

Output coverage: fed_virginia

OK Cancel Help Batch

Clip menu.

Step 4: Overlaying Federal Lands in Virginia with the County Coverage

Now that you have created a coverage of federal land in Virginia, determine the counties that have federal land. These will be the counties your tour company will examine in detail as tourist locations.

The Analysis tool set contains a wizard called the Overlay Wizard. Use this to combine *Fed_Virginia* with *Va_County*. The first screen in the Overlay Wizard is shown in the following illustration.

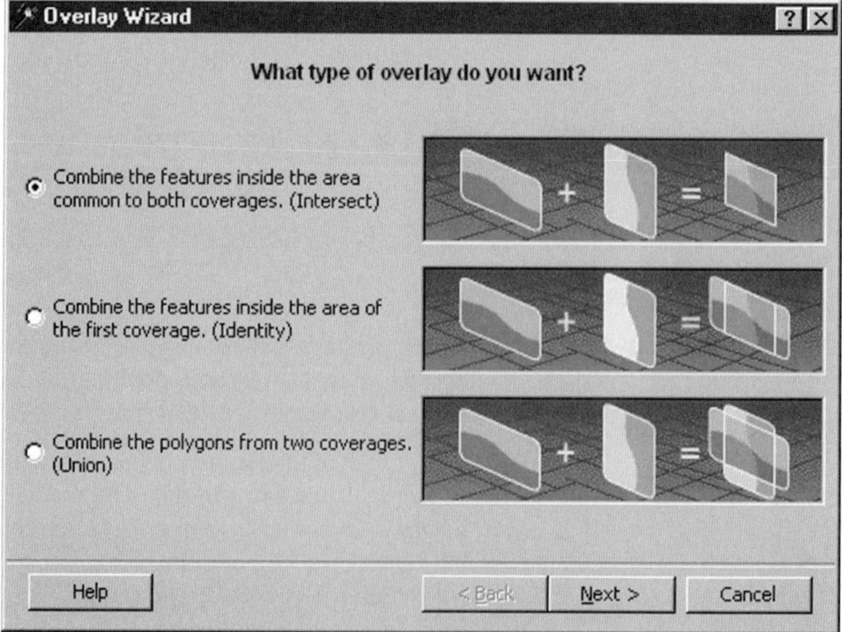

Overlay Wizard.

The Overlay Wizard presents three choices as to the type of overlay you may conduct. Use the mouse to select the first option, "Combine the features inside the areas common to both features."

First, type *fed_virginia* in the text box for the "Input coverage." This will be *Fed_Virginia*. (See the following illustration.)

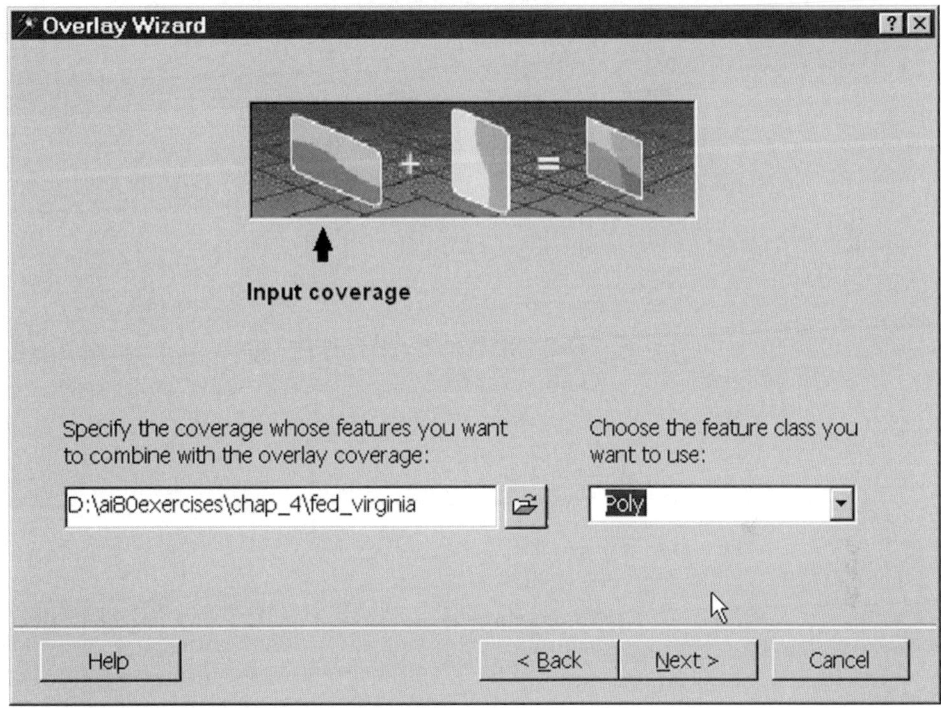

Enter the input coverage in the Overlay Wizard.

Next, type *va_county* in the "Overlay coverage" text box. This is shown in the following illustration.

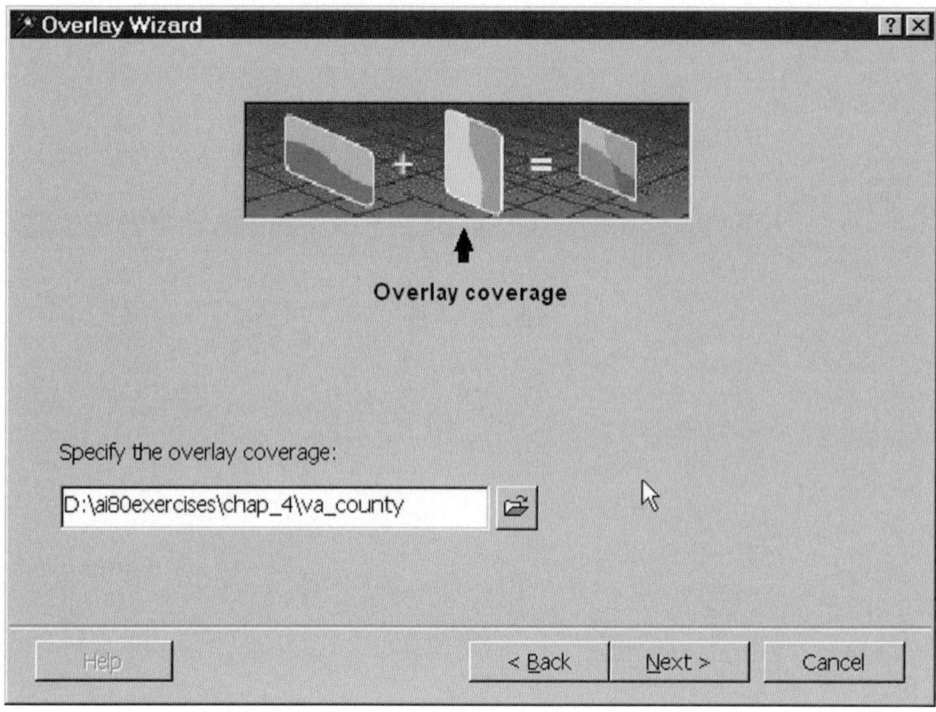

Enter the overlay coverage name.

ArcInfo now asks whether you want to join attributes in the new coverage. This is very important because it means that the new coverage will have all polygons and attributes from both input and output coverages. This will tell you which counties contain federally owned lands. This is shown in the following illustration.

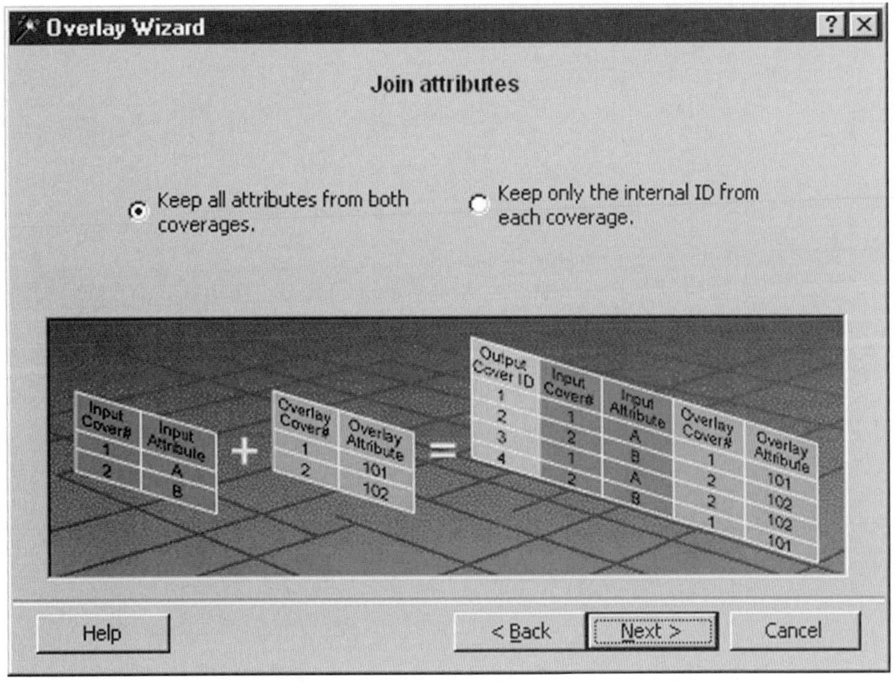

Specify to join attributes in Overlay Wizard.

Now type *fed_county* in the "Output coverage" text box. Your screen should look like that shown in the following illustration.

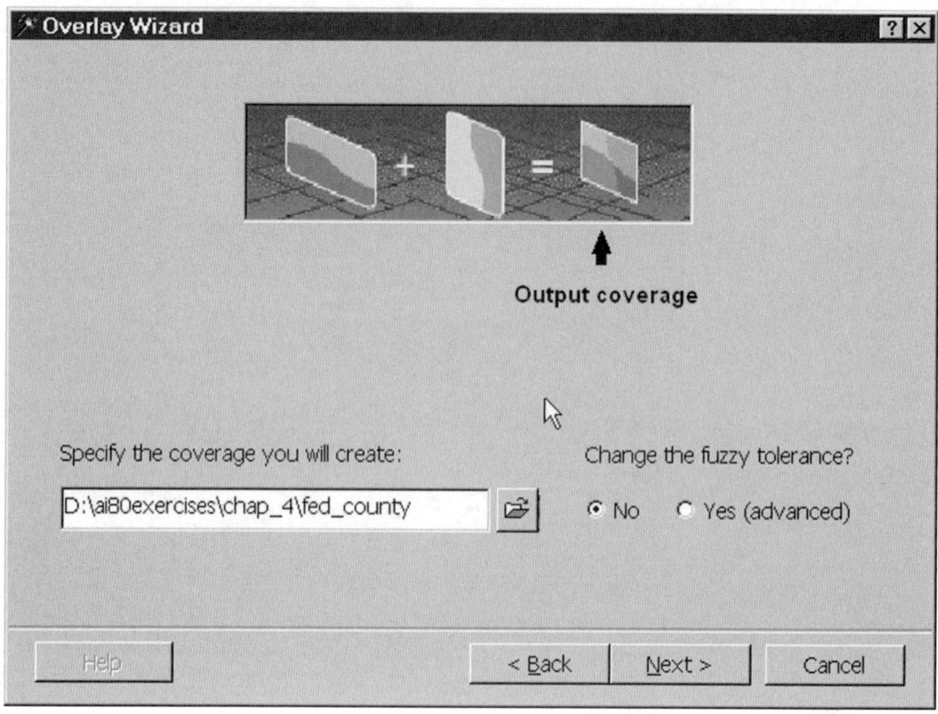

Specify output coverage name.

View the summary screen, shown in the following illustration, to ensure you have entered the correct names and option for the overlay procedure. Click on the Finish button to complete the wizard.

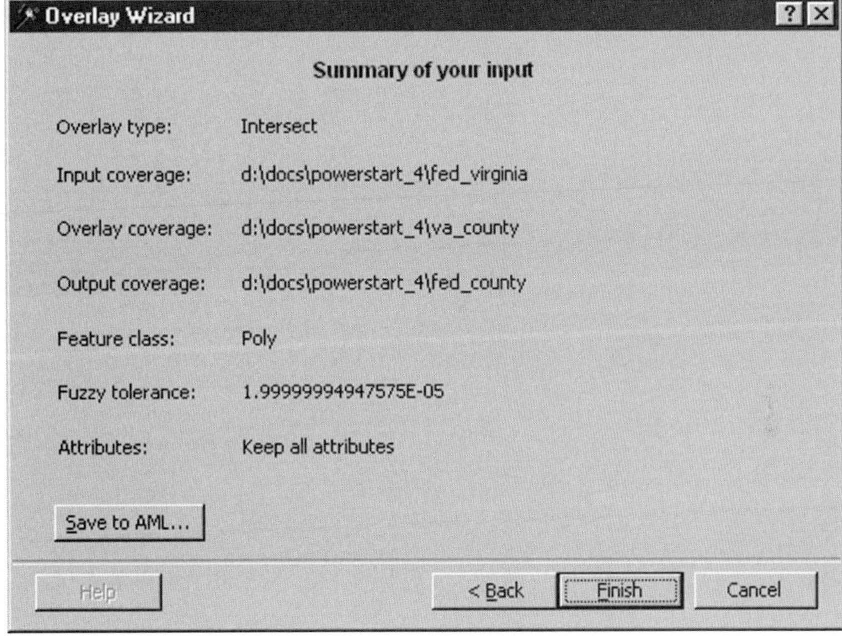

Summary screen in Overlay Wizard.

Step 5: Determining Which Counties Have Federal Lands

You are almost done! Now that you have created a coverage of only those counties in Virginia that contain federal lands, use the Frequency tool to produce a list of these counties. The Frequency tool is found under the Statistics tools in the Analysis tool set. The following illustration shows the Frequency tool menu.

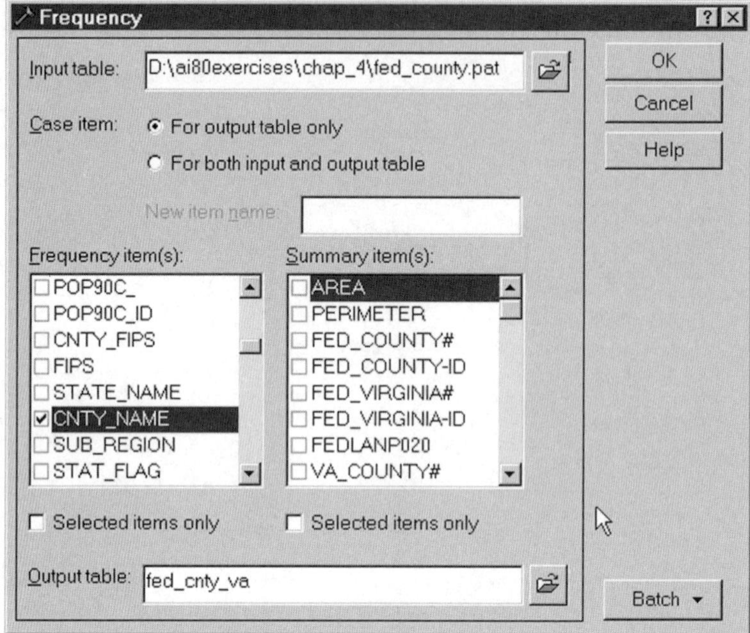

Frequency tool menu.

The letters .PAT added signifies that you are gathering frequency information from the polygon attribute table from this coverage. Select with the mouse two items in the left-hand window, CNTY_NAME and FEATURE1 (CNTY_NAME contains the county name and FEATURE1 contains the type of federal land present in that county). Type *Fed_cnty_va* in the "Output table" text box. Click on the OK button.

Use ArcCatalog to view the frequency table just created (see Chapter 2 to learn how to view tables). As a brief refersher, simply start ArcCatalog, open the *D:\ai80exercises\chap_4 folder*, and select the *fed_cnty_va* data file. The following illustration shows the frequency table using ArcCatalog. Notice that there can be more than one instance of a national park in a county. This might be because there actually may be more than one park in a county. It also might be because a park may have an irregular boundary and it could weave in and out of a county's border. Still, you now have enough information on federal land holdings in Virginia counties to get your tour promoters started on exploring the Virginia landscape!

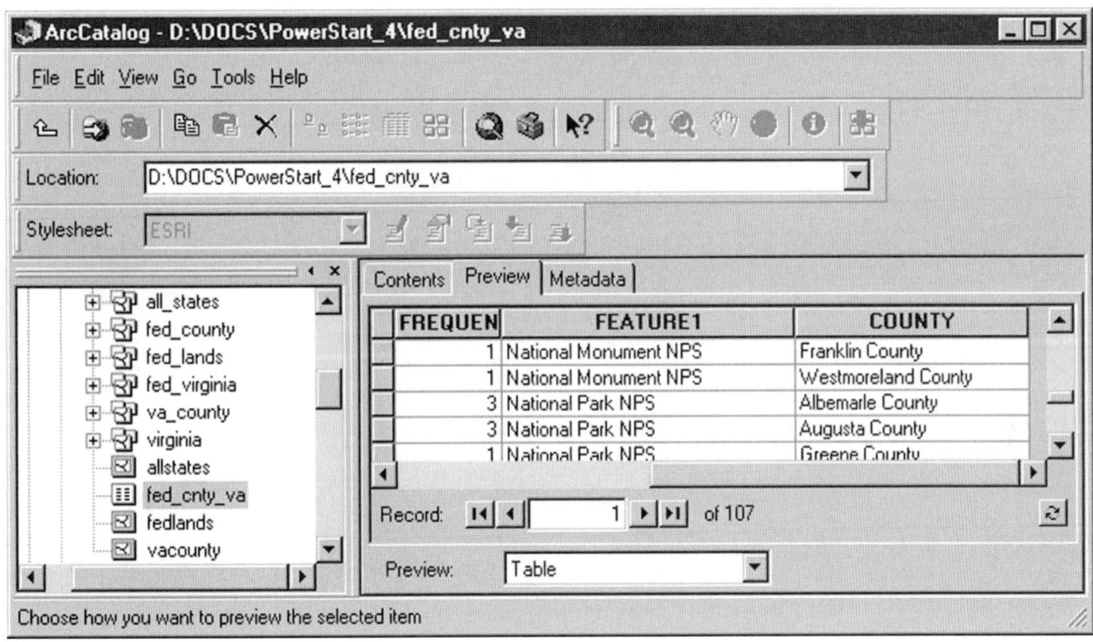

Using ArcCatalog to view the frequency table.

✓ **TIP:** *See Chapter 2 for a discussion of how to use ArcCatalog to view tables.*

A Reference Guide to ArcToolbox

The ArcToolbox main menu, shown in the following illustration, is simple and compact. Four sets of tools are presented (grouped as "Toolsets"), each of which contains many additional tools.

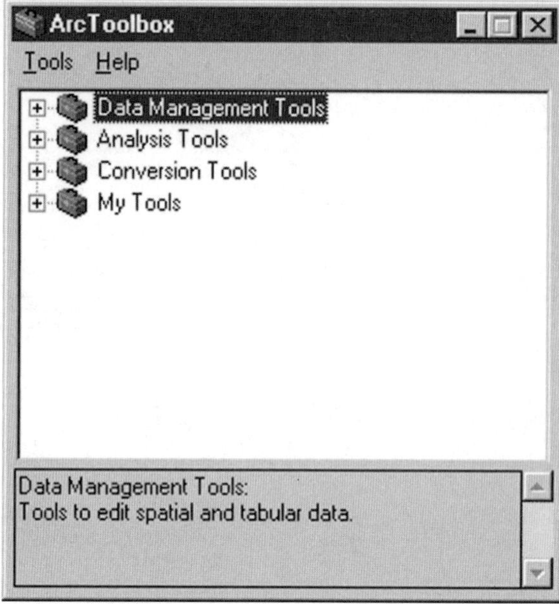

ArcToolbox main menu.

This section documents each of these tools and wizards. Here, you will find an explanation of the tools, and a description of the information needed to use them.

NOTE: *A wizard takes the user step by step through a series of menus in order to run a command. Wizards are often used to make more complicated tasks easier.*

Data Management Tools

The Data Management toolbox contains many of Arc's bread-and-butter tools for building topology, assigning projections, and working directly with attribute data. The sections that follow examine each tool in depth. The Data Management toolbox is shown in the following illustration.

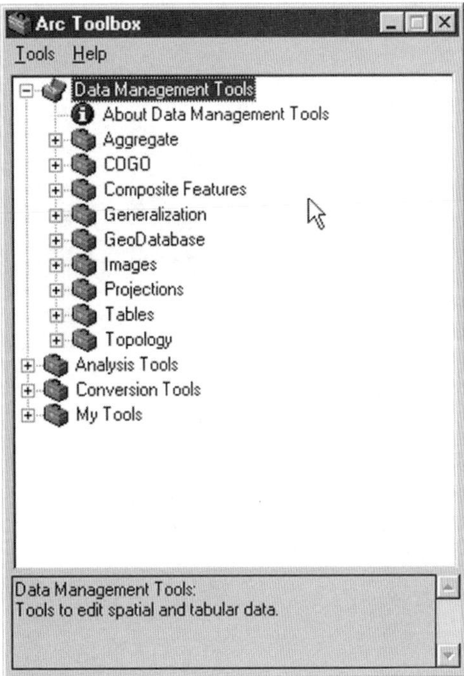

Data Management toolbox.

Aggregate Tool

The Aggregate toolbox is shown in the following illustration.

Aggregate toolbox.

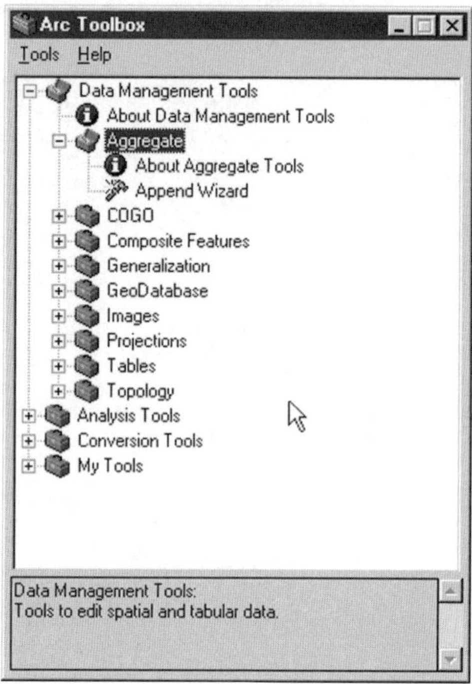

There is just one tool available within the Aggregate menu: the Append Wizard. This wizard invokes a series of menus that let the user supply a list of adjacent coverages to join together into a new coverage. All ArcInfo coverage feature types are supported. That is, you may append line, point, and polygon coverages. You may append coverages with different feature types, but no attributes will be appended. The new coverage will contain spatial data but no attribute data. You may also append annotation, network, and region feature classes. Provided all coverages are of the same feature class and have the same items in the same order in their attribute tables, items can also be appended.

NOTE: *Append does not create topology. You must still use either Build or Clean.*

The following illustration shows the opening screen of the Append Wizard.

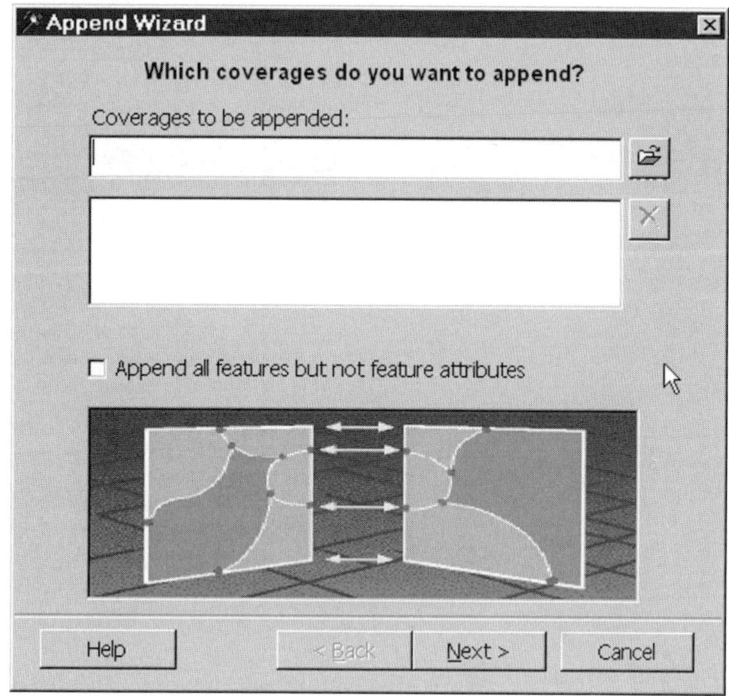

Append Wizard.

COGO Tool

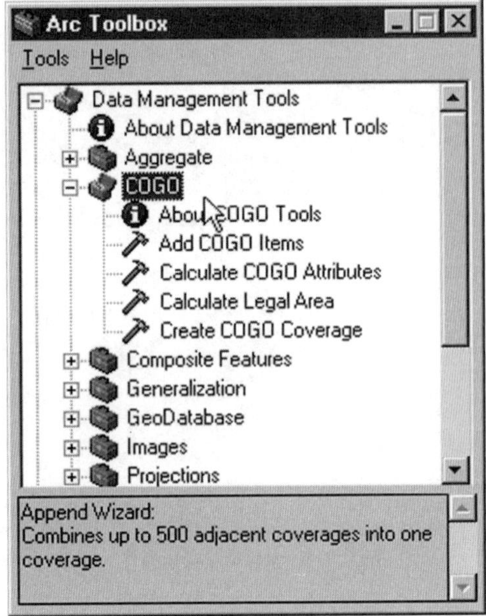

COGO Tool menu.

COGO is an acronym for "coordinate geometry." COGO is used by land record management professionals to create coverages from surveyors' measurements. COGO coverages have special items in their Arc attribute table to hold surveyors' measurements. These items include angle, distance, radius, and tangent, among others. These commands are discussed in the sections that follow.

The Add COGO Items tool, shown in the following illustration, adds the COGO items of the following table to a line coverage (the input coverage). Table 4-1, which follows, summarizes the COGO items added. These items contain no data. To use this tool, simply type in the name of the line coverage in the "Input coverage" text box. For more information on the items added, see the on-line help.

Add COGO Items

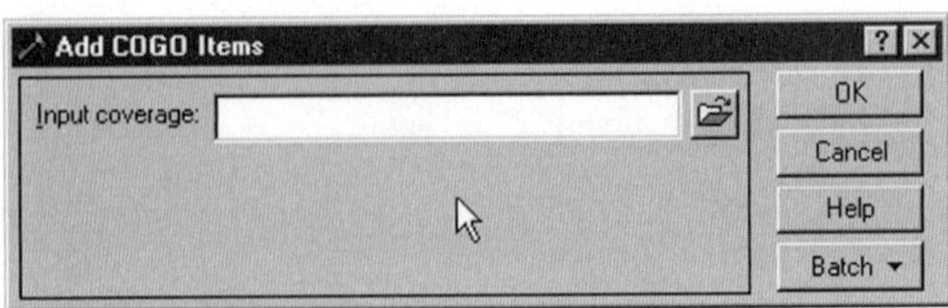

Add COGO Items screen.

Table 4-1: COGO Items

Item Name	Width	Output	Type
Angle	10	10	C
Distance	8	8	C
Radius	8	8	C
Delta	10	10	C
Tangent	8	8	C
Arclength	8	8	C
Side	1	1	C
Radius	2	8	C
Tangent	2	8	C

✓ **TIP:** *If you need help understanding how to view items in a database or understanding what items are in the first place, see Chapter 9 for a discussion of databases.*

Calculate COGO Attributes

The Calculate COGO Attributes command is a tool that calculates COGO data for two-point lines and curves. A new coverage is created. The command screen, shown in the following illustration, contains the following options.

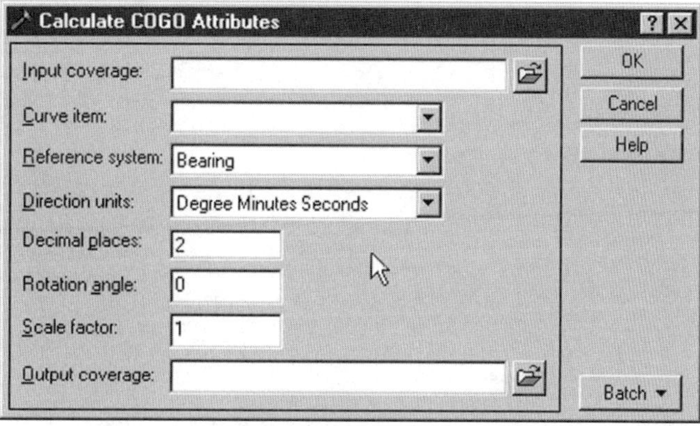

Calculate COGO attributes command.

Input coverage: Coverage whose COGO values you want to calculate.

Curve item: Listing of items in the AAT. Select *Curveitem.*

Reference system: Reference system to use in calculating values. Choose from Bearing, North Bearing, Azimuth, South Azimuth, or Polar.

Direction units: Unit of measure used in distance calculations. Choose from Degrees minutes seconds, Decimal degrees, Grads, or Grads centigrads centicentigrads.

Decimal places: Decimal places used when calculating distance.

Rotation angle: Angle used to change cardinal direction from a local coordinate system to another system such as state plane coordinates. Either positive or negative numbers may be used.

Scale factor: Used to change scale from local coordinate system to state plane coordinate system.

Output coverage: New coverage with COGO values.

Calculate Legal Area

Use the Calculate Legal Area command, shown in the following illustration, to calculate legal area based on survey coordinates in a COGO coverage. This command contains the following options.

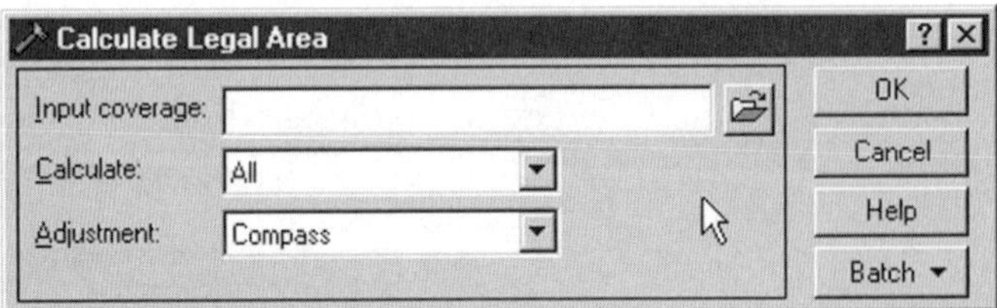

Calculate Legal Area command.

Input coverage: COGO coverage for which you want to calculate legal area.

Calculate: Choose from All, Area, Perimeter, or Closure. Legal areas, perimeters, and lot closures can be calculated separately or together by selecting All.

Adjustment: Choose Compass, Transit, or Crandall. Each is a surveyor's method for handling errors in lot closure.

Create COGO Coverage

The Create COGO Coverage command, shown in the following illustration, creates a new arc or point coverage with COGO attributes and data values. Use this tool to copy COGO attributes from an existing COGO coverage. Once created, the COGO coverage can be edited to match legal deed descriptions of parcels based on survey measurements. This command contains the following options.

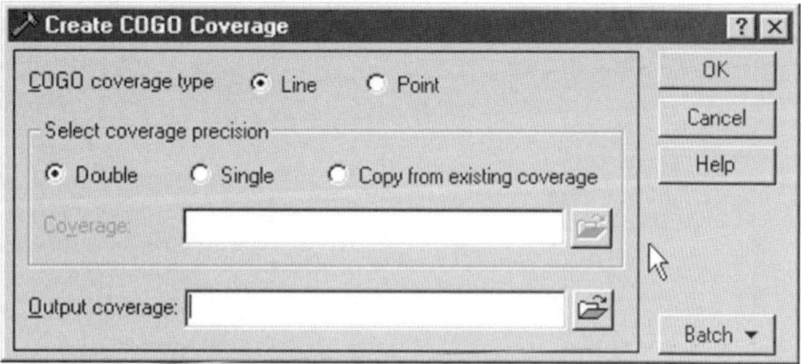

Create COGO Coverage command.

COGO coverage type: Choose either Point or Line.

Select coverage precision: Specify single or double precision. You may also specify a coverage whose precision you want to copy.

Output coverage: New coverage created.

Composite Features Tool

The Composite Features tool provides a means of converting line and polygon coverages to regions. Regions are relatively new to ArcInfo and are very valuable for modeling polygons that overlap or that may be discontinuous. A good example would be the fifty states of the United States on a polygon coverage of all countries of the world. A region could be defined to include the lower 48 states plus Alaska and the islands of Hawaii. Another example would be soils and underground water polygons. They may overlap or even share the same polygon boundaries. When classified as regions, however, they are considered separate entities. The Composite Features tool menu, shown in the following illustration, contains region sections.

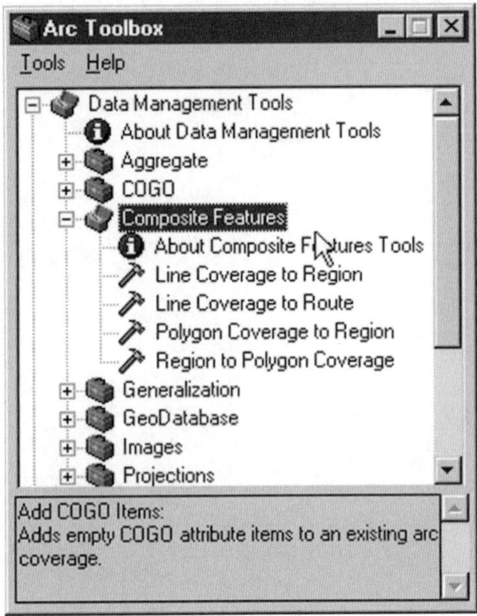

Composite Features tool menu.

Another benefit of defining regions is that it can reduce the number of polygons and lines stored in coverages. A coverage of all counties in the United States can have its state boundaries stored as a region, thus obviating the need to have a separate state boundary coverage. ArcToolbox supplies three tools for working with regions, as well as an additional tool for transforming a line coverage into a route. These are discussed in the sections that follow.

Line Coverage to Region

The Line Coverage to Region menu, shown in the following illustration, creates new regions from an Arc coverage. It may also append new regions to an existing region data set. This command contains the following options.

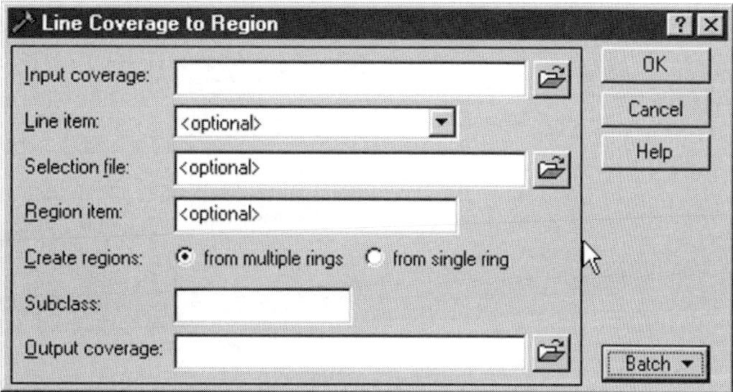

Line Coverage to Region menu.

Input coverage: Coverage whose arc features you want to convert to regions.

Line item: Item whose value will be used to create regions.

Selection file: ArcPlot selection file that contains arcs that will be used to create regions.

Region item: Stores arc values used to create regions. You may specify your own name for the item here.

Create regions: Specify if you want to create regions from multiple or single rings of arcs.

Subclass: Region subclass that will be created.

Output coverage: New coverage created.

Line Coverage to Route

Use the Line Coverage to Route menu, shown in the following illuistration, to create a new route system or to append new routes to an existing network. The input coverage must be a line coverage. This command contains the following options.

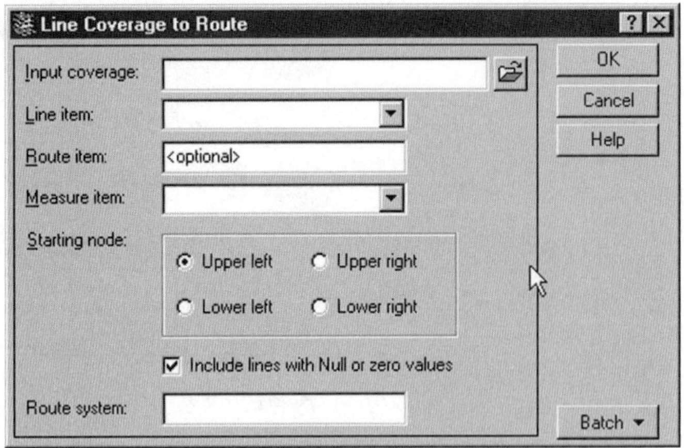

Line Coverage to Route menu.

Input coverage: Arc coverage that will be used to create routes.

Line item: Used to create routes. All arcs with same item value will generate one route.

Route item: Item that stores arc values used to generate routes. You may rename it here (by default, it remains the same as "Line item").

Measure item: Attribute that stores route's total length.

Starting node: Quadrant of coverage that contains the route's starting node (location).

Route system: Route system created. It will be contained in the arc coverage.

Polygon Coverage to Region

The Polygon Coverage to Region menu, shown in the following illustration, creates a new coverage with regions from a polygon coverage. This command contains the following options.

Polygon Coverage to Region menu.

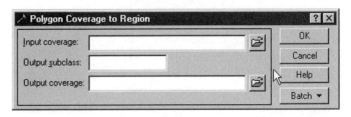

Input coverage: Polygon coverage you will convert to regions.

Output subclass: Region subclass created.

Output coverage: Either new coverage containing region subclass or input coverage with added region subclass.

Region to Polygon Coverage

The Region to Polygon Coverage menu, shown in the following illustration, creates a new polygon coverage from an existing region's coverage. This command contains the following options.

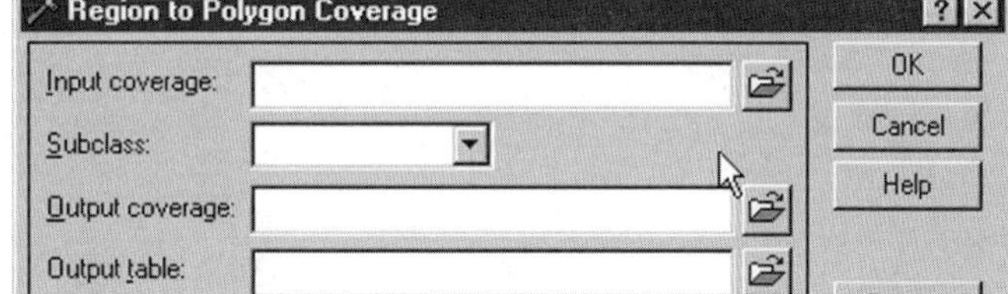

Region to Polygon Coverage menu.

Input coverage: Coverage with regions to be converted to polygons.

Subclass: Region subclass to be converted to polygons.

Output coverage: New coverage created.

Output table: Output table containing errors on overlapping polygons.

Generalization Tools

Arc's generalization tools work to make simpler polygon coverages from more complex ones. Simpler data sets mean that analysis and display operations run faster. Be sure, however, not to simplify too much. You may lose important data in an effort to trim the size of your data sets. The sections that follow discuss the generalization tools. The Generalization tool menu is shown in the following illustration.

Generalization tool menu.

Create Centerlines

The Create Centerlines menu, shown in the following illustration, creates a new coverage of centerlines derived from a coverage of street casings (the edges of roads). This command contains the following options.

Create Centerlines menu.

Input coverage: Coverage that contains the street casings.

Maximum width: How far ArcInfo will look on either side of a potential centerline for a road edge. Be sure to use coverage units.

Minimum width: Least amount ArcInfo will look for a street edge.

Output coverage: New centerline coverage.

Dissolve

The Dissolve menu, shown in the following illustration, merges polygons or lines based on values specified in the "Dissolve item" field. If adjacent polygons or lines have the same value in the Dissolve item, they will be merged into one polygon or line. A new coverage of all merges is created. This command contains the following options.

Dissolve menu.

Input coverage: Coverage to be dissolved.

Feature class: Polygon or line.

Dissolve item: Attribute that will be used to merge polygons or lines.

Output coverage: Coverage containing dissolved polygons or lines.

Dissolve Regions

The Dissolve Regions menu, shown in the following illustration, merges regions based on values specified in the "Dissolve item" field. If adjacent regions have the same value in that item, they will be merged into a single region. A new coverage of all merges is created. The command contains the following options.

Dissolve Regions menu.

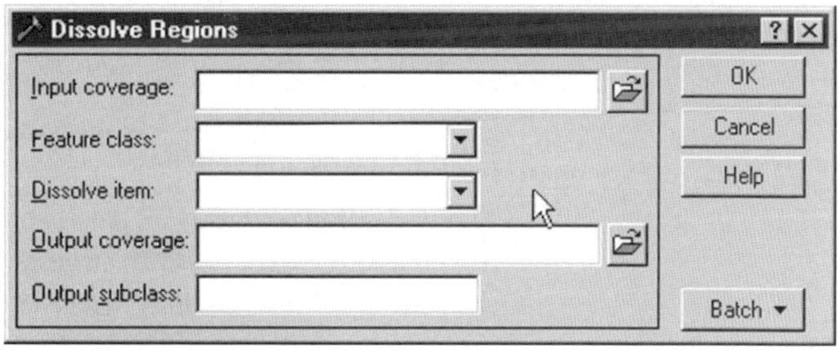

Input coverage: Coverage to be dissolved.

Feature class: Choose polygon or region. Dissolve will be conducted on chosen feature type.

Dissolve item: Attribute that will be used to merge polygons or lines. Select All to dissolve on all items instead of a single attribute.

Output coverage: Displays input coverage name by default. New region will be created within coverage by default.

Output subclass: New region created.

Find Building Conflicts

The Find Building Conflicts menu, shown in the following illustration, creates a new coverage of building footprints that fall within a user-specified distance. This command is appropriate to help find buildings that may be too close together or to weed out small adjoining buildings (sheds, garages, and so on) from a buildings coverage. This command contains the following options.

Find Building Conflicts menu.

Input coverage: Coverage of building footprints.

Distance: Maximum distance in coverage units to search for adjacent buildings. Buildings within this distance will be flagged as conflicted.

Output coverage: Coverage of conflicting buildings identified as being within the search distance.

Generalize Lines

The Generalize Lines menu, shown in the following illustration, removes any extra points or arcs in a line coverage. A coverage of shorelines, for instance, may have every bend of beach front included, which would contain many more arcs than necessary for use as a national boundary. This command contains the following options.

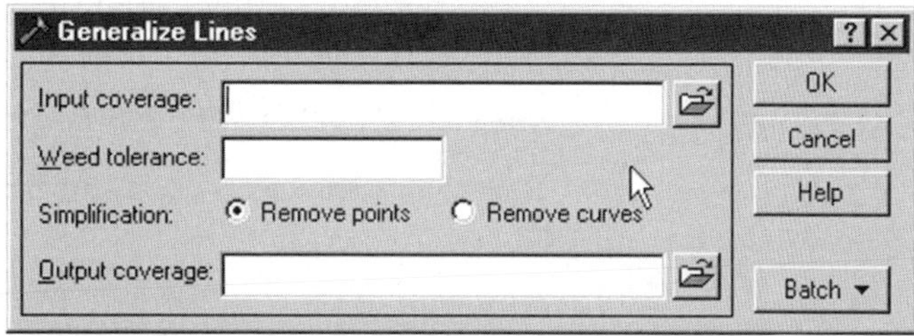

Generalize Lines menu.

Input coverage: Coverage containing arcs to be simplified.

Weed tolerance: Tolerance in coverage units used to remove arcs. It must be greater than 0.

Simplification: Choosing "Remove points" uses the Douglas-Peuker algorithm to weed out unnecessary points. Choosing "Remove curves" looks for unnecessary bends in each line and is often best when high-quality cartographic output is required.

Output coverage: New coverage created.

Simplify Buildings

Often a coverage containing structures will include many unnecessary lines and angles that while important to architects and engineers lend little to geographic analysis. The Simplify Buildings command, shown in the following illustration, simplifies footprints in a coverage composed of buildings. You might want to do this to save storage space or to speed up display or analysis when using this coverage.

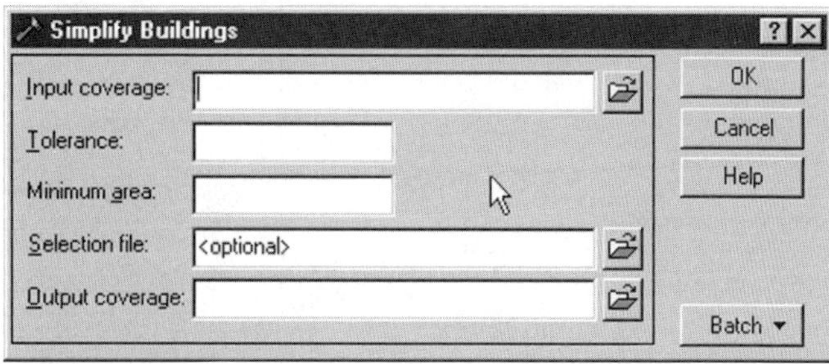

Simplify Buildings menu.

By setting a minimum area, you can even eliminate buildings that have a smaller area. One use for this would be to create a new coverage containing only office buildings or shopping malls. The minimum distance setting would eliminate all smaller buildings. This command contains the following options.

Input coverage: Coverage containing building footprints.

Tolerance: Sets tolerance as to how much buildings will be simplified.

Minimum area: Smallest area for buildings to be included in output coverage.

Selection file: ArcPlot selection file containing arcs you want deleted.

Output coverage: New coverage created.

GeoDatabase Tool

The GeoDatabase tool helps build a geodatabase, a new feature type in ArcInfo 8.0, from a line coverage. Specifically, use the Build Geometric Network Wizard to convert a line coverage to a geodatabase line coverage. A geodatabase line coverage supports complex attributes useful in applications used by the utilities industry. The GeoDatabase tool menu is shown in the following illustration.

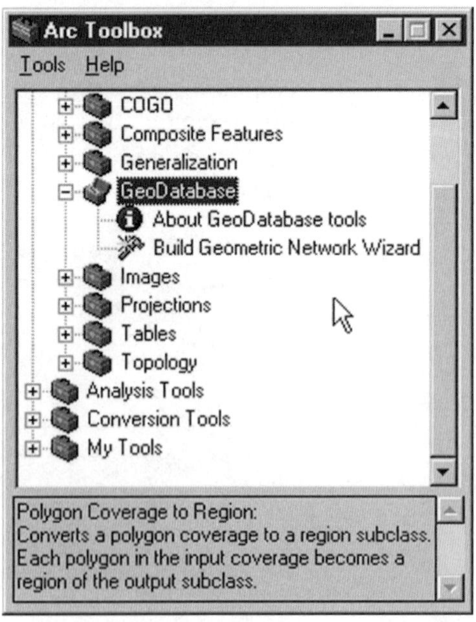

GeoDatabase tool menu.

Build Geometric Network Wizard

The Build Geometric Network Wizard, shown in the following illustration, takes you step by step through converting a line coverage to a geodatabase coverage. Coverages of utilities, for instance, can be converted. The benefit to making them geodatabase network coverages is that you can add features specific to utilities operations, such as complex edges, simple edges, and simple junctions.

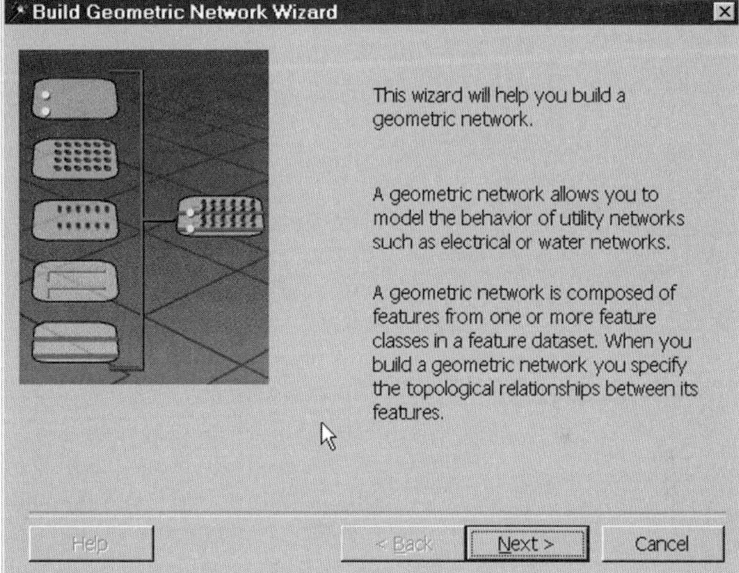

Build Geometric Network Wizard.

Projections Tool

The Projections tool set contains two wizards to aid in both defining projections for a data set and for changing a data set's projection. The Transform tool shifts coverage coordinates using user-specified parameters. The Projections menu is shown in the following illustration. The tools and wizards within this tool set are discussed in the sections that follow.

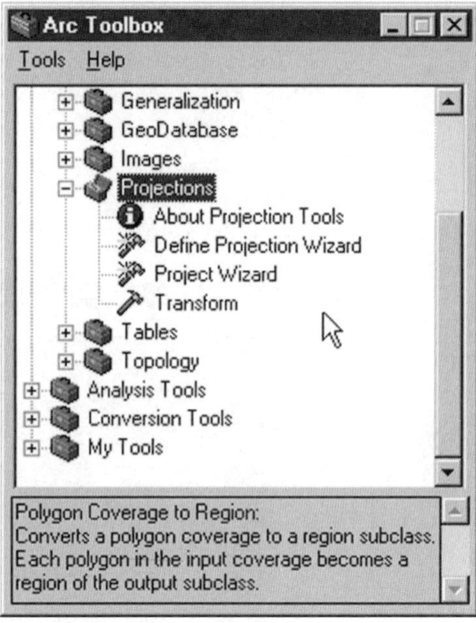

Projections menu.

Projections

Cartographers have for centuries struggled to develop means of representing the earth's surface on a map. Map projections are the means cartographers use to transform locations from 3D earth to a 2D paper map. GIS users are no different; they must pay the same attention to projections as cartographers.

All map projections distort to some degree shape, distance, or area. Some projections, such as the Mercator projection, maintain accurate shape but lose accurate area measurement. The GIS user must make some decisions about what projection properties he or she needs to best complete the spatial analysis. These decisions are governed by what the map is meant to convey (e.g., regions of the world or downtown parking lots) and by the map's scale.

For small-scale maps (maps that show a large area but little detail), pay attention to the projection used if you are concerned about using the map to measure distances or calculate area. Large-scale maps (ones that show a small area and a lot of detail) usually have minimal impact from distortion because the area they protraty is so small. Because these maps cover a very small portion of the earth's surface, projections are usually not an overriding concern.

ArcInfo comes with a rich collection of map projections that can be accessed from ArcToolbox or ArcInfo Workstation. See ArcInfo's on-line help (ArcInfo Concepts ➡ Map Projections) for a listing and discussion of the many map projections available. The on-line help includes information that helps you decide which map projection might be best for your application.

When using ArcInfo, be sure that the spatial coverages you use all have the same map projection. If any do not, they cannot be used with the other coverages for spatial analysis! If you display two coverages you think ought to cover the same physical space, such as a coverage of property boundaries and one of zoning, but they do not line up or do not display together, chances are one is in the wrong projection. Use the Describe command at the Arc prompt in ArcInfo Workstation, or use Map Catalog in Arc Desktop, to view the projection information. See the command reference in ArcToolbox for how to change a coverage's projection.

The Define Projection Wizard, shown in the following illustration, is used to define the projection of a coverage. You can either specify the projection in formation interactively or use the projection parameters from another coverage.

Define Projection Wizard.

Define Projection Wizard

Define the coordinate system of your coverage or grid

Define the coordinate system interactively.

You select a projection and specify its parameters.

Define a coordinate system for my data to match existing data

Matches the coordinate system of an existing coverage or grid.

Help | < Back | Next > | Cancel

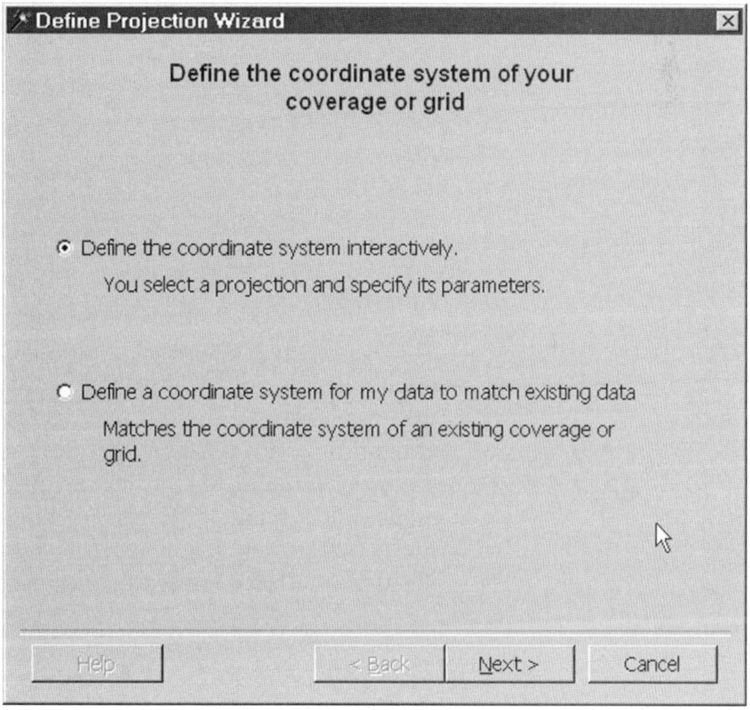

Project Wizard

The Project Wizard, shown in the following illustration, is used to change the projection of a coverage that already has a projection. You can specify the new projection's information interactively or use the projection parameters of another coverage.

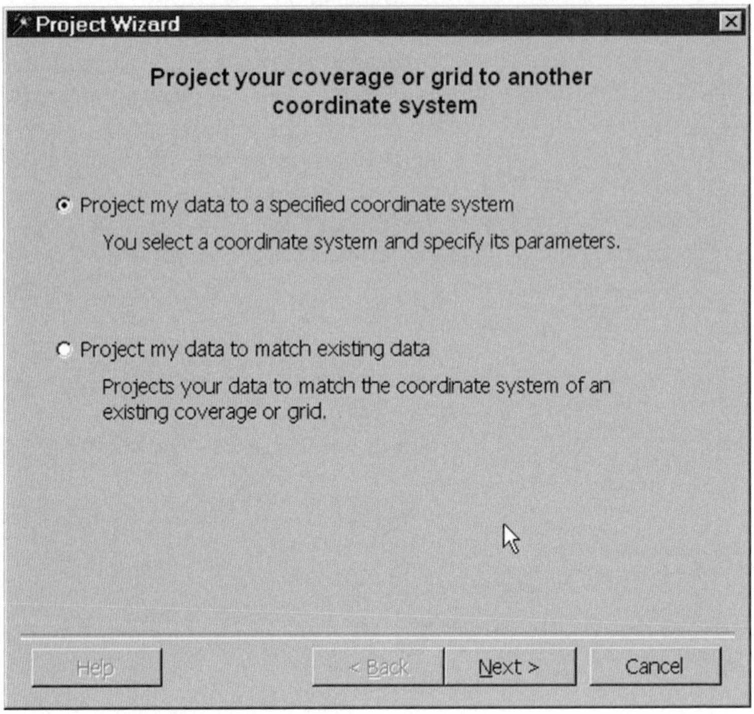

Project Wizard

Project your coverage or grid to another coordinate system

Project my data to a specified coordinate system

You select a coordinate system and specify its parameters.

Project my data to match existing data

Projects your data to match the coordinate system of an existing coverage or grid.

Help | < Back | Next > | Cancel

Project Wizard.

Transform

The Transform tool skews, rotates, scales, and shifts coverage coordinates using one of three transformation options: Affine, Projective, or Similarity. The output coverage must already exist and must contain tics with the new coordinates that are to be matched with the input coverage's coordinates. After completion, the output coverage will contain all features of the input coverage transformed to new coordinates. The Transform menu is shown in the following illustration. This command contains the following options.

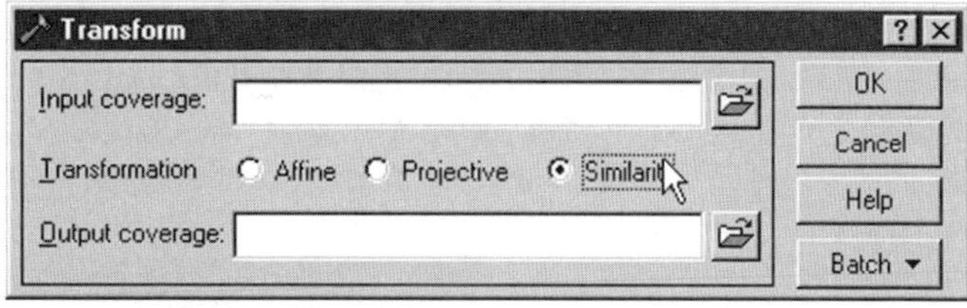

Transform menu.

Input coverage: Coverage to be transformed.

Transformation: Geometric method to be used.

Output coverage: New, transformed coverage.

Tables Tool

The Tables tool set allows access to INFO files for editing, file joining, and file creation. The Tables menu is shown in the following illustration. The tools within this tool set are discussed in the sections that follow.

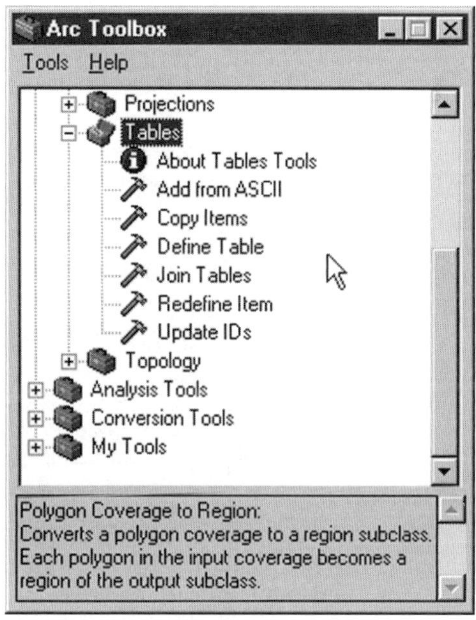

Tables menu.

Add from ASCII

The Add from ASCII tool adds data from an ASCII file to an INFO file. The Add from ASCII menu is shown in the following illustration. This command contains the following options.

Add from ASCII menu.

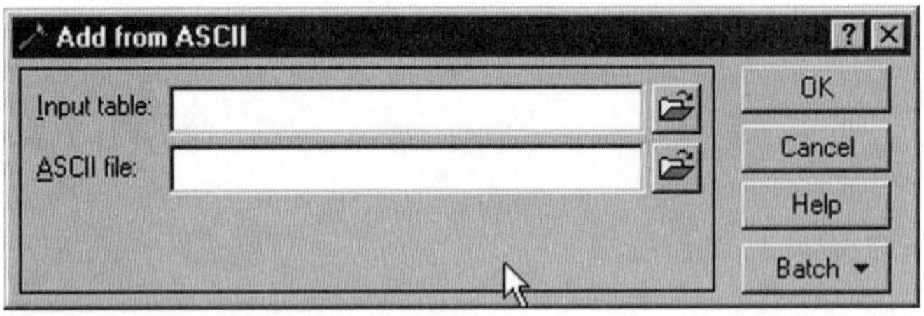

➾ **NOTE:** *Be sure of the following: The number of data elements in each record in the ASCII file must match the items in the INFO database, and the data type for each element in the ASCII file must match the corresponding INFO attribute (i.e., alphanumeric characters cannot be read into numeric or floating point items).*

Input table: INFO file to receive data.

ASCII file: ASCII file to import data from.

Copy Items

The Copy Items tool copies user-selected items from one table to another. The Copy Items menu is shown in the following illustration. This command contains the following options.

Copy Items menu.

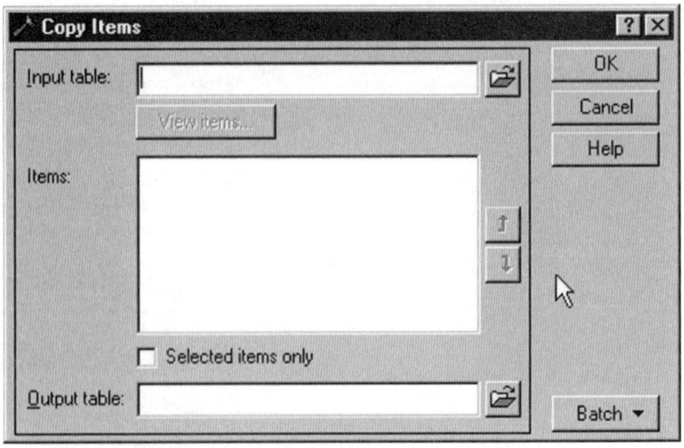

Input table: Table to copy items from.

Items: By default, all items will be copied. To select just some, click on each with the mouse.

Selected items only: Highlight this check box if you want to copy just the selected items above it.

Output table: File to copy items to.

Define Table

Use the Define Table tool to define and create a new table. Because the table is new, it will *not* contain any data. The Define Table menu is shown in the following illustration. This command contains the following options.

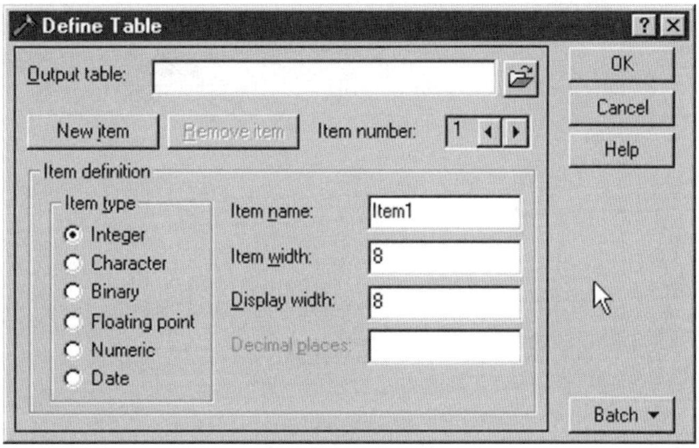

Define Table menu.

Output table: New table to create.

New item: Click on this option to create a new item within the table.

Remove item: Click on this option to delete an item already created.

Item number: Use the left and right arrows to move from item to item in your new table. You may insert or delete items at any time by selecting the "New item" or "Remove item" buttons.

Item type: Specify the item type, such as Integer, Character, Date, and so on.

Item name: Unique name for your item.

Item width: Data width for the item.

Display width: Specifies how many spaces will be used to display the item. You may specify more spaces than the item contains to ensure that blank spaces surround it when its data is displayed on screen or in a report.

Decimal places: Number of decimal spaces for numeric and floating point numbers.

Join Tables

The Join Tables tool is used to join database items from two tables. By default, the items are joined in the first table. You may specify a new, third file to contain the results. When joining two tables, the tables must share a common item, called a relate item. The Join Tables menu is shown in the following illustration. This command contains the following options.

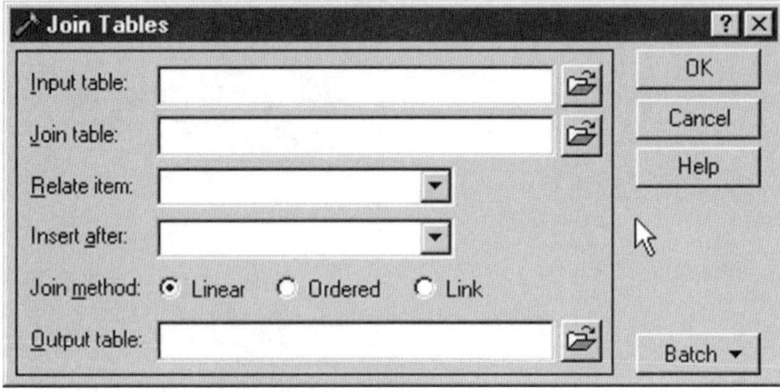

Join Tables menu.

Input table: First table to be joined.

Join table: Second table to be joined.

Relate item: Item each table shares.

Insert after: Items from the second table will be joined to items in the Input table after the item you specify in the Input table. This allows you the flexibility of inserting the new items at any point within the first table.

Join method: Choose Linear, Ordered, or Link. If using Linear, the two tables do not have to be sorted. It will, however, take a while to join the tables. Use Ordered when the join table is sorted on the relate item. This will speed joining. Link is a special sorting method where only the input file must contain the relate item. Here, the relate item holds the record number of the record in the join table to be merged. This method is used most often with INFO data files.

Output table: New table that contains the items of both tables. By default, it will be the Input table.

Redefine Items

The Redefine Item tool lets you change the definition of an item in an INFO data file. Some common uses are to make new, smaller item definitions from an existing, larger item. The existing item definition remains; you simply gain access to some of its constituent parts by breaking it up into smaller segments. A common use would be to break up an item that contained a street address into smaller items such as house number, street name, and so on. Conversely, you may join many smaller items that make up a street address into a single item that displays the complete address. The Redefine Item menu is shown in the following illustration. This command contains the following options.

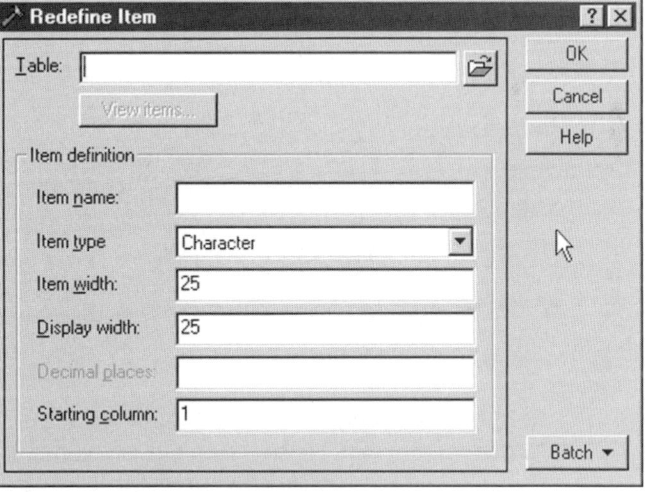

Redefine Item menu.

Table: Table you want to redefine.

Item name: The new name to call the redefined item.

Item type: Item type (Character, Numeric, Integer, and so on).

Item width: Length of the new item.

Display width: Number of spaces to use when displaying the item.

Decimal places: Decimal places allowed.

Starting column: Column in the INFO file where the new item begins. Remember, it can begin anywhere in the file.

Update IDs

The Update IDs tool updates user IDs in a coverage after they have been modified in INFO or another database software. If editing attributes for a coverage and you modify the user IDs, be sure to use this command; it will reestablish the link between a coverage's attribute data and its features. The Update IDs menu is shown in the following illustration. This command contains the following options.

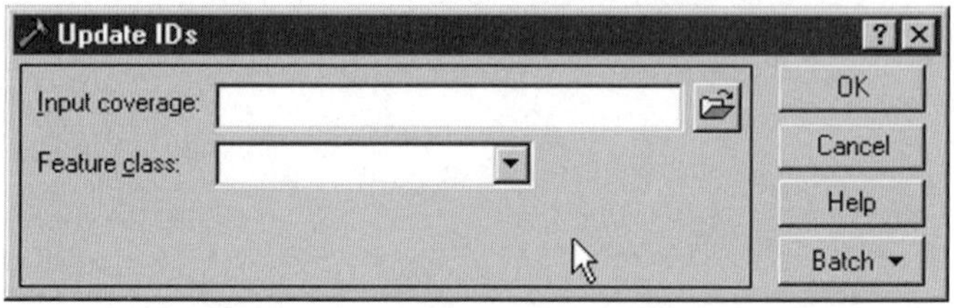

Update IDs menu.

Input coverage: Coverage whose IDs were modified using a database editor.

Feature class: Feature class of the coverage. Choose Point, Line, or Polygon.

Topology Tools

The last of the data management tool sets, Topology, contains some of the most commonly used commands in ArcInfo. These commands create and maintain topology for ArcInfo coverages. See Chapter 1 for a discussion of topology. The new GUI in Version 8.0 makes the commands even more accessible. The Topology tool menu is shown in the following illustration. The sections that follow describe the tools of this tool set.

Topology tool menu.

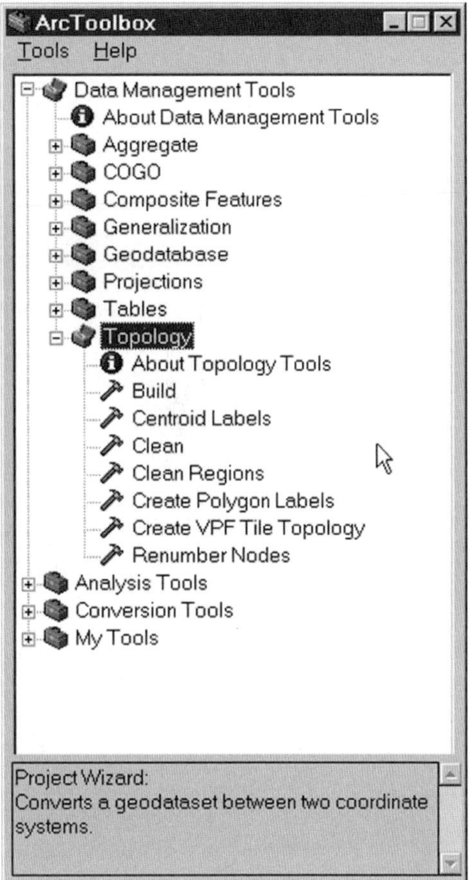

Build

The Build tool reestablishes topology for any coverage that has been edited. Build assumes that all intersecting lines in the coverage have already been identified and that nodes have been created appropriately (if not, use the Clean command). If nodes exist at intersections, the build proceeds. If not, the build will fail. The Build menu is shown in the following illustration. This command contains the following options.

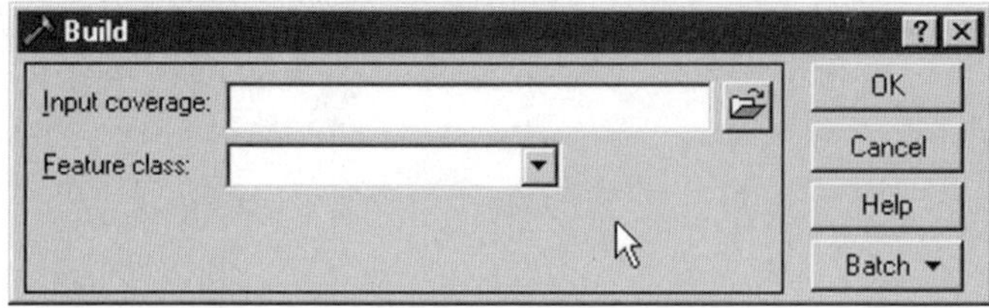

Build menu.

Input coverage: Coverage to have topology rebuilt.

Feature class: Type of coverage. Choose Line, Polygon, or Point. Topology will be built appropriately for the type of feature chosen.

Centroid Labels

The Centroid Labels menu, shown in the following illustration, moves polygon labels to the center of the coverage polygons. However, some irregularly shaped polygons may still have labels outside the polygon. Select "Force labels inside the polygons" to correct this situation.

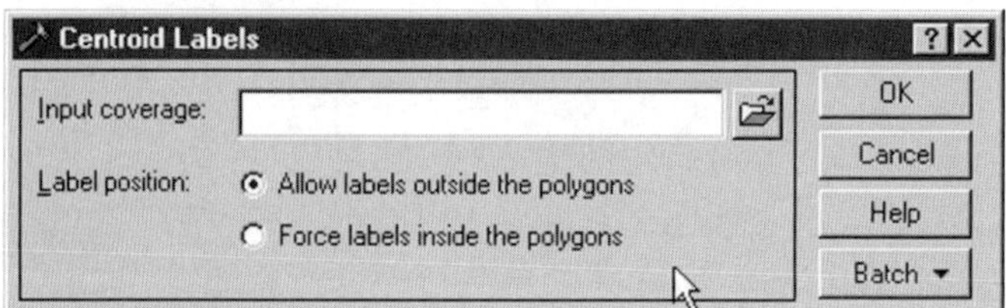

Centroid Labels menu.

Clean

The Clean menu, shown in the following illustration, creates topology for polygon and line coverages. Clean removes dangling arcs and sliver polygons, splits intersecting lines, and adds nodes at intersections. This command contains the following options.

Clean
Input coverage:
Dangle length:
Fuzzy tolerance:
Feature class: Poly Line
Output coverage:
OK
Cancel
Help
Batch

Clean menu.

Input coverage: Coverage to be cleaned.

Dangle length: Minimum length allowed for arcs. Arcs less than this value are deleted.

Fuzzy tolerance: Minimum distance between nodes or vertices.

Feature class: Select polygon or line.

Output coverage: New coverage created by Clean.

Clean Regions

Use Clean Regions, shown in the following illustration, to reduce the size of regions by merging polygons with identical item values. This command contains the following options.

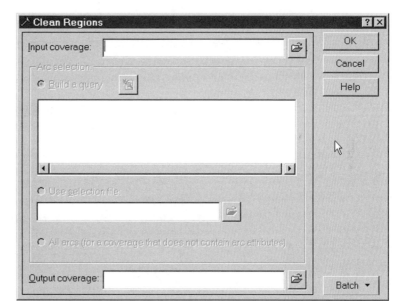

Clean Regions menu.

Input coverages: Coverage whose regions will be cleaned.

Output coverage: New coverage containing cleaned input coverage.

Create Polygon Labels

The Create Polygon Labels tool generates unique labels for each polygon in a polygon coverage. Use this command to place labels in each polygon automatically. The Create Polygon Labels menu is shown in the following illustration. This command contains the following options.

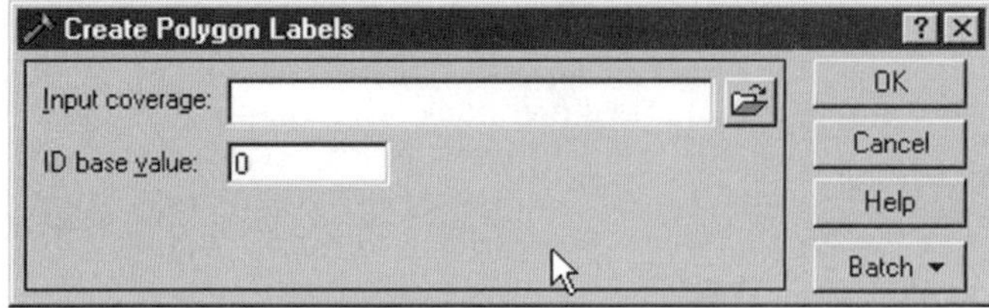

Create Polygon Labels menu.

Input coverage: Polygon coverage.

ID base value: Number with which to begin labeling empty polygons.

Create VPF Tile Topology

VPF tiles are a file storage format used by the military. Use this menu, shown in the following illustration, to create topology in a VPF library. This command contains the following options.

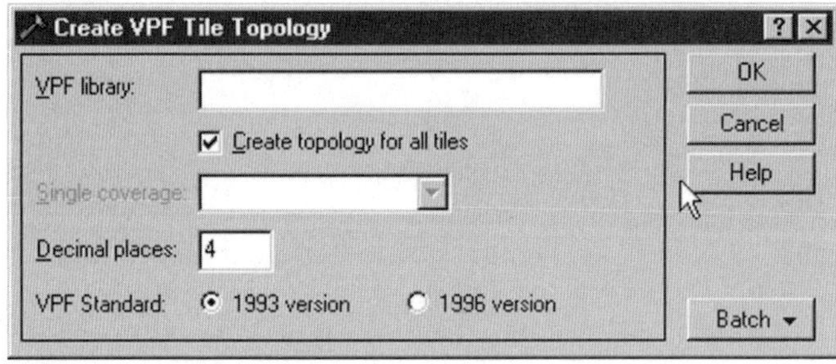

Create VPF Tile Topology menu.

VPF Library: Library for which you want to create topology.

Create topology for all tiles: Used to create topology for all tiles in a VPF library.

Single coverage: Used to specify a single coverage in a VPF library to build topology.

Decimal places: Used to specify more decimal places to create a tighter search tolerance when matching features.

VPF Standard: Select either the 1993 or the 1996 VPF standard.

Delete Feature Class

The Delete Feature Class tool deletes selected feature classes from a coverage. Remember, a coverage can have line, polygon, and point features. The Delete Feature Class menu is shown in the following illustration. This command contains the following options.

Delete Feature Class menu.

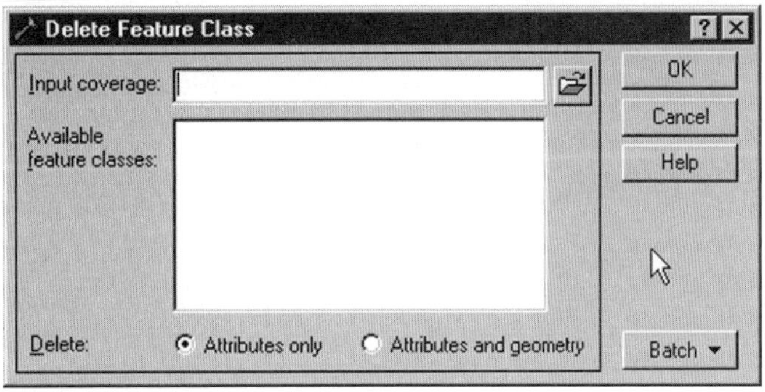

Input coverage: Coverage to delete feature class from.

Available feature classes: A listing of feature classes present in the Input coverage. Choose the one you want to delete.

Delete: Choose either "Attributes only" or "Attributes and geometry." If you choose "Attributes only," the spatial features will remain. For example, you may choose to delete attributes from a road coverage. All roads will remain, but all data items associated with the roads will be deleted. To delete both the physical lines that make up the roads as well as the data, select "Attributes and geometry."

Renumber Nodes

Use Renumber Nodes, shown in the following illustration, to renumber nodes in a coverage. In some instances, arcs may overlap without nodes being present. Use the "From node" and "To node" items to specify features in the coverages as above or below one another. This command contains the following options.

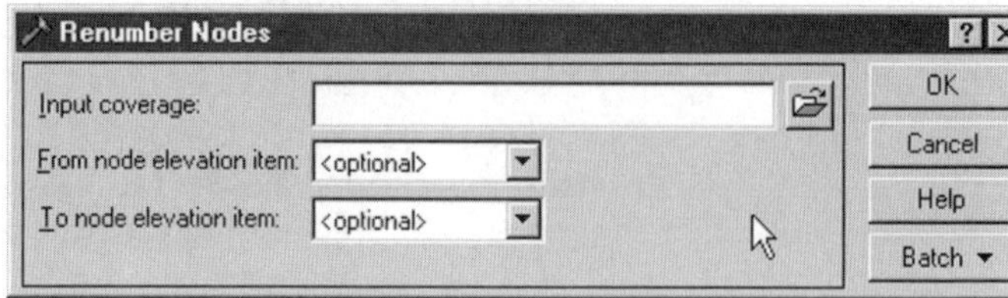

Renumber Nodes menu.

Input coverage: Coverage whose nodes will be renumbered.

From node elevation item: Used to select the item that designates the "From" nodes.

To node elevation item: Used to select the item that designates the elevation of the "To" nodes.

Analysis Tools

Analysis tools contribute toward making GIS more than a CAD program. Divided into five tool sets, these commands provide powerful methods for manipulating spatial data and performing sophisticated analysis. These commands are also available in their command line format at the Arc prompt. The sections that follow discuss the tool sets and commands within this toolbox. The Analysis toolbox is shown in the following illustration.

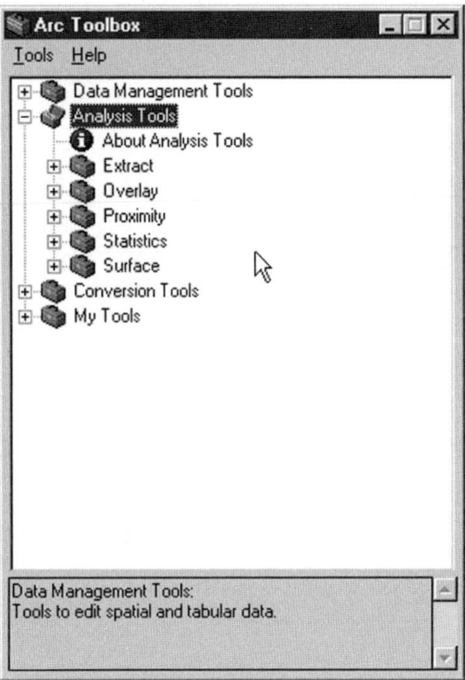

Analysis Tools toolbox.

Extract Tools

The following sections discuss the commands available within the Extract Tools menu, shown in the following illustration. The tools are used to create new coverages by extracting spatial features from other coverages.

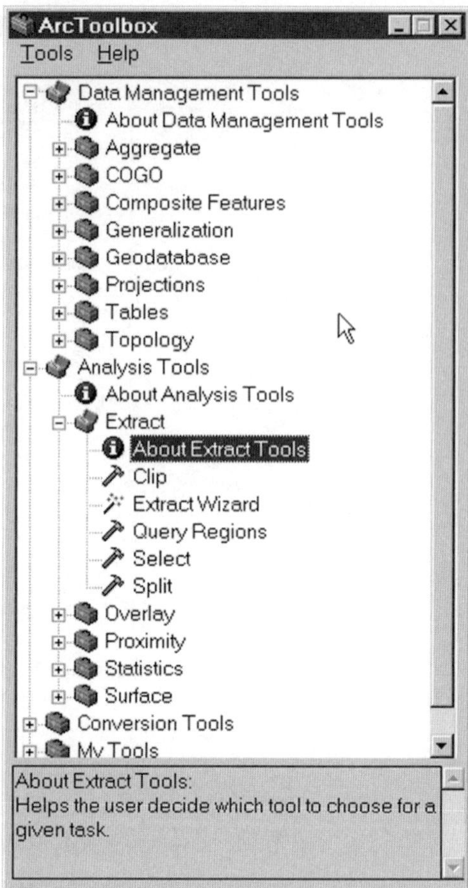

Extract tools menu.

Clip

The Clip tool uses the features in the clip coverage to serve as a "cookie cutter" to extract features from the input coverage. Only those features that fall within the extent of the clip coverage will be extracted. Feature attribute data from the input coverage is maintained. The Clip menu is shown in the following illustration. This command contains the following options.

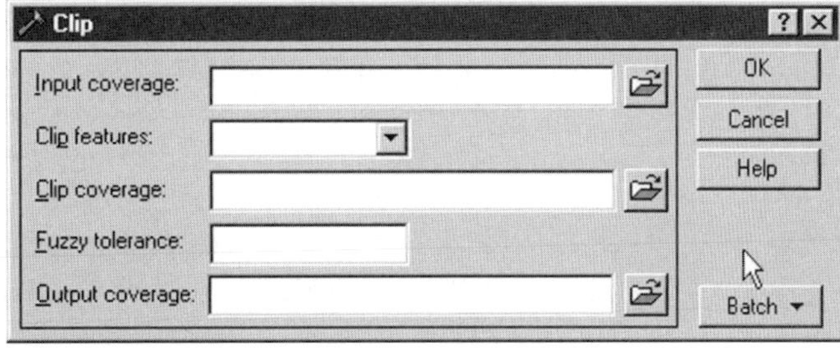

Clip menu.

Input coverage: Coverage to be clipped.

Clip features: Feature type to be clipped. Choose Polygon, Line, Net, Link, or Point.

Clip coverage: Coverage to use as the clip.

Fuzzy tolerance: Minimum distance between coordinates in the output coverage. By default, the tolerance is derived automatically from the input and clip coverages.

Output coverage: New coverage created.

Extract Wizard

The Extract Wizard guides you through a series of screens in which you specify an input coverage, the feature type you want to extract (polygon, line, or point), and the criteria for extracting the feature. The Extract Wizard menu, shown in the following illustration, contains a Query Builder tool for specifying the criteria.

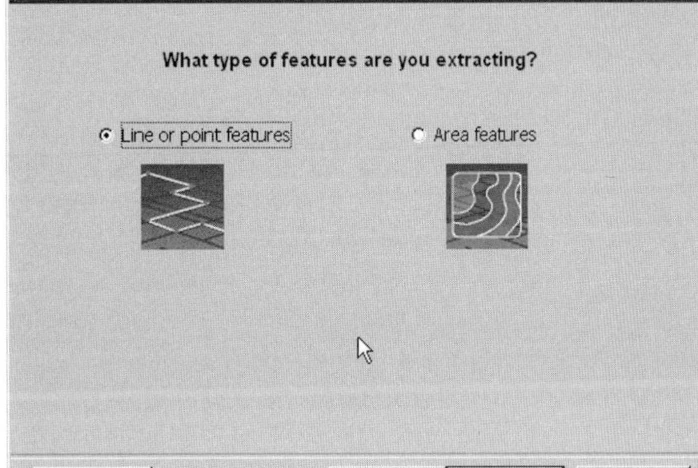

Extract Wizard.

Query Regions

The Query Regions tool, shown in the following illustration, creates new regions based on the attribute values of input region or polygon layers. Query Regions works with polygons and region subclasses in the same coverage. Build a query using logical expressions to select regions and polygons. This command contains the following options.

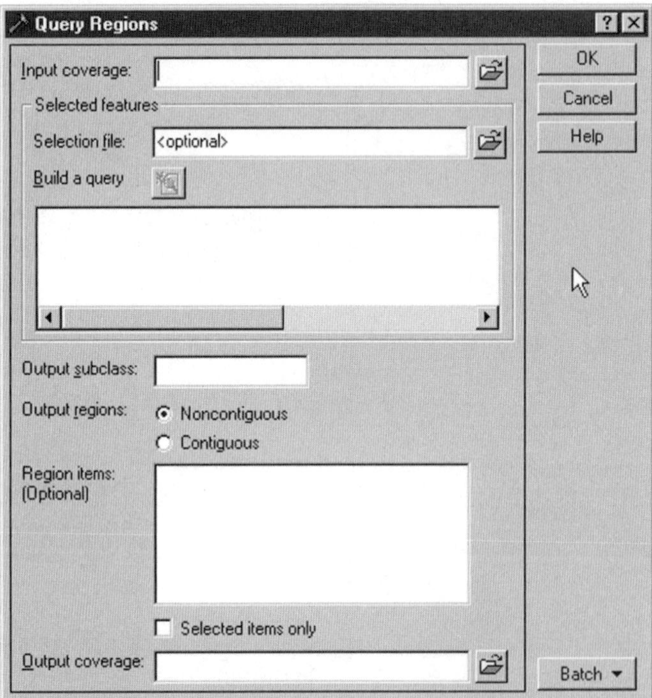

Query Regions menu.

Input Coverage: The coverage whose regions will be queried.

Selection file: Use this option if you have already created a selection file using ArcPlot's Writeselect command, which writes the currently selected set of features in ArcPlot to a file. Specifying a "writeselect" file here uses that file to create new regions.

Build a query: Select this option to open the Query Builder. The Query Builder is used to specify a logical expression used for extracting features.

Output subclass: The name of the output subclass you want to create. It can be up to 13 letters long.

Output regions: Choose either noncontiguous or contiguous regions.

Region items: Specify items from the coverage's feature attribute table you want to dissolve new regions on.

Output coverage: The name of the coverage created by the Query Regions command.

Select

The Select tool is used to extract features from a coverage based on attribute values you select. The Select menu is shown in the following illustration. This command contains the following options.

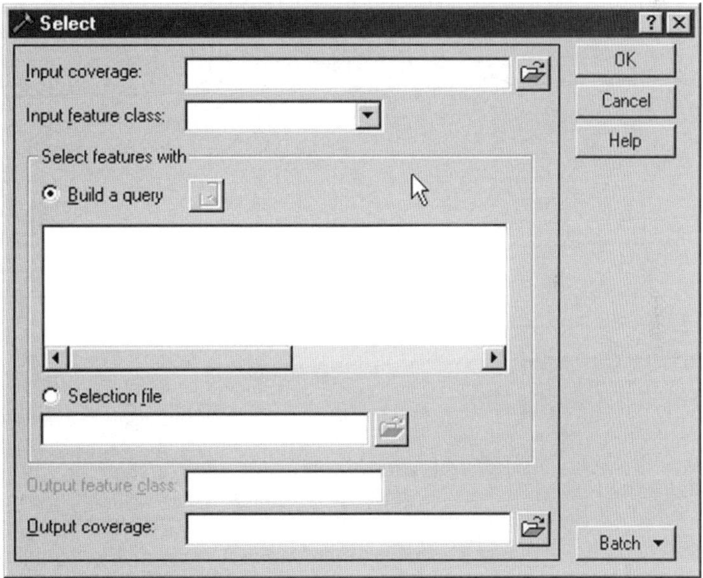

Select menu.

Input coverage: Coverage containing features to be extracted.

Input feature class: Choose the type of feature to extract. Options are Polygon, Line, Point, Anno, Route, and Region.

Build a query: Use Arc's "reselect," "nselect," or "aselect" commands to select features based on their attribute values.

Selection file: Optionally, use this to select features based on an ArcPlot selection file.

Output feature class: Feature type of the coverage to be created after selection.

Output coverage: New coverage containing a subset of features from the input coverage.

Split

Use the Split tool to split a single coverage into multiple coverages. The Split menu is shown in the following illustration. This command contains the following options.

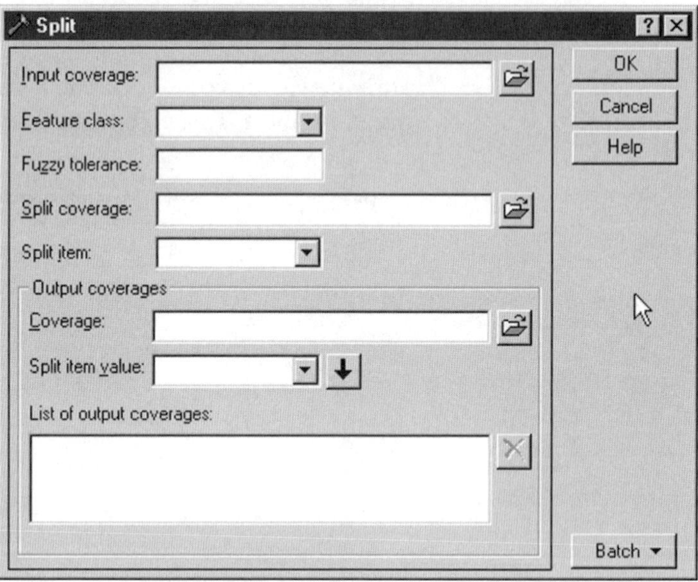

Split menu.

Input coverage: Coverage to be split into multiple coverages.

Feature class: Feature class to be split.

Fuzzy tolerance: Minimum distance between coordinates in the output coverage. By default, it will be automatically determined based on the input and split coverages.

Split coverage: Coverage used to split the input coverage.

Split item: Item in the split coverage used to split the input coverage. Features with the same item value in the split coverage will be used to clip features from the input coverage.

Coverage: Coverage to be created.

Split item value: Specify a value to be used for selecting coverage polygon. Use the down arrow (↓) to add this to the list of output coverages beneath.

List of output coverages: Used to view output coverages created.

Overlay Tools

The Overlay tools provide means of creating a new coverage by combining, erasing, or changing spatial data in two coverages. The sections that follow discuss the commands available within this tool set. The Overlay menu is shown in the following illustration.

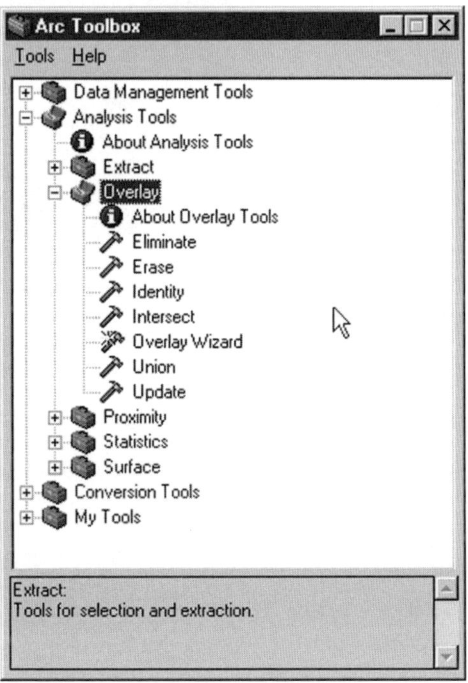

Overlay tools menu.

Eliminate

The Eliminate tool merges adjacent polygons in a coverage based on either the longest shared arc between them or the largest area. Eliminate is most often used to remove sliver polygons. It can also be used with the Line option to merge arcs separated by pseudo nodes (nodes you have manually entered to split an arc into two arcs). Pseudo nodes do *not* occur at intersections of arcs. The Eliminate menu is shown in the following illustration. This command contains the following options.

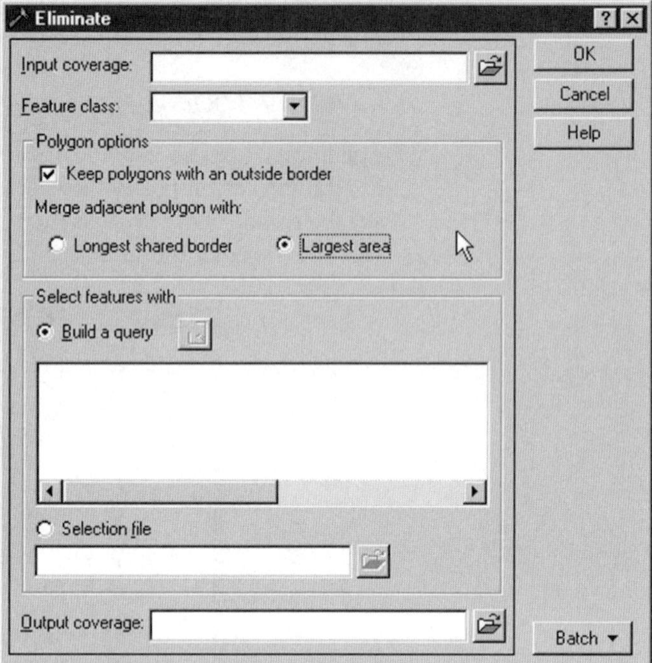

Eliminate menu.

Input coverage: Coverage whose polygons or arcs will be merged.

Feature class: Choose polygons or arcs.

Keep polygons with an outside border: Check this to ensure that polygons on the outside of the coverage maintain their outside arcs.

Merge adjacent polygons with: Choose either "Longest shared border" or "Largest area." If there are more than one polygon a small polygon could merge with, the decision chosen here will govern which polygon is used to merge with.

Build a query: Use Arc's "reselect," "nselect," or "aselect" commands to specify the criteria for which polygons will be eliminated.

Selection file: Optional ArcPlot selection file containing features to be eliminated.

Output coverage: New coverage containing simplified coverage.

Erase

Use the Erase tool to erase features in the input coverage with polygons in the erase coverage. The Erase menu is shown in the following illustration. This command contains the following options.

Erase menu.

Input coverage: Coverage that contains features to be erased.

Feature class: Feature class to be eliminated. Choose Polygon, Line, Net, Link, or Points.

Erase coverage: Coverage whose outer polygon defines the area to be erased.

Fuzzy tolerance: Minimum distance between features in the output coverage.

Output coverage: New coverage created.

Identity

The Identity tool generates a new coverage based on the overlap present between the input and identity coverages. The new coverage maintains all features from the input coverage, as well as features from the identity coverage that overlap the input coverage. This command contains the following options.

Identity menu.

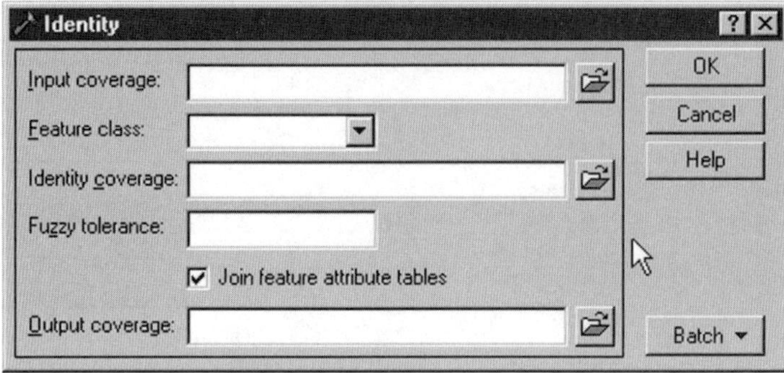

Input coverage: Coverage that will be overlaid on the identity coverage.

Feature class: Feature type in input coverage to be used. Choose Polygon, Line, or Point.

Identity coverage: Coverage containing polygon topology to be overlaid with input coverage.

Fuzzy tolerance: Minimum distance between coverage features in new output coverage.

Join feature attribute tables: Choose this to also merge the attribute tables for each coverage. If you do not choose this, only the coverage internal record number from each coverage will be preserved in the new coverage.

Output coverage: New coverage created from input and identity coverages.

Intersect

The Intersect tool generates a new coverage based on the intersection of two coverages. Only features in the area common to both coverages are preserved in the output coverage. The Intersect menu is shown in the following illustration. This command contains the following options.

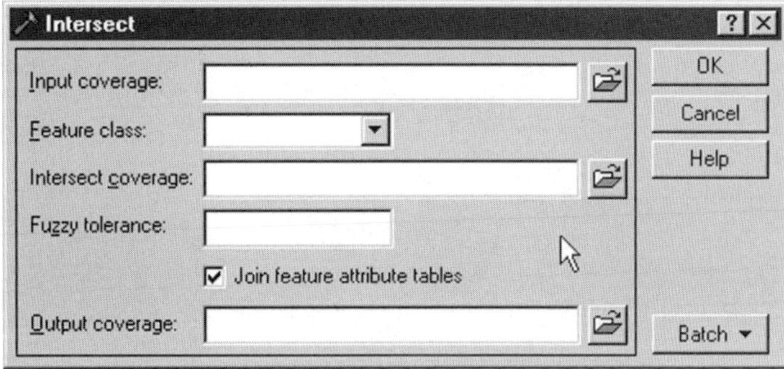

Intersect menu.

Input coverage: Coverage containing polygons, lines, or points.

Feature class: Choose Polygons, Lines, or Points.

Intersect coverage: Coverage with polygon topology.

Fuzzy tolerance: Minimum distance between features in output coverage.

Join feature attribute tables: Choose this to also merge the attribute tables for each coverage. If you do not choose this, only the coverage internal record number from each coverage will be preserved in the new coverage.

Output coverage: Coverage created from input and intersect coverages.

Overlay Wizard

The Overlay Wizard, shown in the following illustration, provides a choice of three types of overlay. The first, Intersect, combines in a new coverage the features in the area common to two coverages. The second, Identity, creates a new coverage by incorporating an area of the second into the first. Union, the third option, combines everything in both coverages into a new coverage. The wizard prompts you for the two coverage names.

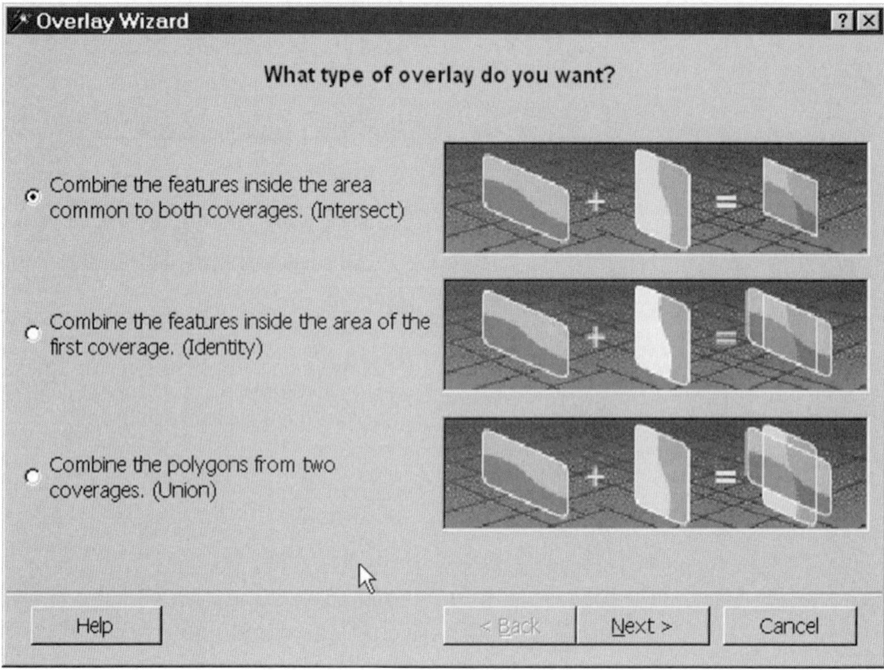

Overlay Wizard.

Union

The Union tool combines all features of the input and union coverage, splitting overlapping features at their intersections, to create a new coverage. Both coverages must be polygon coverages. The Union menu is shown in the following illustration. This command contains the following options.

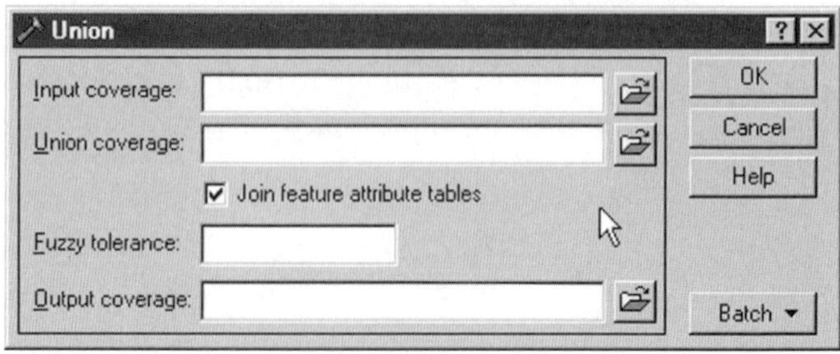

Union menu.

Input coverage: Polygon coverage.

Union coverage: Polygon coverage.

Join feature attribute tables: Select this to also merge attribute tables from both coverages.

Fuzzy tolerance: Minimum distance between new features in new coverage.

Output coverage: New coverage containing all polygons from input and identity coverages.

Update

The Update tool replaces features in the input coverage with features from the update coverage using a cut-and-paste method. It is useful for updating small areas of a larger coverage; for example, replacing a parcel with a new subdivided parcel in a coverage of properties. Attribute information is maintained in the new coverage. The Update menu is shown in the following illustration. This command contains the following options.

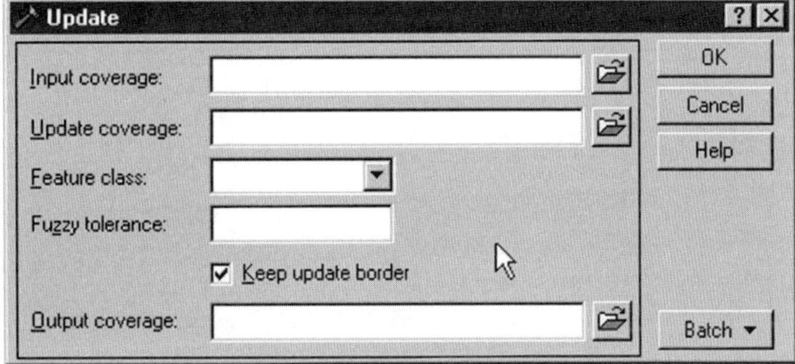

Update menu.

Input coverage: Coverage to be updated.

Update coverage: Coverage containing features to update.

Feature class: Choose Polygon or Net. Choosing Polygon just updates polygon values. Choosing Net also replaces arcs and polygon data.

Fuzzy tolerance: Minimum distance between features in new coverage.

Keep update border: Check this to maintain outer polygon boundary from update coverage in new coverage.

Output coverage: New coverage containing updated features.

Proximity Tools

The Proximity tools are powerful commands for finding features near a given set of features. They include often-used commands such as Buffer and Near. The sections that follow discuss the commands within this tool set. The following illustration shows the Proximity menu.

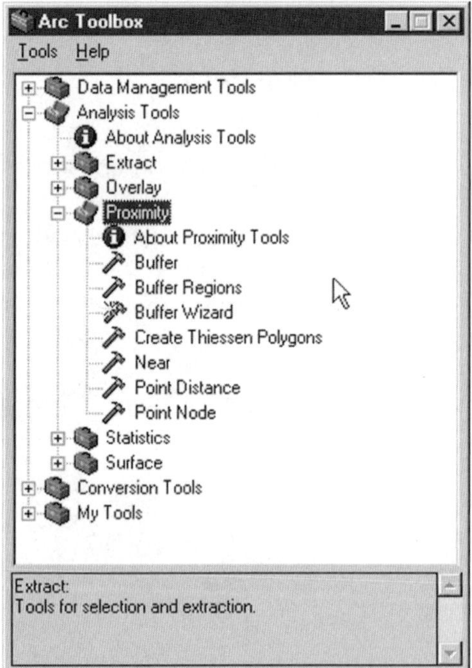

Proximity tools menu.

Buffer

The Buffer command creates a new coverage of polygons based on selected features in the input coverage. Use this command to create buffers around point, line, polygon, or node features. Buffers may be generated for all features in a coverage or just some. The Buffer menu is shown in the following illustration. This command contains the following options.

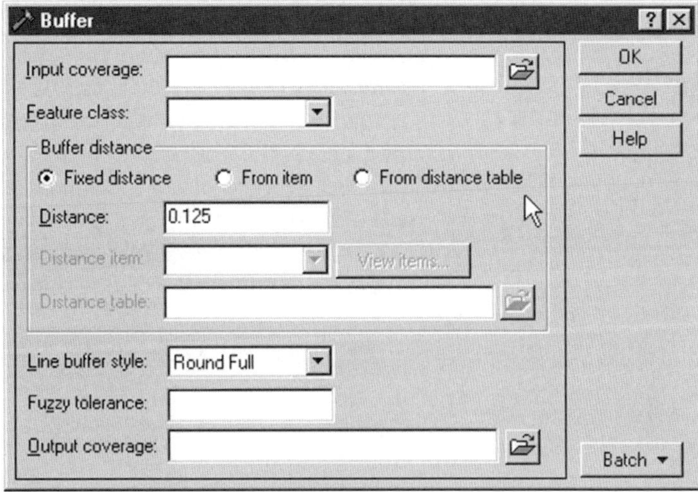

Buffer menu.

Input coverage: Polygon, point, line, or node coverage to be buffered.

Feature class: Choose Polygon, Line, Point, or Node.

Buffer distance: Select "Fixed distance," "From item," or "From distance table." If you select "Fixed distance," you may then enter a distance value in the Distance field. Enter a value in coverage units. If you select "From item," buffers will be generated by the value specified in the "From item" field. If you select "From distance table," enter an item name in the "From item" field and the name of a table in the "Distance table" field. The table specified must have an item named the same as that in the "From item" field, as well as an item called *Dist,* which has the actual buffer distance.

Line buffer style: Choose from Round Full, Round Left, Round Right, Flat Full, Flat Left, and Flat Right. This option controls how your buffer looks when buffering line features, whether it is blocky when choosing any Flat option or has rounded ends when choosing any of the Rounds. Specifying Right or Left means that buffers are created on only one side of a line. Full generates a buffer all around a line.

Fuzzy tolerance: Minimum distance between features in the output coverage.

Output coverage: Name of the new polygon coverage containing buffers.

Buffer Regions

Similar to the Buffer tool, Buffer Regions generates buffers around point, line, polygon, node, or region features. The output is a region feature, either contained in the existing coverage or placed in a new one. Be careful! Be sure to specify a new output coverage if your input coverage is either a point or line feature. Otherwise, your data will be converted into polygons as part of the region creation process. The Buffer Regions menu is shown in the following illustration. This command contains the following options.

Buffer Regions menu.

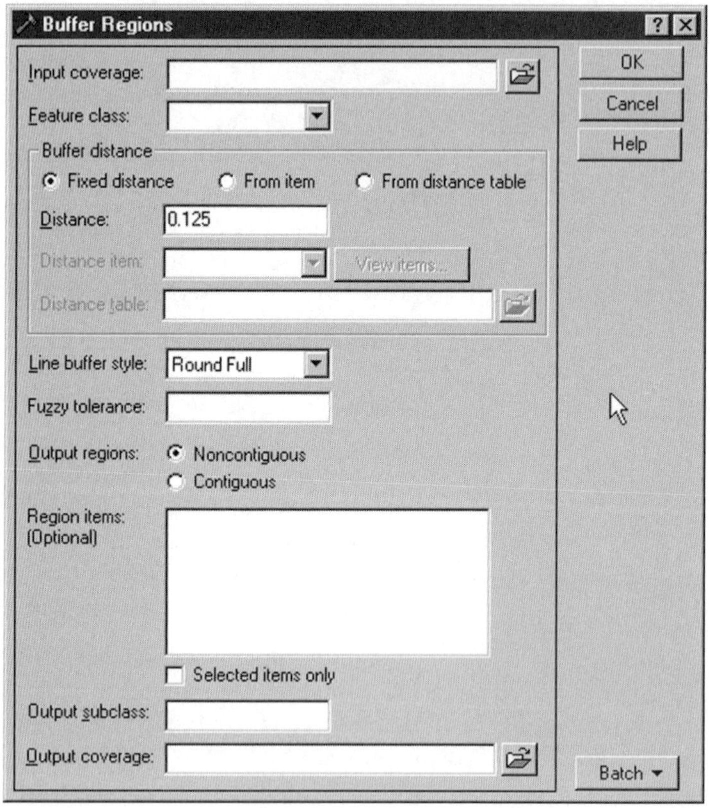

Input coverage: Polygon, point, line, or node coverage to be buffered.

Feature class: Choose Polygon, Line, Point, or Node.

Buffer distance: Select "Fixed distance," "From item," or "Distance table." If you select "Fixed distance," you may then enter a distance value in the Distance field. Enter a value in coverage units. If you select "From item," buffers will be generated by the value specified in the "From item" field. If you select "Distance table," enter an item name in the "From item" field and the name of a table in the "Distance table" field. The table specified must have an item named the same as that in the "From item" field, as well as an item called *Dist*, which has the actual buffer distance.

Line buffer style: Choose from Round Full, Round Left, Round Right, Flat Full, Flat Left, and Flat Right. This option controls how your buffer looks when buffering line features, whether it is blocky when choosing any Flat option or has rounded ends when choosing any of the Rounds. Specifying Right or Left means that buffers are created on only one side of a line. Full generates a buffer all around a line.

Fuzzy tolerance: Minimum distance between features in the output coverage.

Output regions: Choose either Contiguous or Noncontiguous. Contiguous creates only contiguous regions as output, whereas Noncontiguous creates both.

Region items: Specify optional feature attributes you want transferred to the buffer regions in the output coverage.

Output subclass: Region subclass to create.

Output coverage: Coverage you want to create buffer regions in, either the existing coverage or a new one. If a region's class by the same name already exists in an existing coverage, the buffer regions will be appended to it.

Buffer Wizard

The Buffer Wizard, shown in the following illustration, provides a series of menus that walk you through using the Buffer command. The menus relate to the Buffer command previously described and contain the same input options. The wizard is simply another way of running the Buffer command.

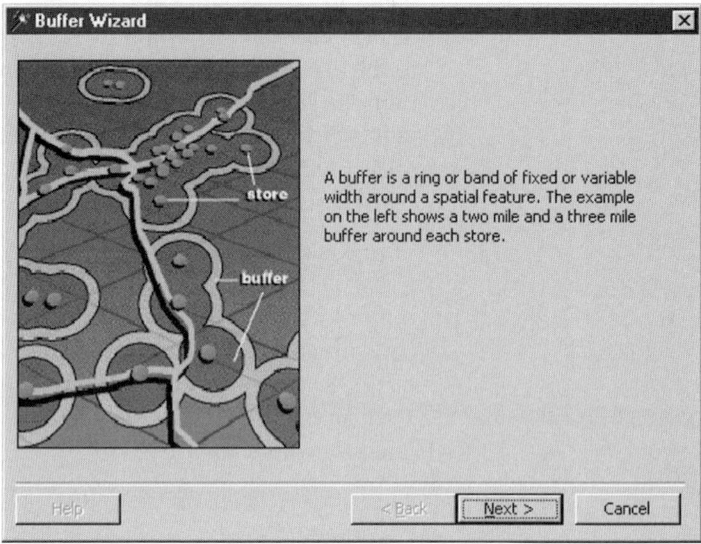

Buffer Wizard.

Create Thiessen Polygons

The Create Thiessen Polygons tool creates a polygon coverage of Thiessen polygons from a point coverage. Thiessen polygons are polygons in which each side is closer to the polygon centerpoint than to any other point. The Create Thiessan Polygons menu is shown in the following illustration. This command contains the following options.

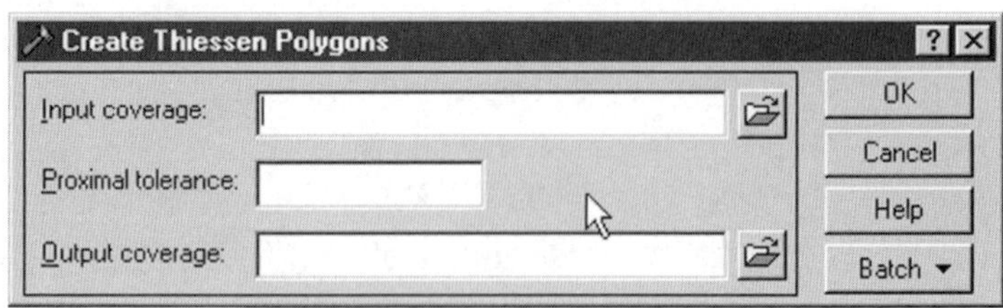

Create Thiessen Polygons menu.

Input coverage: Point coverage.

Proximal tolerance: Distance in coverage units used to weed out points in the input coverage that are too close together.

Output coverage: Coverage with Thiessen polygons.

Near

Use the Near tool to find the distance between any point in the input coverage to any point or line in the near coverage. Distances are added to the PAT of the input file. The Near menu is shown in the following illustration. This command contains the following options.

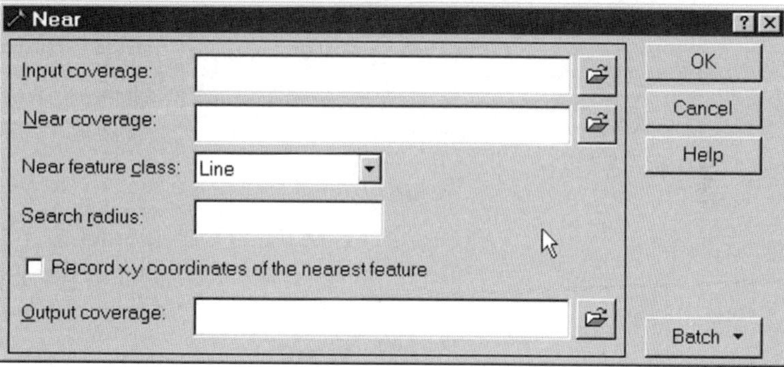

Near menu.

Input coverage: Point coverage.

Near coverage: Line or point coverage.

Near feature class: Feature class of near coverage.

Search radius: Maximum point a feature can be from a point in the input coverage for a calculation to be made.

Record X,Y coordinates of the nearest features: Check this box if you want the X-Y coordinates of the nearest feature to be recorded in the output coverage.

Output coverage: Point coverage that contains the distance calculations. It may be a new coverage or the input coverage.

Point Distance

The Point Distance tool calculates the distance between points in one coverage to points in a second coverage. The Point Distance menu is shown in the following illustration. Caution! The output table containing distances can be quite large, depending on the number of points in each coverage and the search radius specified. This command contains the following options.

Point Distance menu.

From coverage: Point coverage to measure distances from.

Search radius: Maximum distance to the search point in the To coverage.

To coverage: Point coverage to measure distances to.

Output table: INFO table containing distances between all points in the From coverage to points in the To coverage.

Point Node

The Point Node tool transfers the attribute information from a point coverage to a node attribute table. If a node is within the search radius of a point, attribute data is transferred to that node. If more than one node falls within the search radius, the closest node receives the data. The Point Node menu is shown in the following illustation. This command contains the following options.

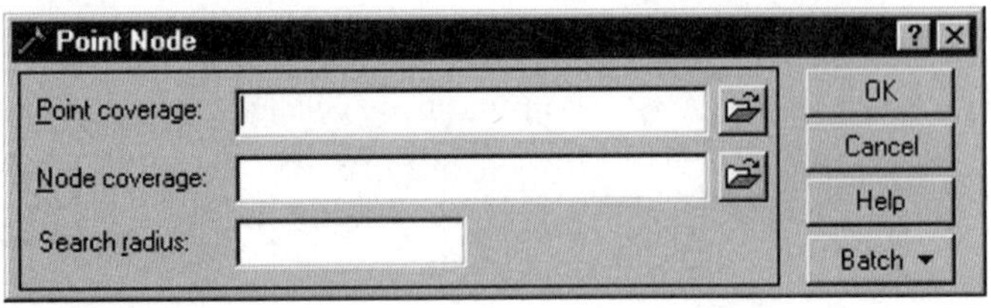

Point Node
Point coverage:
Node coverage:
Search radius:
OK
Cancel
Help
Batch

Point Node menu.

Point coverage: Point coverage whose data will be transferred to the node coverage.

Node coverage: Node coverage that will be updated by the point coverage.

Search radius: Distance in coverage units to search around point locations for nodes.

Statistics Tools

The Statistics tools let you calculate frequency and summary statistics for any feature attributes. These tools create a new table that contains unique instances, as well as the number of their occurrence in the chosen attribute items. The sections that follow discuss the commands available in this tool set. The following illustration shows the Statistics menu.

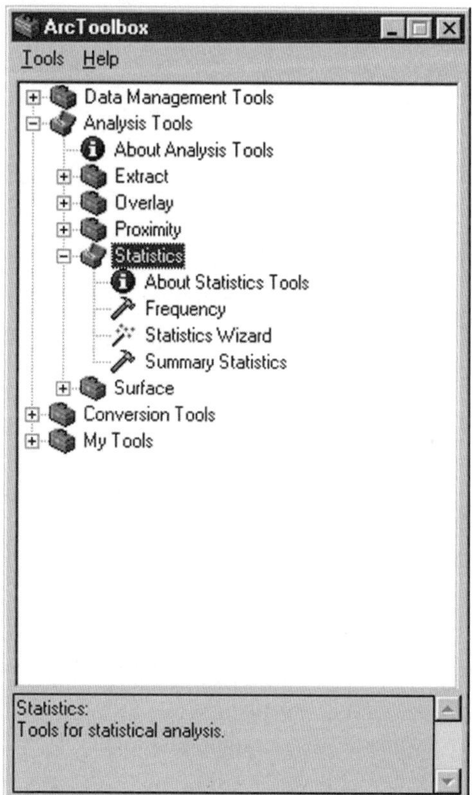

Statistics menu.

Frequency

The Frequency tool lets you calculate how often a value occurs in any attribute. The Frequency menu is shown in the following illustration. This command contains the following options.

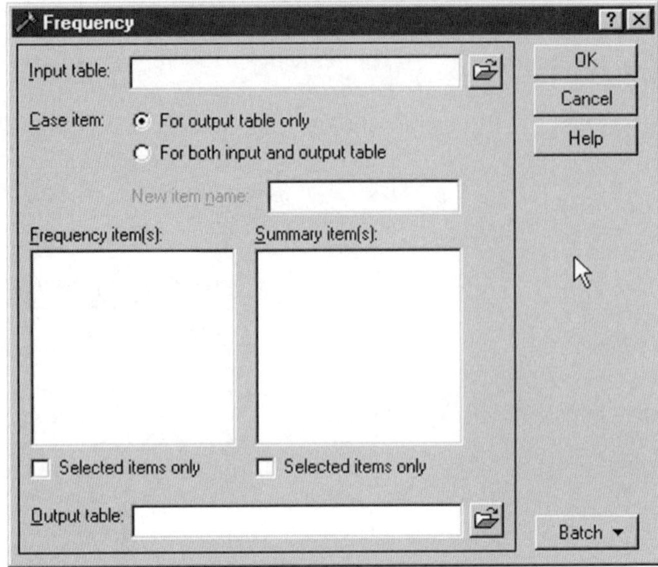

Frequency menu.

Input table: Table that contains attribute(s) to calculate frequency.

Case item: Choose "For output table only" to have an item called CASE# added to it. Choose "For both input an output table" to have CASE# added to both. Choosing the second option allows you to make a link between the two tables.

Frequency items: Choose the attribute items whose unique elements you want to count.

Summary items: Choose the items you want to summarize for unique occurrences of the items chosen in Frequency.

Output table: New table that contains unique values and the number of occurrences of each item.

Statistics Wizard

The Statistics Wizard, shown in the following illustration, contains a series of windows that walk you through using Frequency or summary statistics.

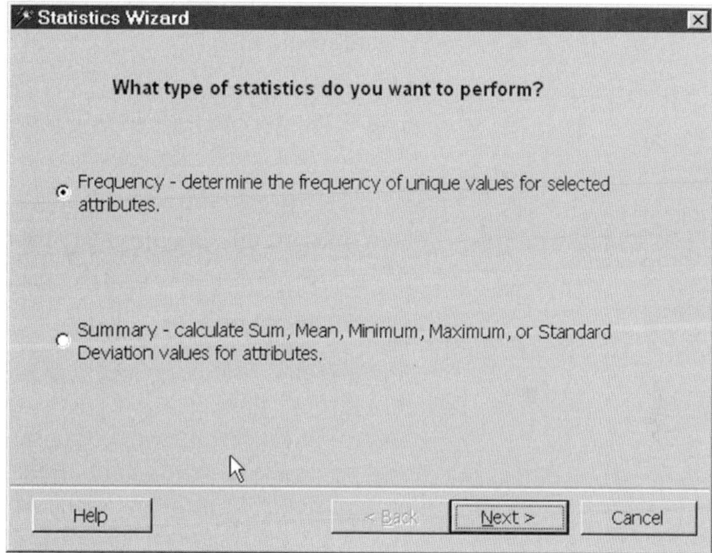

Statistics Wizard.

Summary Statistics

The Summary Statistics tool generates summary statistics for the items you select in an INFO table. The results are produced in a new INFO table. The Summary Statistics menu is shown in the following illustration. This command contains the following options.

Summary Statistics
Input table:
Case item: <optional>
Statistical expressions
Statistics: Sum
Item:
Weight value: None Constant From item
Sum
Statistical expressions:
Output table:
OK
Cancel
Help
Batch

Summary Statistics menu.

Input table: INFO table containing items for which statistics are to be gathered.

Case item: Optional. If you choose a case item, statistics will be gathered for records containing unique values in the case item. Otherwise, statistics will be gathered for all records in the INFO file.

Statistics: Choose the type of statistics to gather. Choices are Sum, Mean, Minimum, Maximum, and Standard Deviation.

Item: Lists the numeric items in your INFO table. Choose those to gather statistics about.

Weight value: You may specify a weight value to use when applying statistics. You can choose None, Constant, or "From item." None means that no weight value will be used. Constant means that the value you specify here will be multiplied by each item you chose in the Item field. "From item" is the same as Constant, except that the value will be retrieved from an item in the INFO database.

Statistical expressions: Lists the equations here for use in creating statistics. You may enter up to 25 expressions. An example might be "SUM area," which will create a summary of all values held in an INFO item called "area." Each time you use the Item and Statistics buttons above, a new equation will be listed here. You may click on the X button to the right of this window to delete an equation.

Output table: Name of the INFO table that will hold the results of the statistical equations.

Conversion Tools

One of ArcInfo's greatest strengths is its ability to use data from many other sources. ArcToolbox contains a full suite of tools to both read spatial data in from other sources and to save ArcInfo data sets in other formats. The following illustration shows the main categories of ways to read spatial data and to export it for use in other programs. Because there are so many file formats ArcInfo can read and write to, the discussion does not attempt to cover the details of each format. The Conversion Tools menu is shown in the following illustration.

Conversion Tools menu.

Export from Coverage

The export tools are useful for converting an ArcInfo coverage into one of a number of common spatial data formats. Some of the formats are very popular, such as the DXF format used by AutoCAD. Others, such as VPF, are formats used by federal government agencies and are less likely to be needed by most users. The Export from Coverage menu is shown in the following illustration.

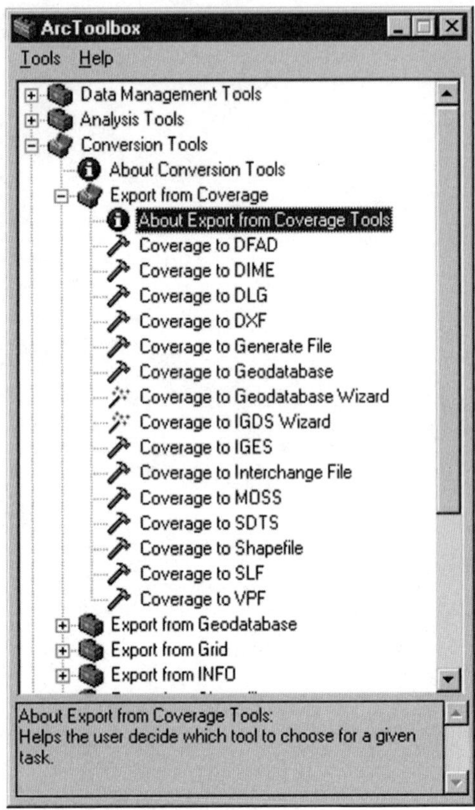

Export from Coverage menu.

Export from Geodatabase

These tools provide a means to export a geodatabase to a coverage, a shapefile (ArcView's native file format), or to an INFO table. They are useful if you must use the geodatabase in another program, or if the geodatabase will be used by an earlier version of ArcInfo that does not support this new data type. The Export from Geodatabase menu is shown in the following illustration.

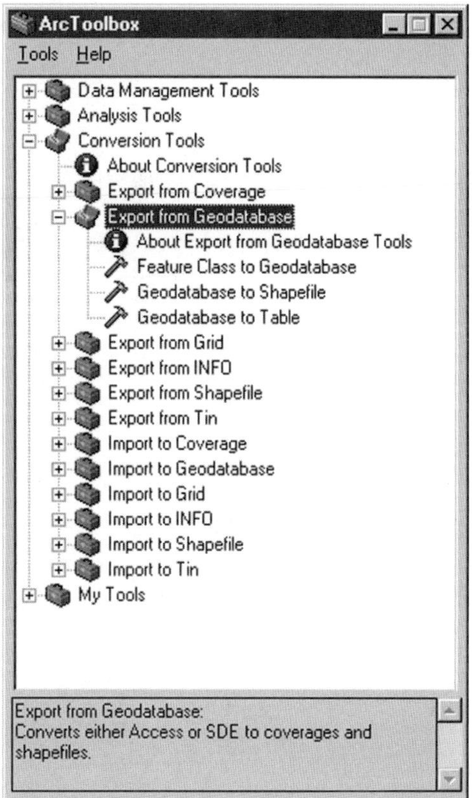

Export from Geodatabase menu.

Export from Grid

These tools convert grid data to other file formats such as ArcInfo polygon or point coverages, ArcView shapefiles, or TINs. You may also export a grid as a DEM (a U.S. Geological Survey format used for elevation data), as an SDTS file (a federal spatial data file standard), or as an interchange file, which can be read by other software programs. The Export from Grid menu is shown in the following illustration.

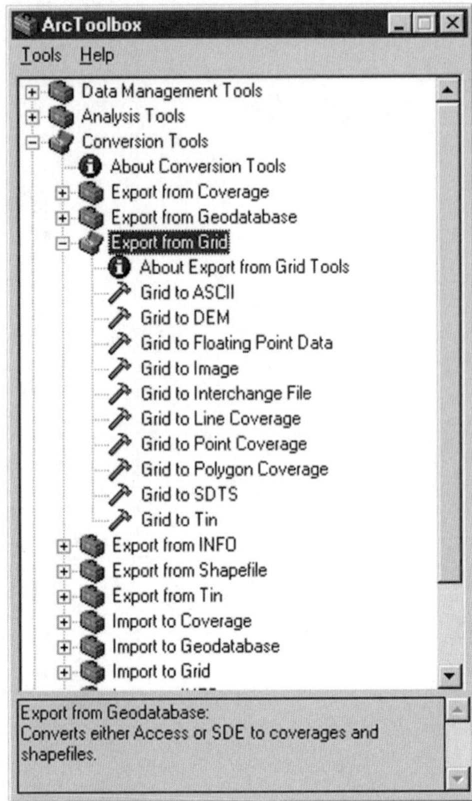

Export from Grid menu.

Export from INFO

These tools are useful for converting an INFO data file to a format used by other database systems. You may convert files to the popular dBase format, to a DBMS format (a third-party database program such as Oracle or Sybase), or to a geodatabase. You may even convert it to a standard interchange file format readable by many programs. The Export from INFO menu is shown in the following illustration.

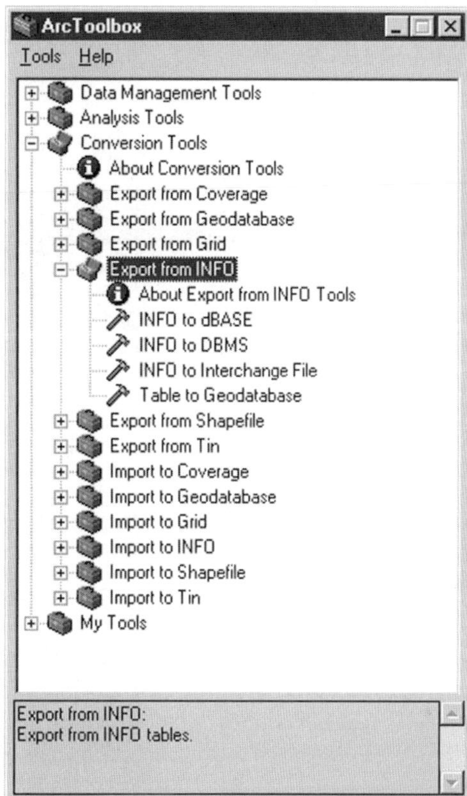

Export from INFO menu.

Export from Shapefile

These tools convert shapefiles to ArcInfo coverages, AtlasGIS files, or to a geodatabase. A shapefile is the file format used by ESRI's ArcView software. The Export from Shapefile menu is shown in the following illustration.

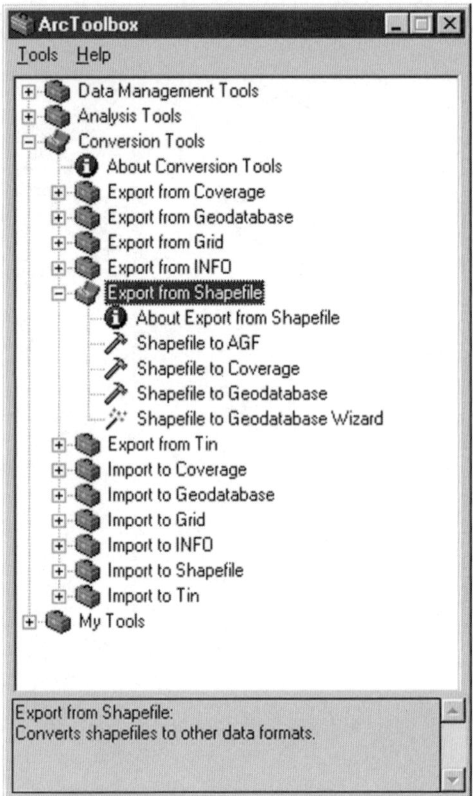

Export from Shapefile menu.

Export from Tin

These tools convert TIN data to ArcInfo coverages or to a common interchange format. TIN data are 3D modeling data used in 3D analysis. The Export from Tin menu is shown in the following illustration.

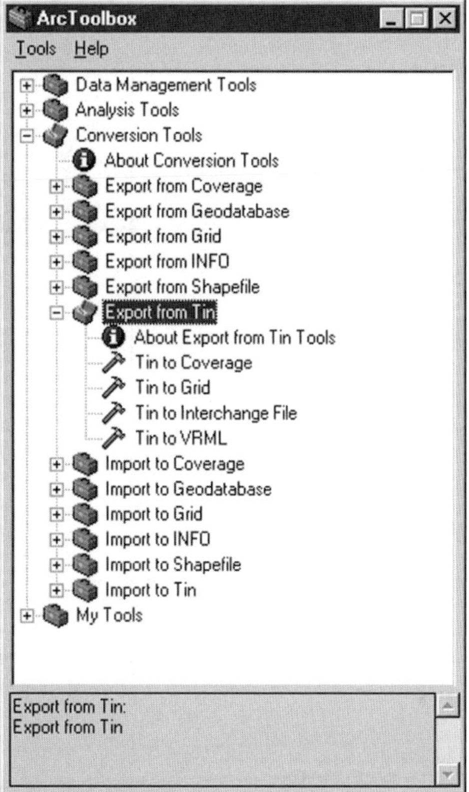

Export from Tin menu.

Import to Coverage

These tools convert a spatial data file into an ArcInfo coverage. There are many formats supported by ArcInfo, from common ones such as DXF (AutoCad), to older, less used ones such as DIME (a U.S. Bureau of the Census file format from the 1980s). Some of the commands, such as "DLG to Coverage Wizard," are wizards that present a series of screens prompting you for input. The Import to Coverage menu is shown in the following illustration.

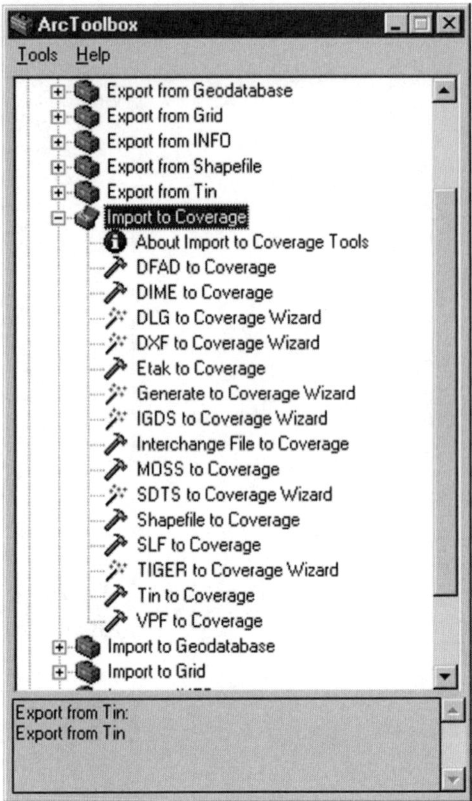

Import to Coverage menu.

Import to Geodatabase

These tools import ArcInfo coverages or ArcView shapefiles into a geodatabase. A tool for converting an INFO table to a geodatabase is also included. The Import to Geodatabase menu is shown in the following illustration.

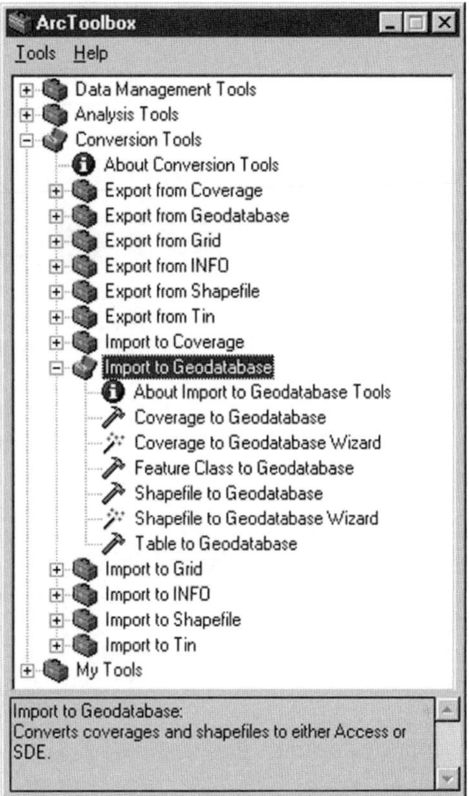

Import to Geodatabase menu.

Import to Grid

These tools convert ArcInfo coverages, shapefiles, DEMs (U.S. Geological Survey's Digital Elevation Model), and so on to a grid. Grid data sets are raster data used by ArcInfo's GRID module, which is outside the scope of this book. The Import to Grid menu is shown in the following illustration.

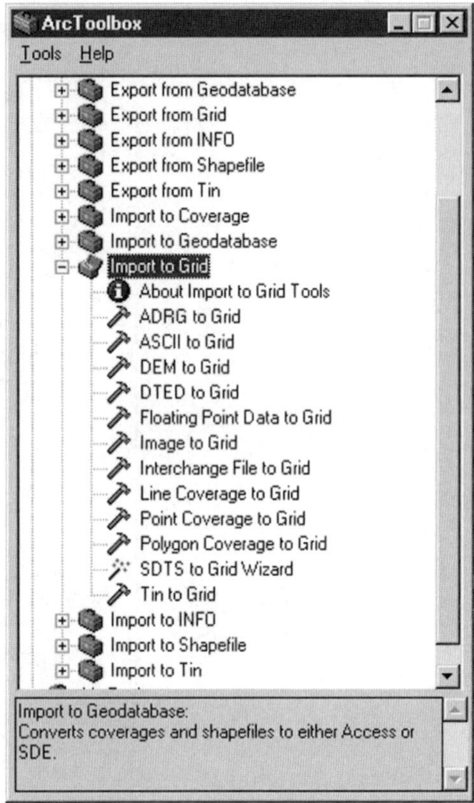

Import to Grid menu.

Import to INFO

These tools are useful for converting a dBase or geodatabase files to an INFO file. A third-party database file such as Sybase, Oracle, or Access may also be converted to an INFO file using the DBMS to INFO tool. The Import to INFO menu is shown in the following illustration.

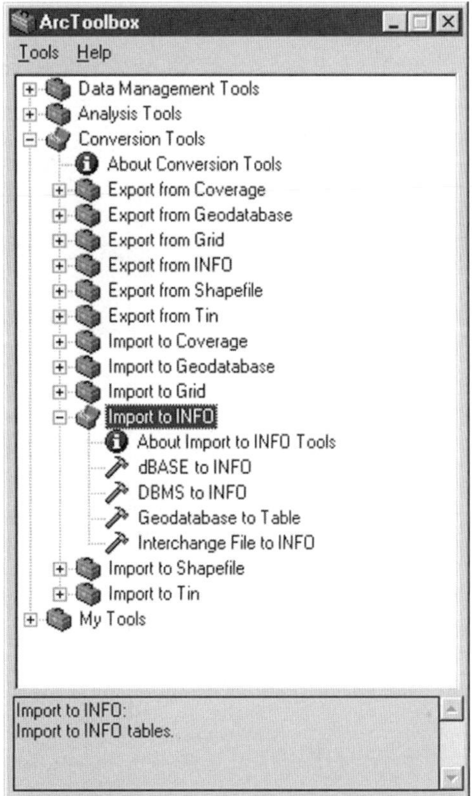

Import to INFO menu.

Import to Shapefile

These tools convert coverages, geodatabases, AtlasGIS files, and MapInfo files to shapefile format. Although shapefiles are the native format used by ArcView, they can also be used in ArcInfo 8.0. The Import to Shapefile menu is shown in the following illustration.

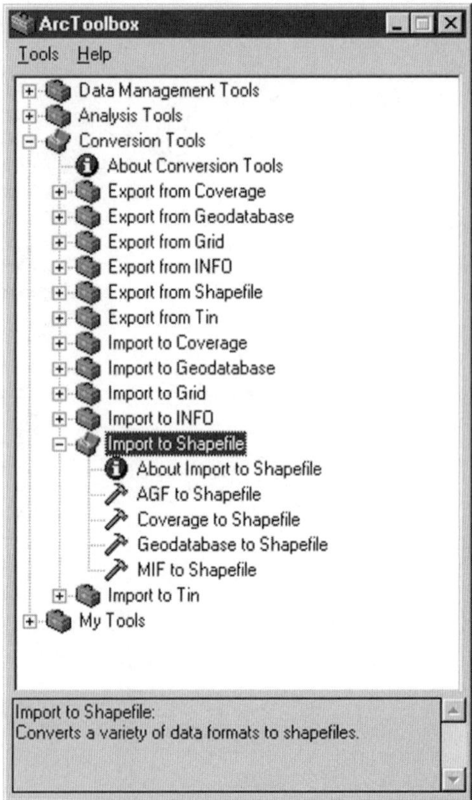

Import to Shapefile menu.

Import to Tin

Using these commands, a TIN may be created from an ArcInfo grid, a standard interchange file, or from a coverage. TIN data sets are used for 3D analysis. The Import to Tin menu is shown in the following illustration.

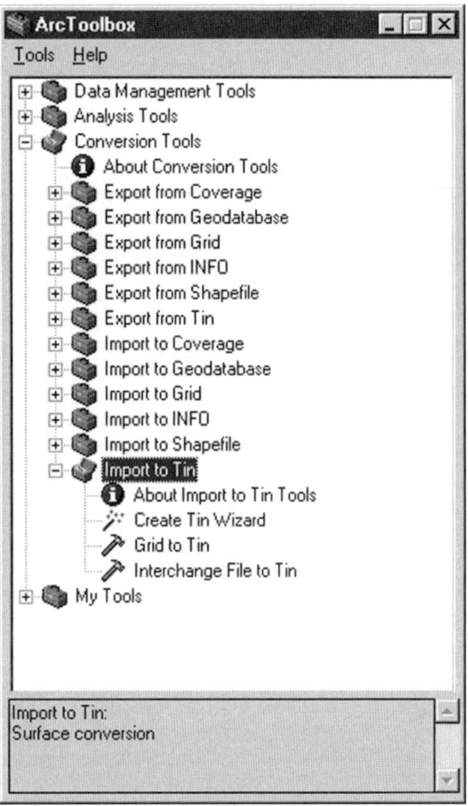

Import to Tin menu.

Summary

ArcToolbox is a welcome addition to the new Arc Desktop. The toolbox provides menus and wizards to make many tasks easier to perform. The graphical user interface is well suited to the ArcInfo newcomer, who will find the pop-up menus convenient and reasonably intuitive. The reference sections in this chapter provided information about each tool and what input each tool requires.

Arc Desktop's three modules (ArcCatalog, ArcMap, and ArcToolbox) represent a move away from the traditional ArcInfo command format. Traditional ArcInfo, or ArcInfo Workstation, is the subject of the remaining chapters of this book.

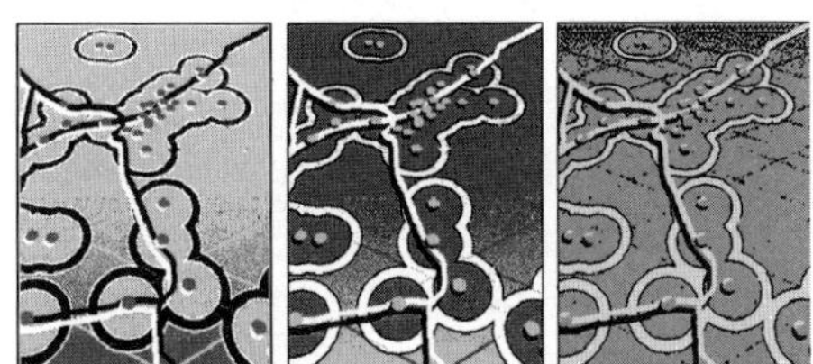

Chapter 5

An Introduction to ArcInfo Workstation

Although the components that constitute Desktop ArcInfo represent a new departure for ESRI and GIS users, ESRI has continued to update the workstation version of ArcInfo. Users familiar with earlier versions of ArcInfo will feel at home with Workstation ArcInfo 8.0, which continues to maintain the traditional command line interface, pull-down menu system, and powerful macro language (Arc Macro Language or AML).

ArcInfo Workstation is available for Windows NT as well as a number of different UNIX systems. The commands and graphics screens are identical regardless of the system used. The screens presented in the following chapters are from ArcInfo running under Windows NT, but UNIX users will see virtually identical results. This chapter presents a general description of ArcInfo, as well as some basics you need to know about ArcInfo's modules before working with them. This chapter covers the following.

- The modules of ArcInfo Workstation
- An Approach to using ArcInfo Workstation commands
- Accessing on-line help

Before jumping into ArcInfo Workstation, you should be familiar with the basic concepts of GIS (if you are not, see Chapter 1). In the following chapters you will explore the capabilities ArcInfo Workstation has to offer. Chapter 6 covers ArcEdit; there you will learn powerful and efficient ways to edit spatial data. Chapter 7 discusses ArcPlot and shows you how to use that module to query and display your spatial data. The Arc module, traditionally used for data management and spatial analysis, is the subject of Chapter 8. Chapters 9 and 10 introduce you to the INFO database and AML, respectively.

The Modules of ArcInfo Workstation

ArcInfo Workstation consists of three main modules, the INFO database, and AML. A number of specialized modules that users license separately from ESRI may also be a part of your organization's ArcInfo setup. The following core modules are covered in this book: Arc, ArcEdit, and ArcPlot. The specialized modules are described briefly, but are outside the scope of this book. The following are the available modules. Descriptions of these modules follow the list.

- Arc
- ArcEdit
- ArcPlot
- Grid
- Network
- TIN
- COGO
- ArcScan
- ArcStorm
- ArcPress

The following modules are covered in this book.

Arc: The module traditionally used for data management and spatial analysis. Some of the other modules (such as ArcEdit, ArcPlot, and Grid) are started from the Arc command prompt.

ArcEdit: The "workhorse" you will use for creating and editing spatial data. It is a full-featured graphics editor that also lets you work with data in associated data files. You may create data interactively using the mouse and keyboard, or you may use a digitizing table or scanner to enter data.

ArcPlot: Use this module for creating map compositions and presentation-quality graphics. Combining strong graphics manipulation capabilities with access to GIS database tables, ArcPlot allows you to compose maps that portray the results of geographic analysis.

The following are separately purchased modules not covered in this book.

Grid: Dedicated to the manipulation of raster-based data. It contains an extensive suite of commands and statistical analysis routines commonly used for working with raster data.

Network: Although a separate module, Network commands are accessed from ArcPlot. This module performs typical network applications such as finding the fastest route between two or more locations, analyzing flow in a network (e.g., traffic or water systems), and so on. Common uses include routing emergency vehicles and finding the best route for school buses or delivery vehicles.

TIN: TIN stands for Triangulated Irregular Network. This module is used for 3D visualization and analysis, and is typically used for viewshed analysis. Its commands are also accessed from ArcPlot.

COGO: An acronym for Coordinate Geometry, COGO is a specialized suite of commands used in ArcEdit. These commands let you enter data using surveyors' coordinates that are then made to fit your existing coverages. For example, some municipalities use COGO to enter boundary line coordinates when parcels are subdivided.

ArcScan: A specialized product used in conjunction with a scanner. A scanner is a device that converts a paper map all at once into a digital file. ArcScan is used to convert the digital map into features ArcInfo understands, such as polygon and line coverages.

ArcStorm: A data management module ideal for maintaining large spatial databases. It provides a means of managing large data sets where there are many simultaneous users. A good example would be a coverage of parcel boundaries in a county. ArcStorm allows many users to access and edit the parcel coverage at one time.

ArcPress: Used to help speed the printing of large graphics files. Many maps you create can become very large files that may take a long time to print. ArcPress speeds up the printing process by optimizing the file to be printed before it is sent to the printer. ArcPress can even load files generated from other software and provide the same speed enhancement.

ESRI provides an Arc command, *productinfo*, for viewing the modules available on your computer. This command also tells you how many licenses per module are available. A license refers to how many computers may run that module at any one time. For example, if you have five ArcPlot licenses on your system, five computers may run ArcPlot at the same time. If you have only one Grid license, for instance, only one computer at a time may run Grid.

Go ahead and view the modules available on your own computer. Start Arc by clicking on Start ➡ Programs ➡ ArcInfo ➡ ArcInfo Workstation ➡ Arc. A blank, black background window appears with software copyright information and the Arc command prompt in it. At the Arc prompt, type *productinfo*. An example screen is shown in the following illustration.

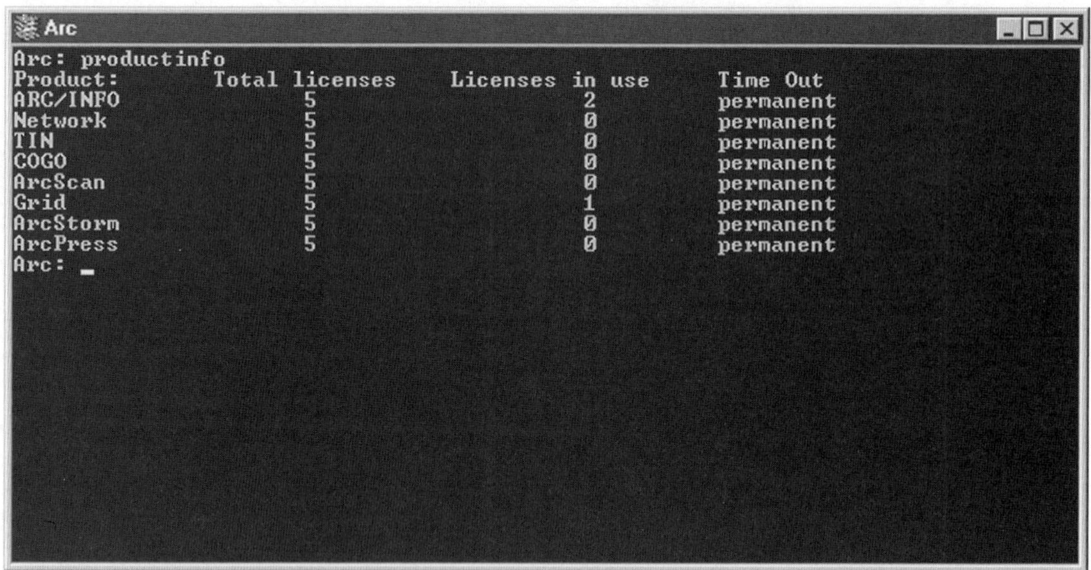

Results of the productinfo *command.*

In the previous illustration you see that *productinfo* lists all the ArcInfo modules, the total number of licenses, the licenses in use, and the time-out date for the license. Here the Time Out column lists “permanent”; on this system, the licenses never expire. A date could also be listed here, and after that date passes the module is disabled and you can no longer use its commands. You can also see in the previous illustration that there are five licenses per module.

Note that it is not necessary to have the same number of licenses for each module. For example, your organization may have five licenses for the ArcInfo Workstation modules, but only two for the Grid module. Using *productinfo*, you can also determine if the modules are already being used by others. In the previous illustration, you can see that two persons are currently using ArcInfo, and that one of them is also using the Grid module.

✓ **TIP:** Productinfo *lists only the modules you have licensed. If you do not have licenses for a particular module, it will not be listed.*

An Approach to Using ArcInfo Workstation Commands

ArcInfo now has well over a thousand commands that may all be typed in from the command line. How do you begin to learn something so large? Keep in mind that once you begin using ArcInfo regularly you will probably use the same sets of commands repeatedly.

In addition, you will probably discover that many of the commands you use could be strung together as macros. A macro is a text file of sequential commands you can access and execute with one command. For example, you may use eight Arc commands to start up when you first log on. Storing them in a text file named *startup* means you could execute all eight commands by just typing *startup* at the Arc prompt.

➾ **NOTE:** *For more information on macros, see the sidebar in Chapter 6 and the section on scripting in Chapter 10.*

There are also many specialized commands you will probably never use. They are included because specific industry or federal government users who find the commands useful requested inclusion of those commands.

In the chapters that follow, you will learn how to use ArcEdit to edit spatial data and ArcPlot to display data and make maps. Discussion includes how to use commands at the Arc prompt for spatial analyses, how to use INFO to view and edit attribute data, and an introduction to AML. First, however, you need to know three things: how to start ArcInfo, how to find available commands, and how to access ArcInfo's extensive on-line help.

How to Start and Quit ArcInfo

Starting ArcInfo can be done in one of two ways. You may use the Start button on Windows NT as previously demonstrated: Start ➡ Programs ➡ ArcInfo ➡ ArcInfo Workstation ➡ Arc. Alternatively, you may open an MS-DOS command window and simply type *arc*. Either way, a new window with a prompt that says "Arc" will appear, ready for you to enter commands.

If you launch ArcInfo via the Start button, a command window will always automatically start in a directory specified by your system administrator when ArcInfo was installed. You may start ArcInfo in a different directory by opening a command window, changing to a new directory, and typing *arc*.

Quitting an ArcInfo session is easy. To quit a session, simply type *quit*. Note that clicking on the upper right-hand X in an attempt to dismiss ArcInfo Workstation windows gives you a message asking you to first quit the application.

Entering Commands

There are two methods of executing many of the ArcInfo Workstation commands. First, you may simply type the command and its parameters at the prompt and then press <Enter>. This is called command line entry. Second, you may access a command from one of the ArcTools pull-down menus.

Say you want to use the Build command, shown in the following illustration, to restore topology to a coverage. Type *build* at the Arc prompt and press the <Enter> key to view the command's syntax.

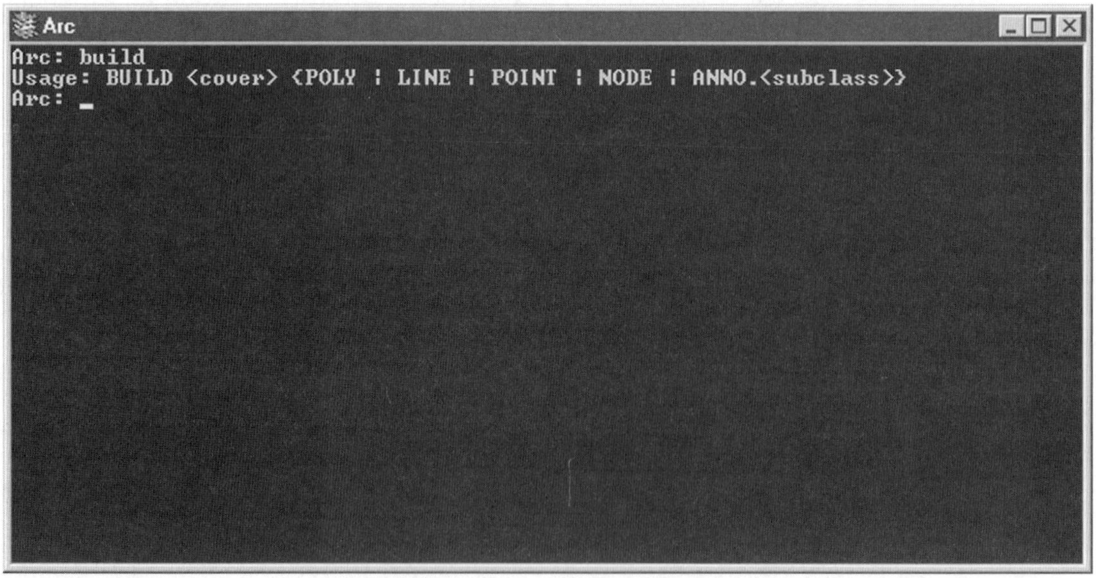

```
Arc
Arc: build
Usage: BUILD <cover> <POLY | LINE | POINT | NODE | ANNO.<subclass>>
Arc: _
```

Build command.

ArcInfo returns the usage for the Build command. The syntax used for its usage follows.

```
Usage: BUILD <cover> {poly | line | point | node |anno.<subclass>}
```

The items enclosed within <> are required parameters when issuing the command. In this case, it means that a coverage name must be supplied after typing the command's name. The feature types that follow (within brackets) are arguments that could be supplied as well. For instance, there are five feature types listed. If you do not type a feature name after the coverage name, the Build command will default to use the first feature in the list, which in this case is *poly*. The " | " means "or," meaning that you could use the *Build poly* command on a coverage or use the *Build line* option. Your command might look as follows.

```
BUILD mycover line
```

This would run the Build command on a coverage called *mycover* using the *line* option. If you had left out a feature option, *line*, Build would have defaulted to the *poly* option.

This example is representative of all ArcInfo commands. To find out what parameters are required by a command, just type the command and press the <Enter> key. A full list of all information the command may use will be displayed.

The second way to issue many of the commands is to use ArcInfo Workstation's ArcTools menu systems. ArcTools is a menu system that provides a menu interface to many of ArcInfo's commands. ArcTools has menu systems that work with Arc, ArcEdit, and ArcPlot, and each is discussed in the following chapters. For now, refer to the following illustration, which shows an example of an ArcTools menu, specifically the Build command's menu.

Build command menu.

Examine the Build command menu closely. You can see that all parameters you can specify using Build at the Arc prompt are available as menu choices. Using the ArcTools menu is also a good way of seeing what options the Build command parameters have. Simply enter the necessary information in the menu and click on the Apply button.

✓ **TIP:** *Sometimes ArcInfo Workstation menus are so long they will not entirely fit on the screen. You can fix this problem by changing your screen's resolution. Simply right click on the Windows Desktop (your Windows background) and select Properties. Click on the Settings tab and use the Desktop Area slider bar to change to a higher resolution (from 800 X 600 to 1024 X 768, for example).*

ArcInfo is very flexible. You can work with ArcInfo by entering commands at the command prompt or by using ArcTools. You can even switch between the two methods of entering commands. As you gain experience using the software, you will develop a working style that is best for you.

Finding Commands

An easy method of obtaining a list of all commands available in any of the ArcInfo modules is to type the word *commands* at the prompt. This is shown, at the Arc prompt, in the following illustration.

Typing commands at the Arc prompt.

The screen that appears shows all available commands, in alphabetical order. Notice that the last line says "Continue?" This is because there are more commands to list than can appear on one screen. Press the <N> key to stop the display of commands, or press any key to see more of the list.

ArcInfo lets you use wildcards to view just some of the commands. For instance, say you want to see all commands that start with the letter B. You would type *commands B**. This is shown in the following illustration. Now you see all commands that begin with the letter B.

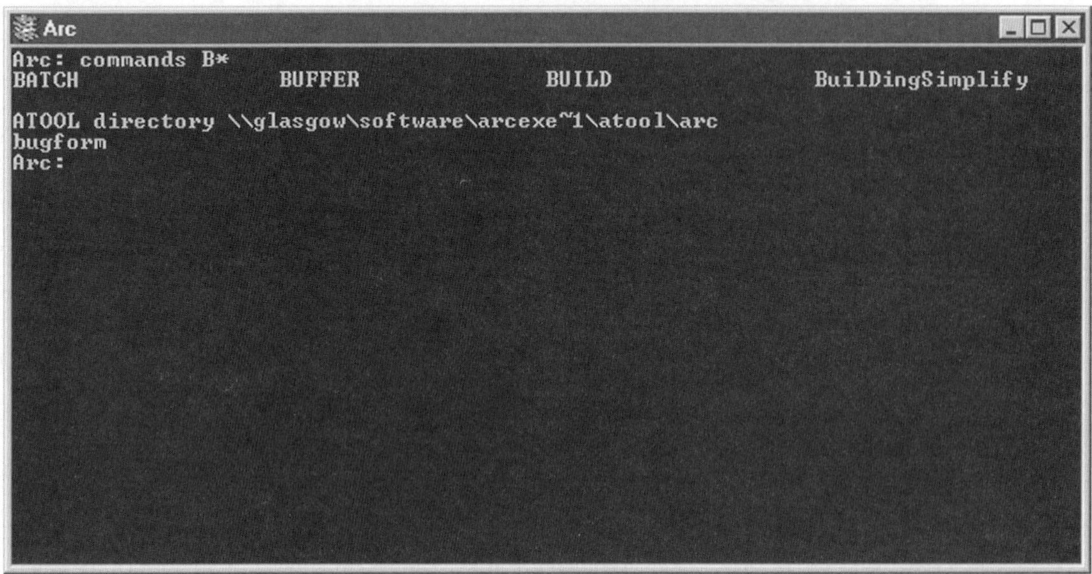

Typing commands B* *at the Arc command prompt.*

There is also something else to take note of. Four commands are listed, as well as a command in an ATOOL directory (which is where you place frequently used AMLs for easy access; see Chapter 10). The first three commands are all in capital letters, and the last is a mix of capitals and small letters. ArcInfo accepts abbreviations for many long commands.

Some commands are short enough that they do not need abbreviation, such as BUILD, BATCH, and BUFFER. However, it is another matter to have to type in *BuilDingSimplify* each time. Instead, ArcInfo lets you type *BDS*, making it much easier to do. The BDS abbreviation for *Buildingsimplify* is indicated by the capital letters among the small letters.

✓ **TIP:** *Although the commands are listed using capital letters, ArcInfo does not require that they be typed using capitals. You may use small letters at any time.*

You may use wildcards in one other way. If you type *commands *ld* you will get a list of Arc commands that end with the letters *ld*. This is shown in the following illustration.

Finding commands that end with ld.

As you can see, ArcInfo provides a lot of flexibility in finding the names of available commands.

Accessing On-line Help

ArcInfo comes with an extensive on-line help system. The information is organized by topic, but it can also be searched by keywords. Note that ArcInfo Workstation commands are listed, by module name, both functionally and alphabetically.

To start the Help system, type *help* at the Arc command prompt. You will see the introductory screen, shown in the following illustration.

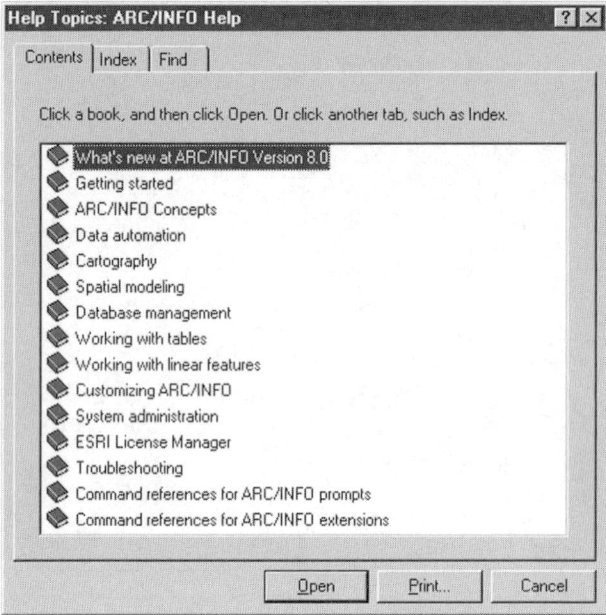

Help introductory screen.

ArcInfo uses the Windows NT built-in help facility as the front-end for its help system. There are three tabs along the top: Contents, Index, and Find. Contents displays headings much like chapters in a book. Click on a heading and additional topics will be displayed. Index lets you type a word and finds references to it within the on-line help. Use this to quickly access information without having to navigate through the many headings in the Contents tab. Find is similar to Index but makes a more in-depth search.

Clicking on the book icon next to "Command references for ARC/INFO prompts" opens it to reveal module names. This is shown in the following illustration.

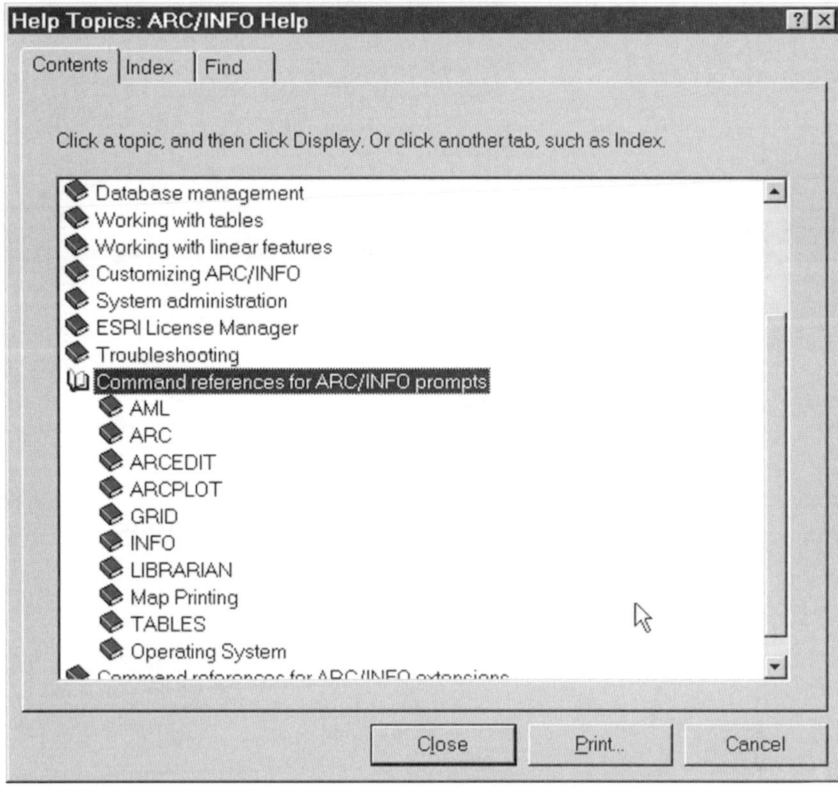

Command references for ArcInfo modules.

Clicking on a book icon next to a module name gives you a choice between the functional list of commands and the alphabetical list. This is shown for the Arc module in the following illustration.

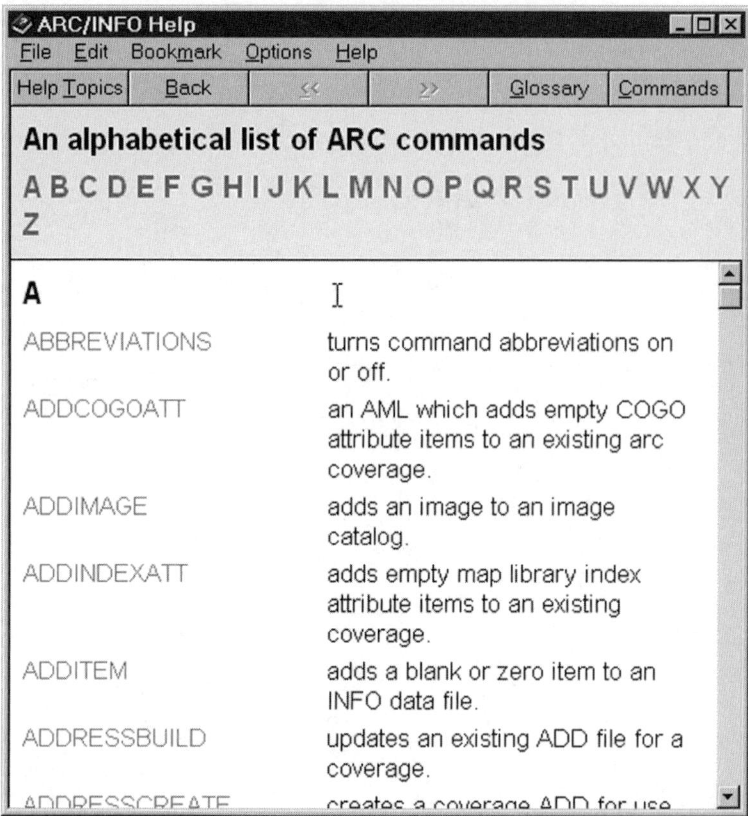

Arc command options.

Choosing the functional list, shown in the following illustration, gives you a menu of command function categories. This is most helpful when you are unsure of a command's name but know what you want to do.

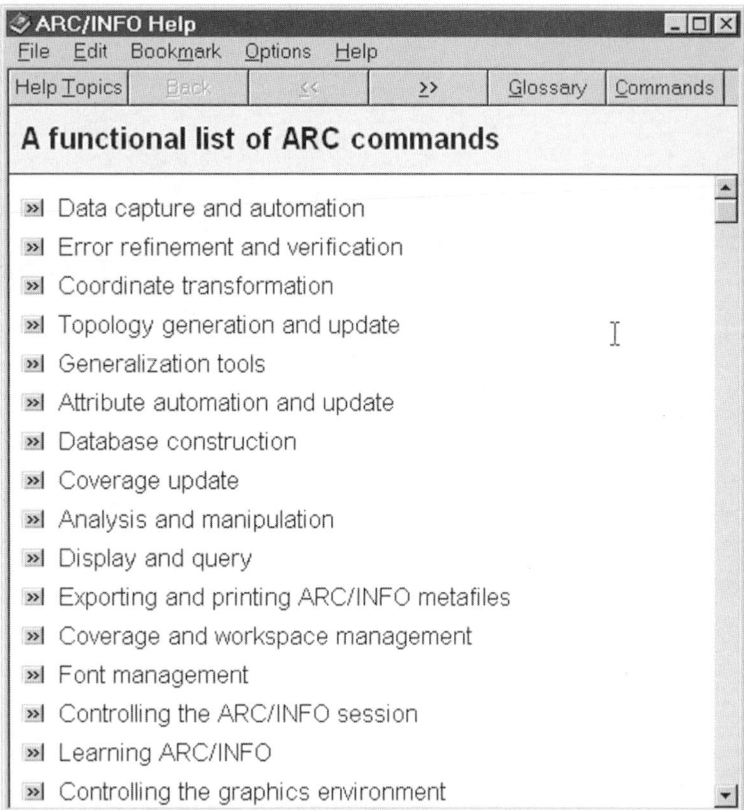

Functional list of Arc commands.

Choosing a letter under the alphabetical list gives you all commands beginning with that letter. This is helpful for finding detailed information about a particular command, such as for the B command shown in the following illustration.

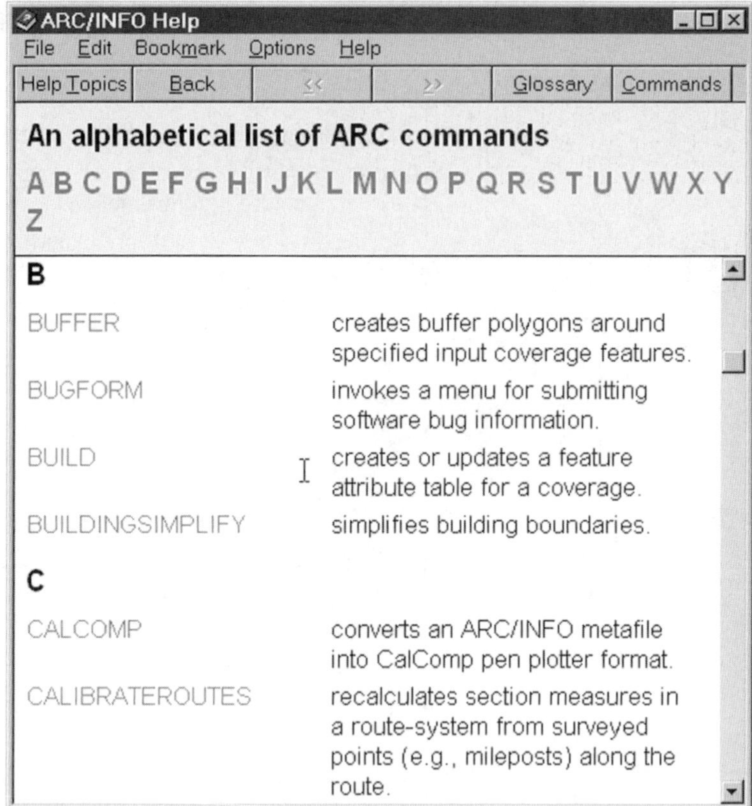

B commands in Arc.

This section introduced you to the organization of the on-line help system. You will find it a convenient and substantial source of information about ArcInfo Workstation commands and processes.

Summary

In this chapter you have learned about ArcInfo Workstation's modules, an approach to using commands, and how to access the on-line help. The next five chapters discuss putting ArcInfo to work. Separate chapters are devoted to using ArcEdit, ArcPlot, Arc, the INFO database, and AML, ArcInfo's scripting language. It is through these chapters that you will master the basics of ArcInfo Workstation.

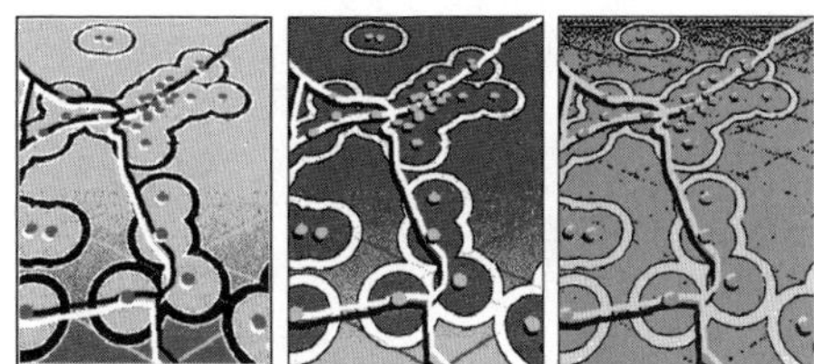

Chapter 6

ArcEdit

ArcEdit is ArcInfo's coverage editing module. You can modify, add, or even create coverages in ArcEdit. Coverages are ArcInfo's unique way of storing geographic data. They are sets of polygons, lines, and/or labels linked to data in a relational database.

ArcEdit is a powerful tool for modifying coverages, and is also one of the original ArcInfo modules. It includes some simple pull-down menus, but the interface, because it was not originally designed for Microsoft Windows NT, is not as sophisticated as Version 8.0's new modules ArcCatalog, ArcMap, and ArcToolbox. Because ArcEdit is one of the original modules, you can also use command line entry of editing commands, and Arc Macro Language programs (AMLs) can contain and execute ArcEdit commands (Arc Macro Language is introduced in Chapter 10).

As the software's name implies, arcs are the graphic foundation of ArcInfo. They are the fundamental feature type and consist of nodes (end points) and vertices (points at which the arc changes direction) connected by arcs (line segments). Neither nodes nor vertices can exist without arcs. In fact, nodes are the only feature type that cannot exist without another feature type (namely, arcs).

Arcs also make up polygons. Polygons are sets of arcs that enclose an area. The beginning point of the first arc and the end point of the last arc have the same geographical coordinates.

Although a lot of the functions of Arcplot and Arc (two other traditional modules of ArcInfo) have been replaced by ArcMap and ArcToolbox, respectively, ArcEdit is still its own powerhouse of editing capabilities. It is here that you will digitize, edit, or type in coordinates to create and modify coverages. ArcEdit contains dozens of commands. It is not the purpose of this chapter to provide details on every command. The more common commands are demonstrated to give you a feel for ArcEdit's capabilities. Detailed information on commands can be found in the on-line help documents. This chapter covers the following.

- Setting up work environments
- Exercise 6-1: Exploring ArcEdit
- Editing arcs, nodes, and polygons
- Editing labels
- Annotation

Before jumping into the exercises in this chapter, you should be familiar with GIS concepts (review Chapter 1). In addition, be sure the Chapter 6 exercises are loaded from the companion CD-ROM to your workspace (assumed to be *D:\ai80exercises\chap_6*).

The exercises in this chapter use fictional data from an integrated pest management (IPM) project. The "project" conducts widespread trapping of the mythical blue-eyed corn moth and looks for instances of high populations. The purpose is to prevent the insect's spread by spraying an insecticide across small areas of higher insect populations rather than across large areas of small populations. Your group is responsible for the data from five Ohio counties.

Exercise 1 contains two parts. The first part uses the pull-down menus to walk you through setting up a work environment and performing some simple arc editing. The second part repeats the steps in Part I using command line entry. You will get a feel for both interfaces. This familiarity will enable you to interact more efficiently with ArcEdit. That is, you can use the interface that will most quickly accomplish the task at hand. Before beginning, be sure to load the Chapter 6 exercise data to your workspace (*D:\ai80exercises\chap_6*).

NOTE: *Exercise 6-1 may appear lengthy. It is in reality a short exercise that includes detailed discussion of the pull-down menus you will be using.*

✓ **TIP:** *Sometimes ArcInfo Workstation menus are so long they will not entirely fit on the screen. You can fix this problem by changing your screen's resolution. Simply right click on the Windows Desktop (your Windows background) and select* Properties. *Click on the settings tab and use the Desktop Area slider bar to change to a higher resolution (from 800 X 600 to 1024 X 768, for example).*

ArcEdit Environments

When using ArcEdit you will be working in several ArcEdit environments. The word *environment* used here means a general category into which certain setup commands can fall. Before actually starting your edits or data additions you must first specify these environment settings. The characteristics of the environments must be set each time you start an ArcEdit session. This is easier than it sounds, and the settings can be stored in a time-saving macro (see sidebar), which can be executed with a single command. The ArcEdit environments are described in the following material.

The *edit* environment names the coverage and what feature category in that coverage (arcs, polygons, and so on) you will be editing. You can only edit one type of feature at a time. That is, for example, if you are editing arcs you cannot edit points without first changing your edit feature. The edit environment also includes settings for snap and edit tolerances. It could also include digitizing settings.

The *draw* environment specifies which attributes will appear on the screen. You could, for example, have only label points appear, or both arcs and label points. You can also specify which symbol sets to use and at what scale you want the information displayed, as well as how you want items labeled.

The *background* environment contains coverages that appear behind your edit coverage (usually in different colors so that you can easily distinguish them), and determines which features (polygons, arcs, and so on) in those coverages will appear. Background coverages can be useful guides for editing.

✓ **TIP:** *Saving these environment settings is a good means of maintaining consistency from session to session and person to person. You can also set up environments specific to coverages.*

Exercise 6-1: Exploring ArcEdit

Part I: Getting Started with the Pull-down Menus

To start ArcEdit, simply select Start ➡ Programs ➡ ArcInfo ➡ Workstation ArcInfo ➡ ArcEdit. A blank, black background window pops up containing the copyright information and the ArcEdit command line prompt. This window is used to execute editing commands (you may see a message in it stating "Warning, the map extent is not defined"). Map extent refers to the geographic area you want to display. The default is the extent of the edit coverage.

A few moments later another blank, black background window appears. This is your display canvas, where the coverage graphics will appear. You can resize and move these windows to suit your working habits and needs. One working method is to make the command window long and narrow at the bottom of the screen, with the canvas window a large rectangle above it. This is how the windows will be shown in the exercises for this chapter, shown in the following illustration.

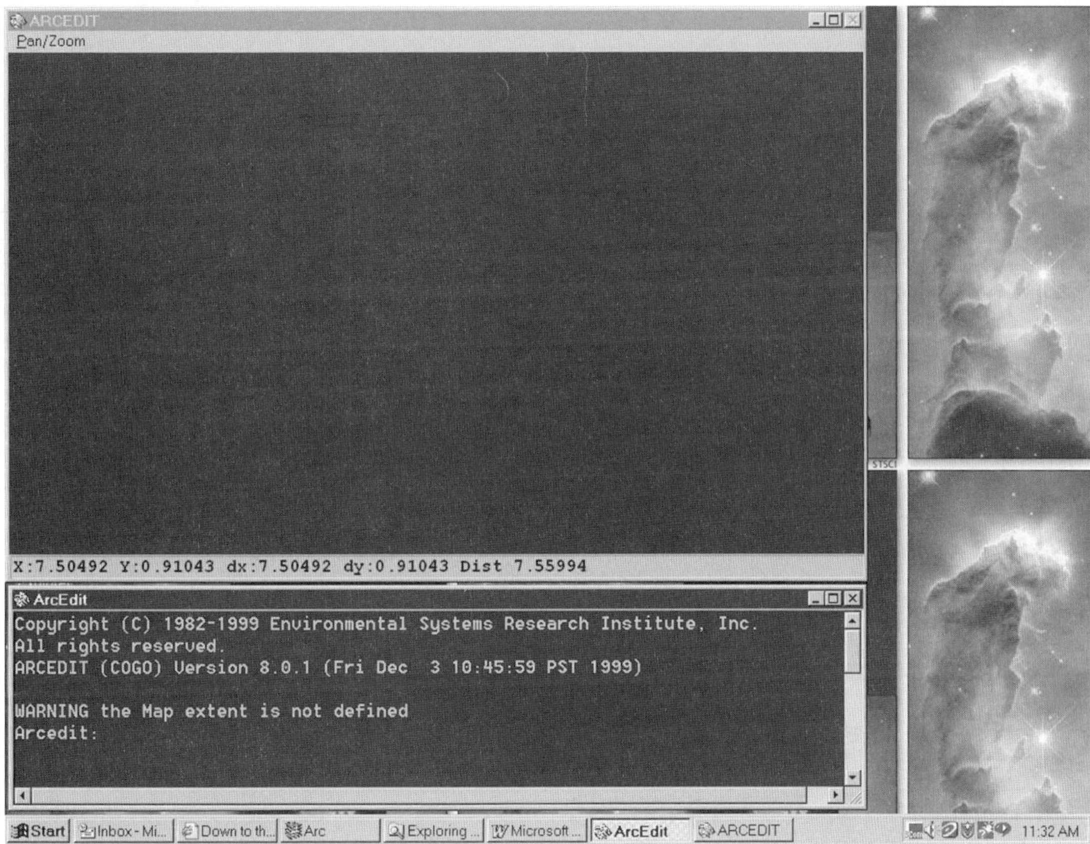

One way of arranging your ArcEdit command and canvas screens.

Establishing Environment Settings

This exercise takes you through setting an edit environment using the pull-down menus, and performing some basic edits. To start the menus, type *arctools* at the ArcEdit prompt.

The Arc Tools menu, shown in the following illustration, pops up in the upper center of your screen. It lists several menu choices. Highlight "editools" and click on the OK button.

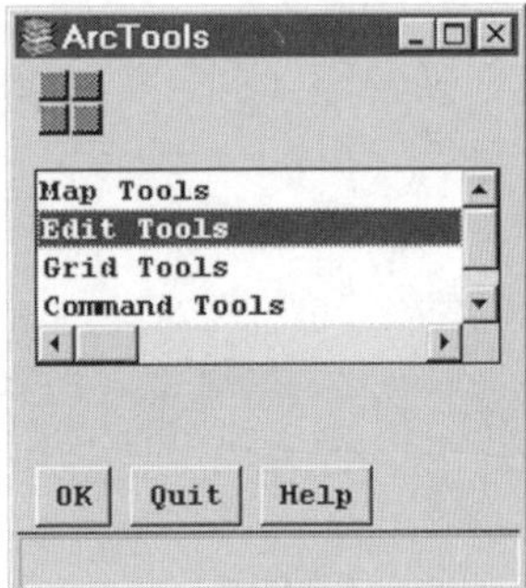

A narrow menu bar named Edit Tools appears in the upper left of your screen. Again, you may move these menus around to suit your needs by simply clicking in the blue bar area and dragging them to their new location. Your screen should now look like that shown in the following illustration.

The initial Arc Tools menu.

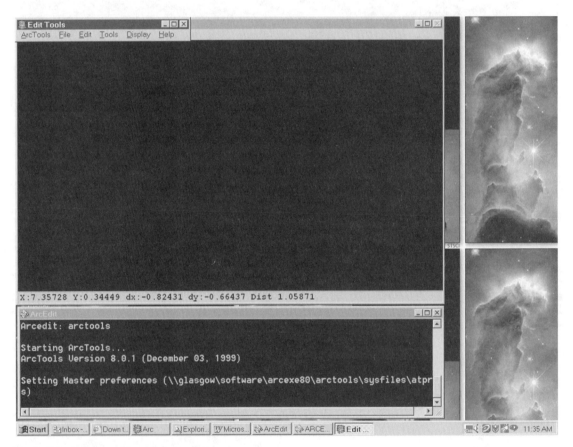

The ArcEdit windows and the pull-down editing menu.

Now specify a coverage to be edited. Select File ➡ Open. Use the large arrow and subdirectories listing to navigate to *D:\ai80exercises\chap_6*. Once there, you should see the coverages listed, shown in the following illustration.

The Open Coverages menu with Chapter 6 coverages listed.

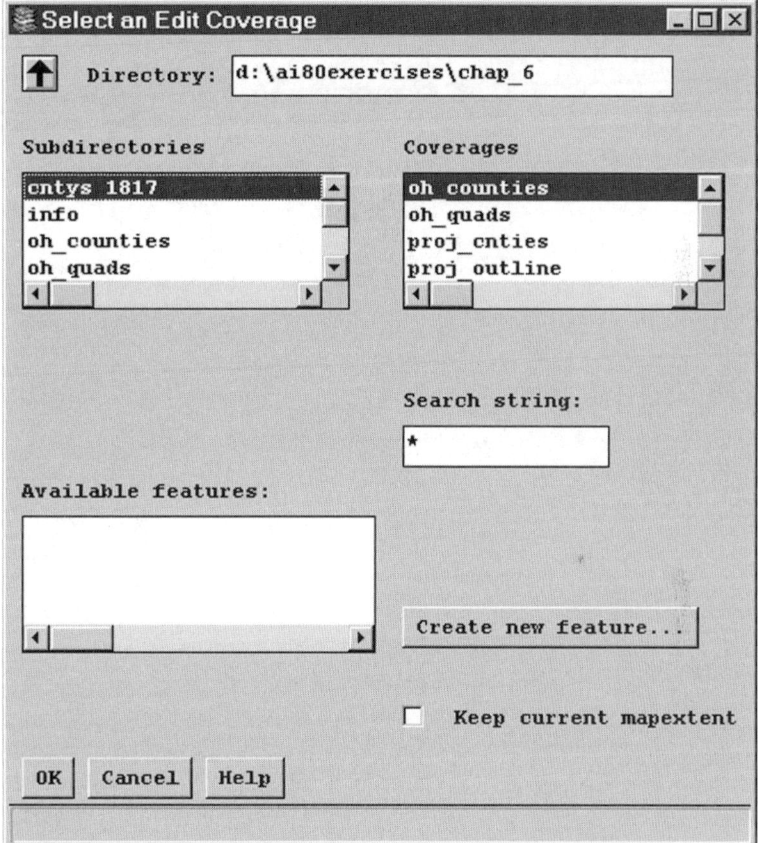

Now specify the edit coverage, and what feature type in that coverage (the edit feature) you will be editing. Highlight the coverage named *proj_cnties*, which contains the five Ohio counties for which your group is responsible. Under "available features," select "arcs," and then click on OK. The arcs of the five Ohio counties will appear in your canvas window. Two menus with commands specific to editing arcs will also appear. For now, minimize the two editing windows (you will maximize them again later). Your screen should look like that shown in the following illustration.

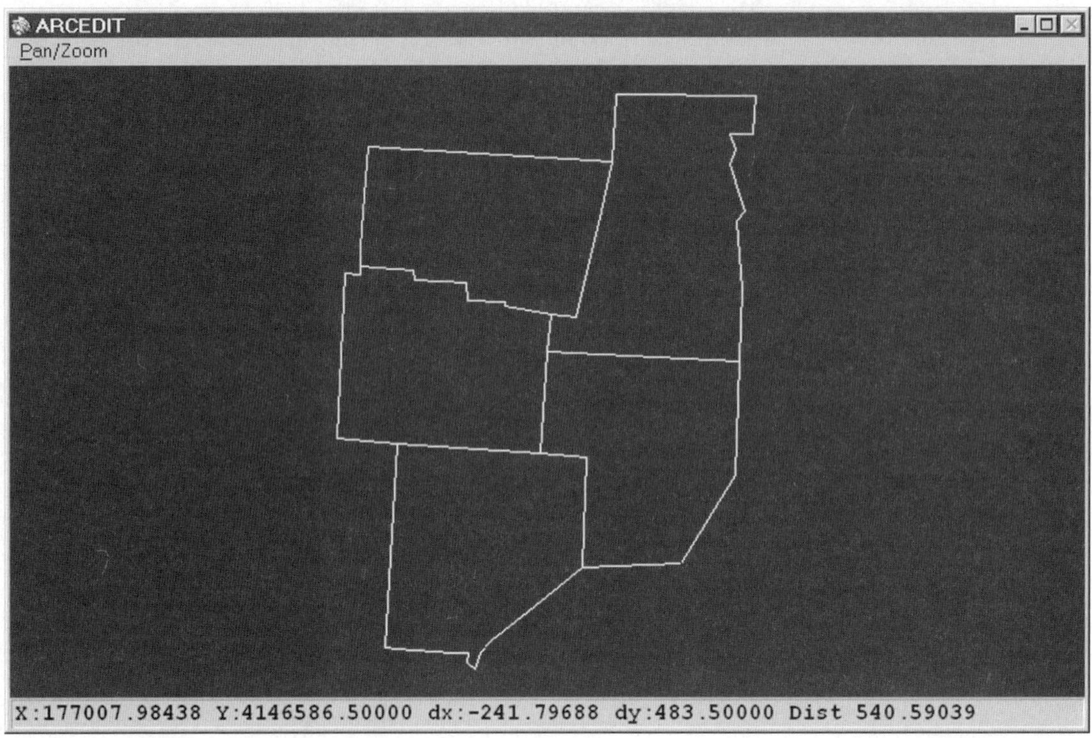

The coverage proj_cnties.

You can specify up to ten edit coverages, but only the most recently specified one will be active. That is, you can edit only the most recently opened coverage. To remove edit coverages, select Edit ➡ Remove from the Edit Tools menu. A pop-up screen allows you to select which edit coverages you want to remove.

In addition to displaying edit coverages, you can also display background coverages. Background coverages appear behind your edit coverage. You can assign a different color to your background coverage to distinguish it from your edit coverage. Add one now by selecting Display ➡ Back Env General from the Edit Tools menu. A new menu pops up. It shows information on your current background environment. The top row shows that many different items can be used as backgrounds. Select "Cover" from the top row and type *oh_counties* in the "Add back" box.

Background coverages cannot be edited. If you need to edit a background coverage, simply specify it to be your active edit coverage.

Your screen should now look like that shown in the following illustration.

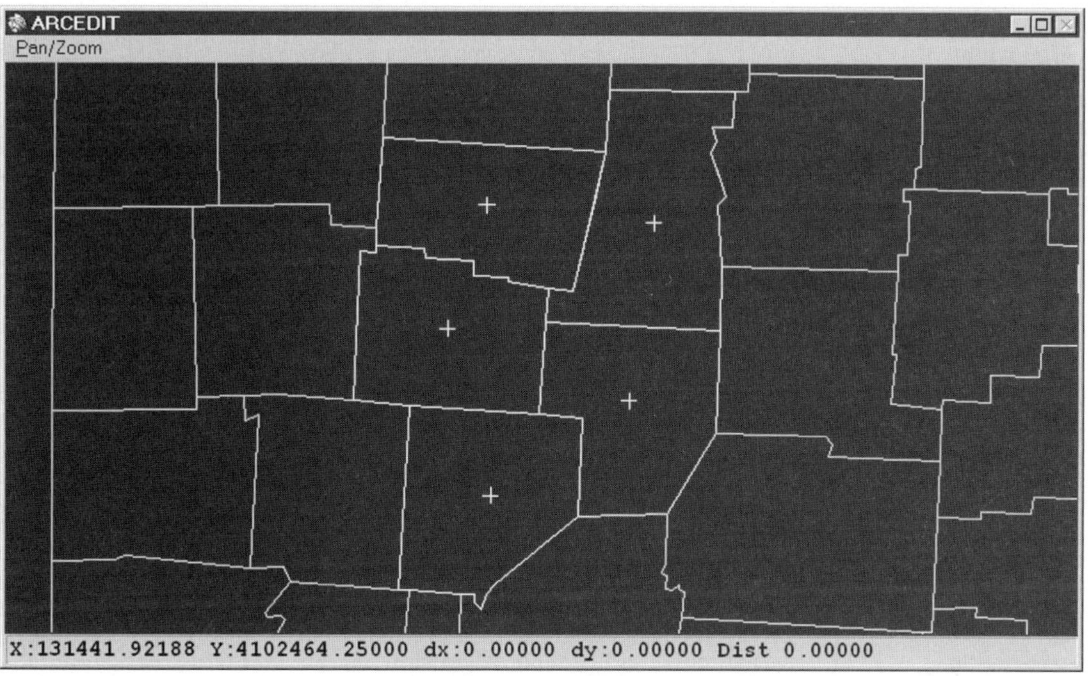

The edit coverage proj_cnties *with the background coverage* oh_counties.

✓ **TIP:** *You can specify a map extent that is different from your edit coverage. For example, if your edit coverage were* oh_counties *and your edits were only in the project area, using the map extent from* proj_cnties *would "zoom" in to just that extent.*

To complete your editing environment setup, set your snap environment and edit tolerances. The snap environment adjusts added features to exactly match a coverage's existing features. For example, if your coverage contains three roads that meet in a T intersection (and you are digitizing a new, fourth road that meets the T and makes it a "+"), setting a snap tolerance can correctly position the new road. Snapping causes added arcs to meet other arcs. From the Edit Tools menu, select Tools ➥ Snap Environment to activate the snapping menu.

Setting your edit tolerances consists of choosing your edit distance, grain, and weed tolerance. The edit distance specifies how close your cursor must be to the item you are attempting to select before the item is actually recognized as being selected. The default value is 1/100 of the coverage height or width, whichever is greater. You can also assign a distance in coverage units or by using the mouse to specify the distance on screen. For this exercise, accept the default edit distance.

The grain and weed tolerances specify the closest allowable distance between adjacent arc vertices. Use grain when adding curved arcs or using commands such as "spline"; set weed tolerance for adding new arcs or using commands such as "generalize." These tolerances are more specific and you may not need to set them for your edit sessions. You will not use them for this exercise.

At this point you have completed your environment settings. You have specified the edit coverage (the coverage to which you will make changes), the edit feature (you will be editing polygons), the drawing environment, the background coverage and background environment, the snapping distances, and the edit tolerances. Now you can begin editing.

Performing Basic Edits

First, create an outline of your project area. Do this by deleting the arcs between the counties in the *proj_cnties* coverage. You will also have to delete the labels. Maximize the two editing menus now and arrange your windows to your preferences.

Remember that you have already specified your edit feature as being arcs. That is, you have told ArcEdit you will be selecting and performing operations on arcs. To select an arc, first click on the New circle (you are creating a new selected set). Then select the diagonal arrow icon from the Feature Selection menu. This icon is the same as the Select Many command and lets you select as many features as desired, one at a time. A crosshair, which is operated by your mouse, appears in your canvas window. Use your mouse to select the arcs between counties. Once selected, the arcs will turn yellow. When you are finished, press the <Ctrl> key and the right mouse button (or the <9> key) to exit.

✓ **TIP:** *Yellow is the default color that represents selected items. If you want a different color to represent your selected set, click on the Prefs button on the Feature Selection menu. There you can type in another color name.*

Notice that the Feature Selection menu keeps track of how many arcs are currently selected. Near the Remove button, you should see that 7 of 23 arcs are in the selected set. If you make a mistake while selecting, simply quit the selecting process and click on the Remove button. Then click on the diagonal arrow icon again. Now the arcs you select turn white to show that they are being dropped from the selected set. To continue adding to the set, click on the "Add to" option and then the diagonal arrow icon.

To the right of the arrow icon are five additional selection buttons. Spend a few moments exploring their differences. You can "lasso" several features at once with a rectangle, an irregular polygon, or a circle. You can also select only those arcs appearing on the screen or all of the arcs in the coverage. You will find these selection methods very useful for more complex coverages.

Once you have the seven arcs between your project counties selected, select the Delete icon (the second button, top row, under "General arc editing" from the Edit Arcs & Nodes menu). The chosen arcs now disappear. Save your results as a new coverage by selecting File ➥ "Save as" from the Edit Tools menu. Name your new coverage *proj_outline*. It should look like that shown in the following illustration.

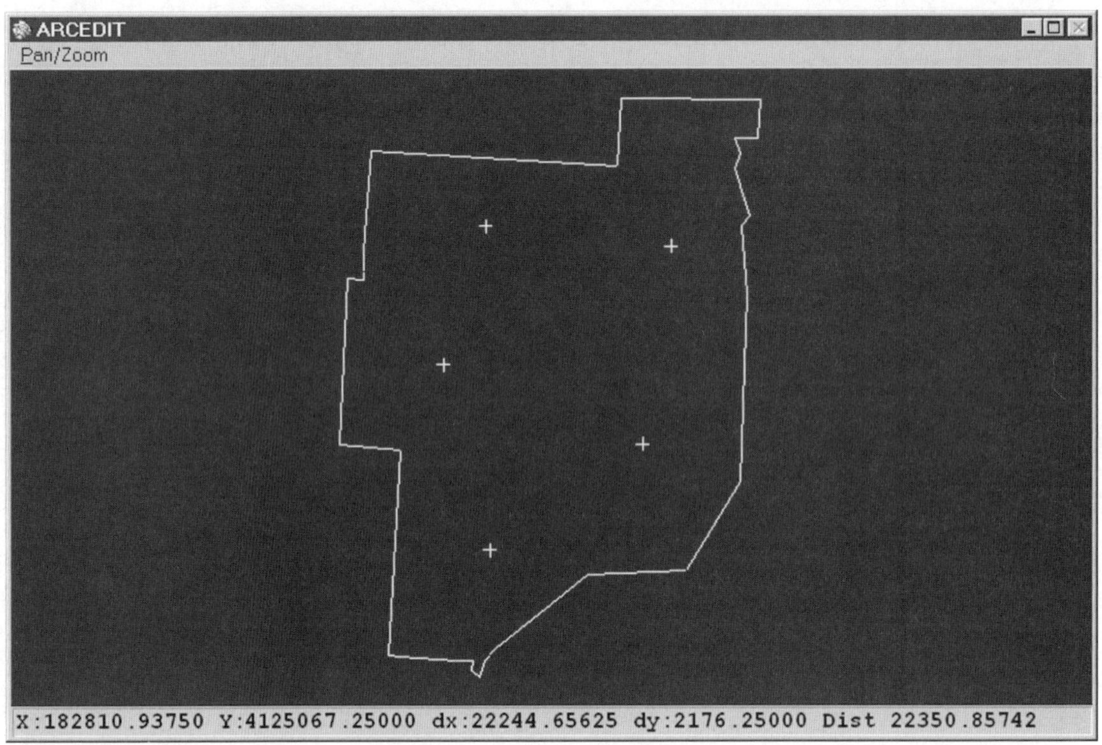

Proj_outline *coverage: an outline of the five counties.*

✓ **TIP:** *Save your edits often! In addition, make copies of your coverages and keep the copies up to date. This is the best method of avoiding "arc angst": lost time, lost money, and the repetition of tedious work.*

Now return to deleting labels. As a rule, each polygon in a coverage should have only one label. First, change your edit feature to Label by selecting Edit ➥ "Change edit feature" and selecting Label.

You will notice that the Feature Selection menu stays the same but a new Edit Lab menu appears. Select all of the labels (five) and then delete them. Now add a single "fresh" label to the center of your project outline. Do this using the Add icon in the Edit Lab menu. Your project outline coverage should now look like that shown in the following illustration.

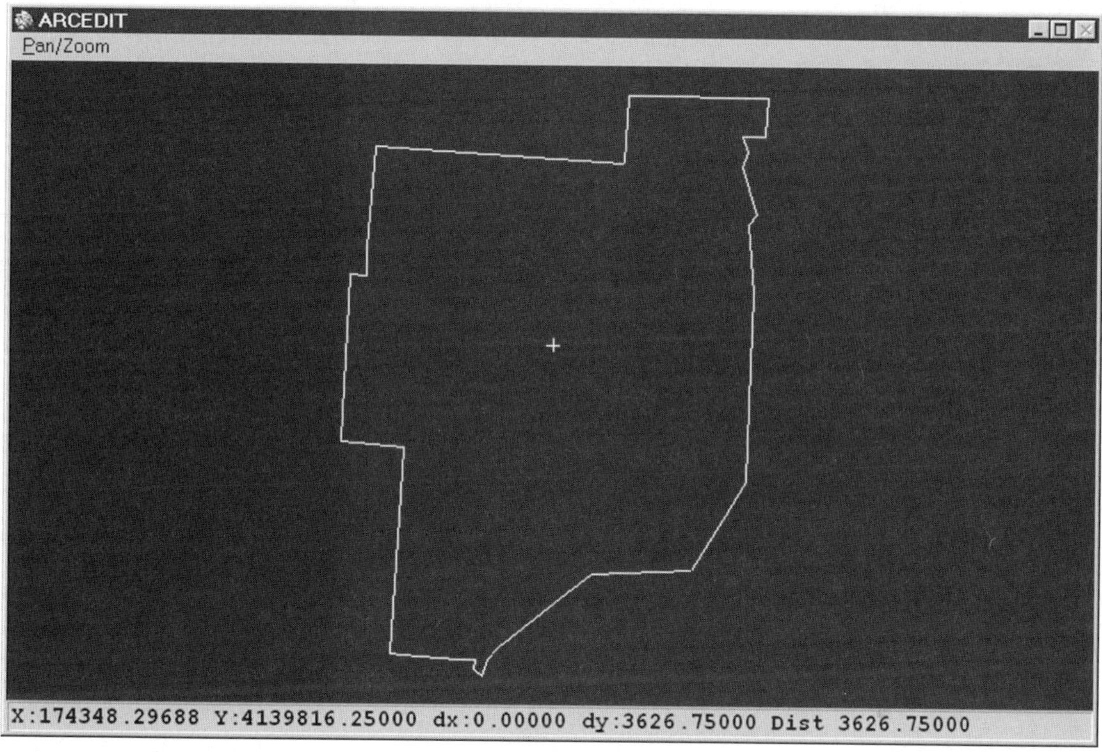

Revised project outline coverage.

Great! Now congratulate yourself on a job well done. In this exercise you have created a working environment, jumped into arc and label editing, and even created a new coverage for your IPM project.

The second part of this exercise repeats the steps in Part I. However, this time you will use command line entry instead of the pull-down windows. Part II gives you a look at the actual commands used in each of the steps. Try it this way to see which interface works best for you. As you become familiar with ArcEdit's commands, you will find that using the command line might be a faster method for some operations.

Part II: Command Line Entry

Start ArcEdit by selecting Start ➡ Programs ➡ ArcInfo ➡ Workstation ArcInfo ➡ ArcEdit. Use the mouse to arrange the windows to suit your preferences. You will be typing in your commands in the ArcEdit prompt window.

Set up your working environment by typing in the commands that follow. Do not type the word *ArcEdit*, which in the following represents the command line prompt.

```
ArcEdit: editcoverage proj_cnties

ArcEdit: editfeature arcs

ArcEdit: drawenvironment arcs labels

ArcEdit: draw
```

With these four commands you have specified your edit coverage and edit feature, and have set your canvas to display arcs and labels. The *draw* parameter tells ArcInfo to do just that: draw your current specifications on the canvas. You can issue the Draw command anytime you need to refresh your canvas.

You can also set a background environment if you want. Type the following commands to do so.

```
ArcEdit: backcoverage oh_counties 3

ArcEdit: backenvironment polys

ArcEdit: draw
```

Now you have set your background environment. The *3* after *oh_counties* in the "backcoverage" command is the color parameter; it means that the background coverage will appear in red. There is no need to specify an edit distance because for this exercise the default is acceptable. With the exception of the background coverage appearing in red, your screen should look like that shown in the following illustration.

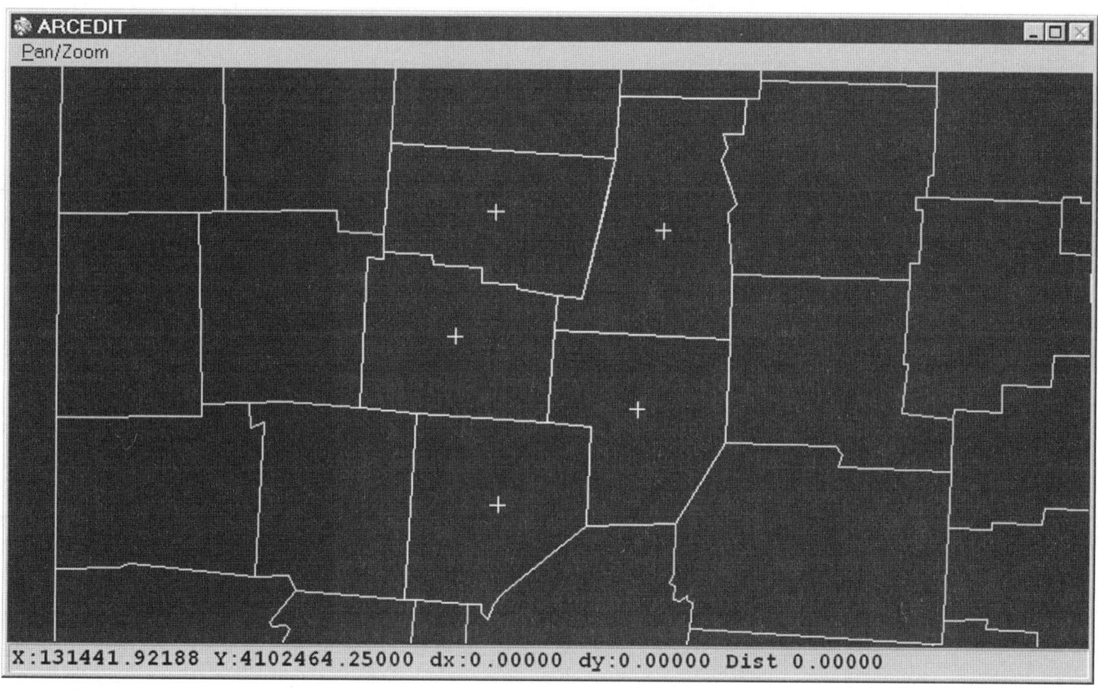

Proj_cnties *as the edit coverage and* oh_cnties *as the background coverage.*

To begin editing, type *select many* at the ArcEdit prompt. The crosshair cursor will appear on the canvas screen. Use your mouse to select the seven arcs between the counties, just as you did previously. Press the <Ctrl> key and the right mouse button (or the <9> key) to end your selection process. Now, click the mouse in the command line window: this tells ArcEdit that the input is changing back to the keyboard from the mouse. At the prompt, type the following.

```
ArcEdit: delete

ArcEdit: save proj_outline2
```

Typing a name after the "save" command saves your edited coverage as a new coverage. Now, delete the labels and add a single, new label using the following input.

```
ArcEdit: editfeature label

ArcEdit: select all
```

```
ArcEdit: delete
ArcEdit: save
ArcEdit: add
```

The crosshair cursor once again appears in the canvas window. Use it to place your new label. When finished, again click the mouse back in the command line window to notify ArcEdit that input has changed back to the keyboard. Finally, do not forget to save your edits and restore topology. Do this with the following input.

```
ArcEdit: save
ArcEdit: build
```

You are done! Again, as you become familiar with ArcEdit's commands you may want to switch between these two methods (using pull-down menus and command line entry) for your work. In addition, by using FormEdit (discussed in Chapter 10) you can even create customized editing tool menus to help you work more efficiently. Both AML programs and macros can eliminate repetitive editing tasks. Keep these ideas in mind as you become proficient with ArcEdit's commands and functions.

Scripting

Scripting is one method ArcInfo uses to let you string together a list of commands to execute, one after another. If you enter the same sequence of commands repeatedly, create a script to make entering these commands faster and easier.

Scripting is actually a simple form of AML programming. AML stands for Arc Macro Language, and is ArcInfo's programming language. Although AML contains a rich set of programming commands and functions, scripting is just executing many commands one after the other.

Look at an example of how a script can save you some typing. Say that every time you start an ArcEdit session, you always have to change workspaces to get to a particular folder, and then you always start ErcEdit. Thus, at the Arc prompt you are always typing the same three commands, which follow.

`Arc: workspace d:\ai80exercises\chap_6` [Move to a new folder]

`Arc: &stat 9999` [Initialize the graphics display]

`Arc: arcedit` [Start ArcEdit menu system]

Instead of typing these each time you start Arc, use a text editor to put these three commands into a text file named *start.aml*. This AML (script) has the following three lines.

```
workspace d:\ai80exercises\chap_6

&stat 9999

arcedit
```

Note that these commands are identical to what you would type. Now, by simply typing *&r start* ("&r" is the abbreviation for "&run") you execute all three commands.

Use a word processing utility such as Windows NT Notepad to write and edit your AMLs. You can also use word processing software (such as Microsoft Word or WordPerfect), but be sure to do two things: always save your AML using the *.aml* file extension, and save your file as a text file, *not* a Word or WordPerfect file. ArcInfo only recognizes AML prorams if they are text files and if they end with the *.aml* extension.

✓ **TIP:** *If using word processing software to write AMLs, be sure to save the files as text files with the* .aml *extension.*

The *&r* command is used to execute all AMLs. It tells ArcInfo to execute the commands in the file specified. If you forget to type *&r* before the file name, ArcInfo will interpret the file name as a command, and return the "Unknown command" error message. Although AMLs must be named with the *.aml* extension, you can reference them with *&r* by their prefix. For example, instead of typing *&r start .aml* you can type *&r start.*

NOTE: *See Chapter 10 for a more in-depth presentation of AML.*

Editing Arcs, Nodes, and Polygons

Exercise 6-1 introduced you to starting ArcEdit and activating the pull-down menus. It described the Feature Selection menu and guided you in making some choices from the Editing Arcs & Nodes menu. In addition to an exercise, this section contains a reference section for that family of menus, showing each button and detailing its functions.

Exercise 6-2 introduces you to some advanced methods of editing that should make your editing sessions even more efficient. The exercise goes beyond the basics of the simple editing operations such as adding, deleting, and copying.

Arcs and Nodes Menu Reference

This reference section supplies descriptions of the editing buttons provided in the Editing Arcs & Nodes menu. Each menu button corresponds to one or more commands. The commands are listed in parentheses next to each button's name. This helps you learn the commands behind the buttons, which will be especially helpful if you choose to use the command line entry method rather than the pull-down menus. This also tells you which command to reference in the on-line help if you need more information.

Use this section as a reference when using Edit Tools menus to edit arc and node feature types. First, familiarize yourself with the descriptions and functions of the buttons, starting with the main "arcs and nodes" editing menu, shown in the following illustration. It may be helpful to follow along on screen: just start an ArcEdit session, specify an edit coverage, and click on the editing buttons as you read through this reference section. Then, explore in depth some arc, node, and polygon editing with exercise 6-2.

Arc Tools menu for editing nodes and arcs.

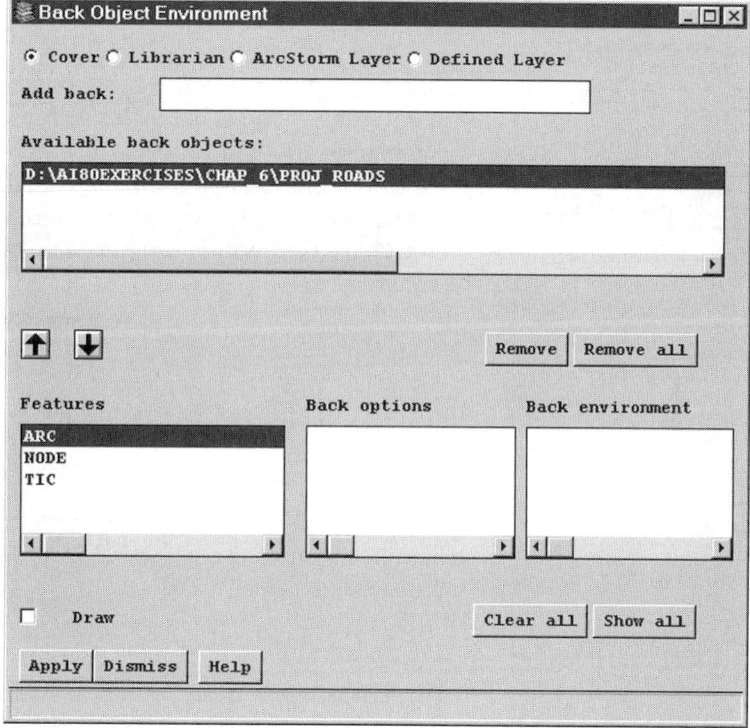

You will notice that the Edit Arcs & Nodes menu shown in the following illustration at left, contains four groups of commands: "General arc editing," "Editing nodes," Error Correction, and Attributes. Each group is discussed in the material that follows.

General Arc Editing

The "General arc editing" menu buttons, shown in the following illustration, represent the most commonly used arc editing commands. This menu is not inclusive but does contain the basic editing commands you will probably use regularly. It also contains a couple of buttons designed to access editing tools specific to the special-purpose modules ArcScan and COGO. (Both ArcScan and COGO are beyond the scope of this book.)

"General arc editing" buttons.

The following sections describe each command. Remember, the "commands behind the button" are given in parentheses after the button name. Use them for command line entry or for looking up more information on individual buttons and commands in the online help.

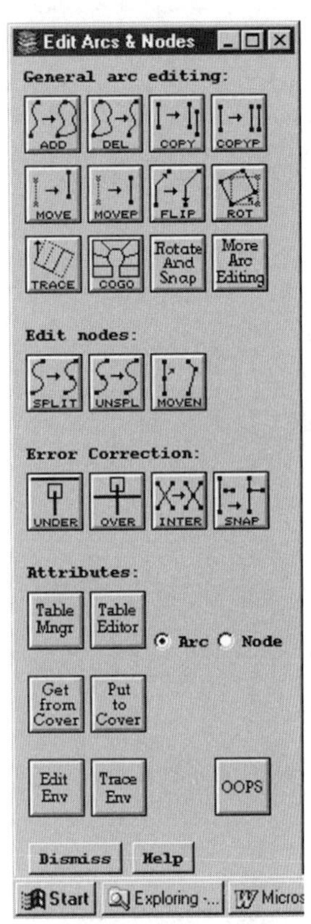

General arc editing command groups.

Add Arcs (ADD)

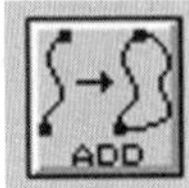

Add Arcs button.

Use Add Arcs (shown at left) to add new arcs to a coverage. When you click on this button, a crosshair appears in the graphics window (the window containing the edit coverage) and a text menu is displayed in the ArcEdit command window. The text menu displays nine options, described in the following material.

Vertex: Use to indicate where a vertex should be. Vertices (if any) are between the arc's starting node and ending node.

Node: Arcs must begin and end with nodes.

Curve: This arc forms a curve between two points you add.

Delete vertex: Use this to delete the last added vertex.

Delete arc: Deletes the added arc.

Spline on/off: Splining adds a mathematical curve between the vertices as you add them.

Square on/off: Draws a square based on two points you specify with the crosshairs. Click on two points and they form the opposite corners of a square.

Digitizing options: Presents a few methods of customizing the addition of arcs. The following nine options are presented.

- *New user ID:* Allows you to specify a new user ID for an arc.
- *New symbol:* Use this to specify a new symbol for drawing newly added arcs. You can change the color of new arcs to red, for instance.
- *Autoincrement off:* Turns off the automatic numbering of new arcs' IDs.
- *Autoincrement resume:* Used to automatically generate a new sequential ID number for all new arcs.
- *Arctype line:* Use this to specify that all new features are lines.
- *Arctype box:* Specifies that all new arcs will automatically create boxes.
- *Arctype circle:* Used to specify that new arcs will draw circles.
- *Arctype centerline:* Use this to draw centerlines that will automatically have parallel lines drawn (one on either side of the arc you just added).
- *Quit:* Returns you to the main text menu for adding arcs.

Quit: Stops the "add arc" command.

To add an arc, first position the crosshairs on its beginning point and click on option 2 (node). Then position the crosshairs at each point the arc "bends," and add a vertex with option 1. Correct vertex mistakes with option 4. Correct arc errors with option 5. Finish the arc by adding another node (option 2) as its last point. The arc is complete once you have specified two nodes, with or without vertices between them. You will use the other options (splining, squaring, digitizing) as appropriate. Select option 9 to quit the arc-adding process.

✓ **TIP:** *The default "shape" for adding arcs is a line, but you could also add arcs as circles, boxes, or centerlines. The "centerlines" option lets you add, for example, street centerlines from maps* *that show a road's edges. The Arctype command changes the shapes. This command takes the following format.*

```
Arctype <line | box | circle | centerline>
```

Delete Arcs (SELECT MANY and DELETE)

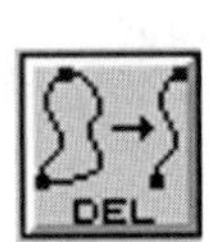

Delete Arcs button.

Use the Delete Arcs button (shown at left) to select and delete arcs. Upon pressing this button, crosshairs appear in the graphics window and the following text menu appears in the ArcEdit window. Descriptions of the options appear in brackets.

`1) Select.` [Use this command to select an arc (or arcs) using the crosshairs in the graphics window.]

`2) Next.` [Deselects your selected arc and instead adds the next nearest arc to your selected set.]

`3) Who.` [Gives you the number and User-ID of the arc just selected.]

`9) Quit.` [As soon as this is selected, the selected arcs are deleted and control returns to the editing menu.]

Use the crosshairs to select the arcs you want to delete. You will see that the selected arcs turn yellow as you select them. Use care in selecting arcs, as all of the arcs in the selected set will be deleted. If you delete arcs by accident, use the Oops button to undo your deletions.

Copy (COPY)

Copy button.

The Copy button (shown at left) is used to select and copy arcs. The following text menu appears.

`1) Select.` [This command is used to select arcs you want to copy.]

`2) Next.` [Selects the next arc nearest the one you selected with the crosshairs.]

`3) Who.` [Makes all arcs currently selected blink, helping to visually identify them.]

`9) Quit.` [Quits the selection process.]

Once you have selected arcs and have pressed the <9> key, the crosshairs become a pointer. Holding down the left mouse button, move the pointer to the location you want the arcs copied to, and release the mouse button. The arcs are then copied.

Copy Parallel (COPY PARALLEL)

Copy Parallel button.

Copy Parallel (shown at left) copies selected arcs to a new position, but copies them in parallel, meaning that they are parallel to the original arcs. When you select this button, you are prompted to point to a new position to copy the arcs. Using the crosshairs, click on a new location. All selected arcs will be copied there, and will maintain a parallel orientation with the original arcs.

Move (MOVE)

Move button.

Use the Move button (shown at left) to move arcs from one location to another. Upon selecting the button, you are prompted to select the arc to be moved. Use the crosshairs to select an arc. You are then prompted to specify a new location. Again, use the crosshairs to indicate a new position for the arc. If a NODESNAP distance has been specified, moving arcs or nodes to a position within that distance of another node will cause the arc or node to snap to the other node.

Move Parallel (MOVE PARALLEL)

Move Parallel button.

Move Parallel (shown at left) works just like the Move button, except that the arcs maintain a parallel direction to the original arc.

Flip (FLIP)

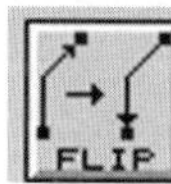

Flip button.

The Flip button (shown at left) does not physically flip an arc, but rather changes its directionality by flipping its "FROM-TO" direction. Use this to correct directionality errors, for example, in editing network coverages when the direction of flow (streams, roads, and so on) is important. If you flip arcs, be sure to update any direction-dependent attributes the arcs may have.

*Rotate (ROTATE * option of ROTATE {* | angle})*

Rotate button.

Use the Rotate button (shown at left) to interactively rotate an arc around a point you specify with the crosshairs. If you select an end point of an arc, you can drag the crosshairs and rotate the arc around that end point. If you choose a midpoint along an arc, the arc will rotate around that point.

Trace

Trace button.

The Trace button (shown at left) is intended for use with ArcScan, a separately purchased ArcInfo module used for editing scanned images. Its usage is outside the scope of this book. ArcScan requires its own license. If you click on this button without having ArcScan installed, the message "No ArcScan license available" will appear.

COGO

COGO button.

The COGO button (shown at left) starts a COGO session for using surveyor's coordinates (coordinate geometry) to enter property boundaries. COGO is a separately purchased module of ArcInfo, requiring an additional license, and a discussion of its capabilities is outside the scope of this book. If you click on this button without having COGO installed, the message "No COGO license available" will appear.

Rotate and Snap Menu Button

Rotate and Snap button.

Selecting the Rotate and Snap button (shown at left) launches the Rotate & Snap Tools menu, shown in the following illustration, which contains options for rotating and snapping arcs. These options (Arcs, including Rotate) contain command options described in the material that follows.

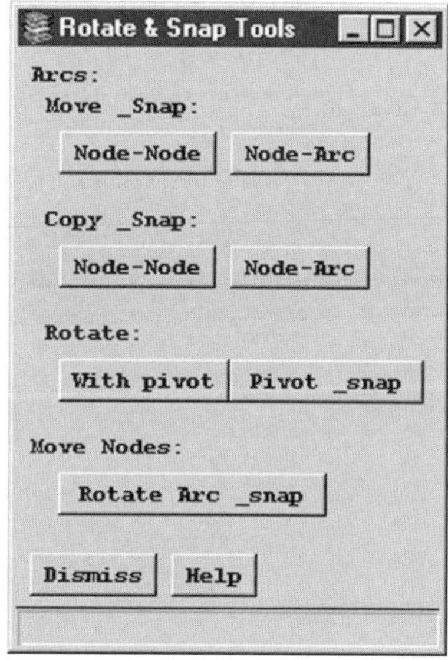

Rotate & Snap Tools menu.

Arcs Option Commands

The following commands are found under the Arcs option of the Rotate & Snap Tools menu.

Move and snap: This command moves selected arcs and controls how they intersect with other arcs in a coverage. The Node-Node option means that the arcs moved must meet the other arcs at a node; the Node-Arc option means that selected arcs will meet other arcs anywhere along the arc.

Copy and snap: This command is identical to "Move and snap," except that the selected arcs are copied instead of physically moved.

Rotate Option Commands

The following commands are found under the Rotate option of the Rotate & Snap Tools menu.

With pivot: Arcs are rotated around a pivot, which is a node interactively selected by the user with the crosshairs.

Pivot and snap: Selected arcs are rotated so that an arc in the selected set snaps to an arc in the non-selected set.

Move Nodes Option

Rotate arc and snap: Allows the user to move a node to a new location and controls whether a node snaps to an existing node.

More Arc Editing

The More Arc Editing button brings up a few additional, not as frequently used, ArcEdit commands. The More Arc Editing menu is shown in the following illustration.

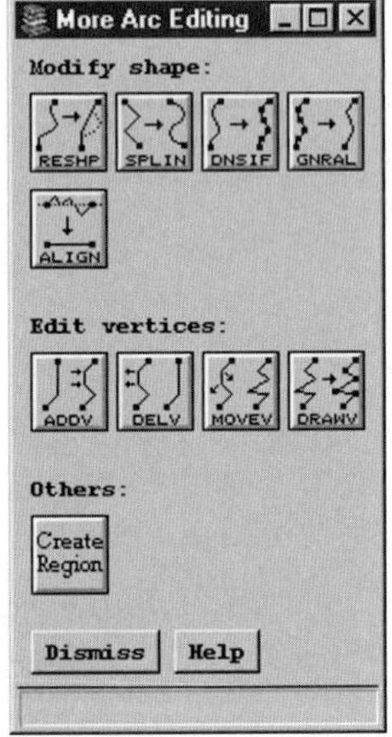

More Arc Editing menu.

Reshape (RESHAPE)

Reshape button.

The Reshape button (shown at left) redigitizes part of a selected arc. First, you use the crosshairs to digitize the new shape of the arc, making sure that the new shape crosses the existing arc at least once. Press the <9> key to end the operation.

Spline (SPLINE)

Spline button.

Select the Spline button (shown at left) to smooth curves in selected arcs. This button is useful for removing the jagged appearance that sometimes occurs when adding new arcs.

*Densify (DENSIFY {DEFAULT | * | distance})*

Densify button.

The Densify button (shown at left) adds new vertices to selected arcs, but does not change their shape. Densify adds new vertices based on the distance set by the current Grain setting (the default), interactively by using the crosshairs to specify a distance, or by entering a number in map units. New vertices are added, spaced per the specified distance, along the arc.

*Generalize (GENERALIZE {DEFAULT | * | distance})*

Generalize button.

The Generalize button (shown at left) removes vertices from selected arcs and generalizes their shape. It is useful if there are too many vertices in an arc that contribute little to accurately portraying the arc. Generalizing is a good means of cutting down on the computer storage requirements of a large arc coverage. Generalize removes vertices that fall within the distance specified by the current weed tolerance (the default), interactively using the crosshairs, or by a distance in map units that you type in.

Align (ALIGN)

Align button.

Use the Align button (shown at left) to snap the vertices of selected arcs to a line you will add after clicking on the Align button. You will be prompted to draw a line using the crosshairs, and this line will be used for snapping.

Add Vertices (VERTEX ADD)

Add Vertices button.

Use Add Vertices (shown at left) to manually add new vertices to an arc. You may have only one arc selected at a time.

Move Vertices (VERTEX MOVE)

Move Vertices button.

Move Vertices (shown at left) lets you use the crosshairs to move existing vertices from one location on an arc to another.

Draw Vertices (VERTEX DRAW)

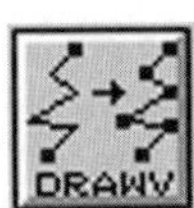

Draw Vertices button.

Draw Vertices (shown at left) draws the vertices of the currently selected arc. It will only work if just one arc is selected. It is useful for displaying vertices prior to adding or moving additional ones.

Create Region (CREATEFEATURE REGION)

Create Region button.

Create Region (shown at left) creates a new region from the currently selected set of arcs.

Editing Nodes

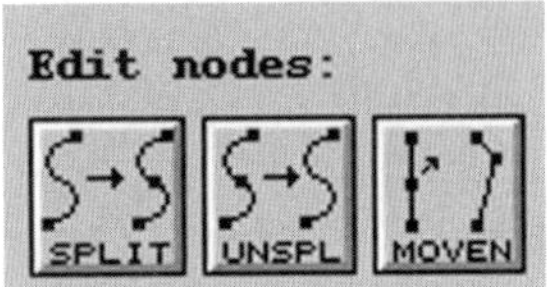

"Edit nodes" buttons.

The three "Edit nodes" buttons (shown at left) are described in the following sections.

Split (SPLIT)

Split button.

The Split button (shown at left) allows you to interactively split an existing arc into two arcs. You are prompted to select where on an arc to add a new node. Use the crosshairs to select the location and click with the mouse. The arc will be split in two.

Unsplit (UNSPLIT)

Unsplit button.

The Unsplit button (shown at left) removes all nodes from selected arcs that are not at the intersection of at least three lines. The two lines that do meet at a node are joined into one arc and the node is removed. It is the opposite of the Split command.

Move Node (MOVE)

Move Node button.

Move Node (shown at left) lets you use the crosshairs to move a node to a new location. The arcs that meet at that node are also moved.

Error Correction Buttons

Error Correction buttons.

The Error Correction buttons (shown at left) are used to correct arcs that fall either a bit short or a bit too far past arcs they are to meet with. Use these buttons, described in the following sections, to correct these mistakes.

Correct Undershoot

Correct Undershoot button.

The Correct Undershoot button (shown at left) extends the selected arcs by a distance you specify with the crosshairs. If a selected arc meets another arc within that distance, it snaps to that arc.

Correct Overshoot

Correct Overshoot button.

Similar to the Correct Undershoot button (shown at left), Correct Overshoot lets you use the crosshairs to specify a distance to which selected arcs will be snapped to non-selected arcs. The difference is that this command is used for arcs that extend past another arc. It deletes that portion of the selected arc that falls past the arc it meets.

Intersect Overlaps

Intersect Overlaps button.

Intersect Overlaps (shown at left) creates nodes wherever selected arcs cross existing arcs.

Snap (SNAP)

Snap button.

The Snap button (shown at left) snaps selected arcs to other arcs. Be sure to set the snap features to "arc" in the Editing environment.

Attribute and Other Buttons

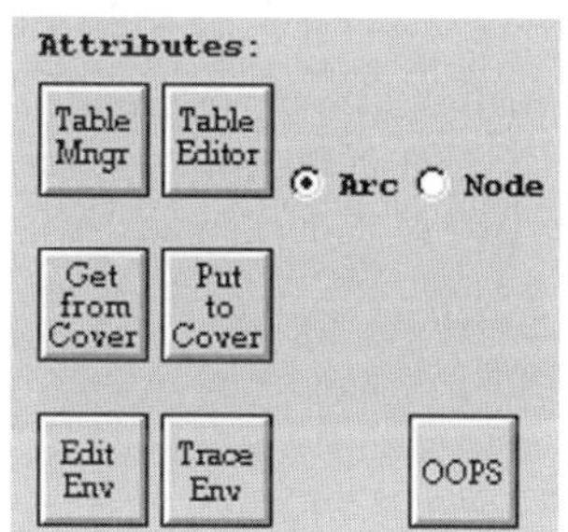

Attributes buttons.

The Attributes buttons (shown at left) are used to manage the feature attribute table for the current edit coverage. You may select the Arc setting or Node setting to open the appropriate feature attribute table. The sections that follow describe the various attribute buttons, as well as some related environment buttons.

Table Manager

Table Manager button.

Table Manager (shown at left) displays the feature attributes for the edit coverage. You may add a new item to the feature attribute table, but you cannot change any data values in the table. For this purpose, use the Table Editor.

Table Editor

Table Editor button.

Table Editor (shown at left) lets you add or change a value for an item for a selected feature. If you have an arc selected in a highway coverage, for instance, use this tool to add a street name for this arc in the feature attribute table.

Get from Cover (GET)

Get from Cover button.

Use Get from Cover (shown at left) to copy all arcs from another coverage into your edit coverage. You are prompted for the name of the coverage whose arcs you will copy.

Put to Cover (PUT)

Put to Cover button.

The Put to Cover button (shown at left) is very useful for taking selected arcs and putting them into a new or existing second coverage. You are prompted to specify the name of the new (or "target") coverage.

Edit Environment

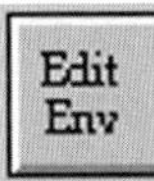

Edit Environment button.

Use Edit Environment (shown at left) to set the arc edit environment for your session. The following illustration shows the Arc Environment Properties dialog's default settings for moving and snapping arcs.

Arc Environment Properties dialog.

You can type in values that specify the distance in map units for all tolerances that control moving and snapping. The default is to specify the distance interactively using the crosshairs.

Trace Environment

Trace Environment button.

Trace Environment (shown at left) sets the environment when using ArcScan, which is outside the scope of this book.

Oops (OOPS)

Oops button.

The Oops button (shown at left) "undoes" the last action undertaken in ArcEdit. You can use it repeatedly to roll back your editing actions. Oops can "undo" actions back to the last time you saved a coverage. Once you save a coverage during an ArcEdit session, Oops cannot "undo" an edit.

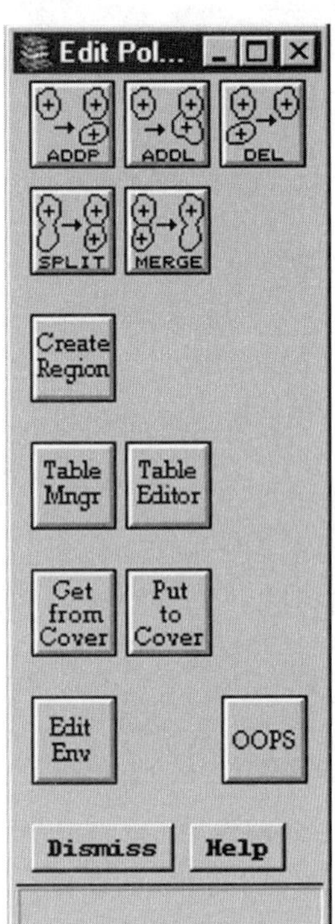

Polygon Feature menu.

Polygon Feature Menu

Polygons, like arcs and labels, can also be edited through the button commands in the Edit Tools menus. This section serves as a reference to the polygon feature commands available under Edit Tools. The Polygon Feature menu is shown in the first illustration at left.

Add Polygon (ADD)

Add Polygon button.

Add Polygon (shown in the second illustration at left) adds new polygons to a coverage. The following text menu is displayed in the ArcEdit dialog window, and crosshairs appear in the graphics window.

`1) Add polygon.` [Press this key to add the first and subsequent polygon boundary points.]

`2) End Polygon.` [Ends the current polygon addition and automatically connects this point with the first point. This connection closes the polygon. It is very important to specify end points; an "open" polygon is really just an arc and will lack the properties of a polygon.]

`4) Delete last point.` [Deletes the last point added.]

`5) Delete last Polygon.` [Removes the last polygon you created. It serves as a mini "oops" command.]

`8) Digitizing Options.` [If you are digitizing your polygons, select number 8 of the foregoing text menu to customize your editing session. The following options are presented.]

> `1) New User-ID.` [Select this to specify a new ID number for the next polygon you create.]
>
> `2) New symbol.` [This selection lets you specify a new symbol for the lines used to outline the polygon you create.]
>
> `3) Autoincrement off.` [Turns off the automatic incrementing of new ID numbers for new polygons.]
>
> `4) Autoincrement on.` [Turns on the automatic incrementing of new IDs for new polygons.]
>
> `9) Quit.` [Returns you to the polygon "Add" menu.]

9) `Quit.` [Ends your polygon addition session.]

Add Lines

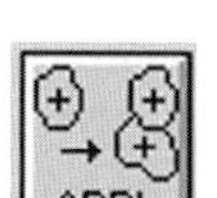

Add Lines button.

Add Lines (shown at left) is used to add lines that are inside a polygon (adding a "donut" hole, for instance) or lines that are outside an existing polygon. It does not create a new polygon label; it simply adds new arcs to a coverage. It is very similar to adding an arc when the edit feature is set to Arc and not Polygon.

As when adding arcs, you must begin and end each new line by adding a node (pressing the <2> key). Press the <1> key or the left mouse button (which acts as a "1") to add vertices. The following text menu, similar to the one seen when adding arcs, will appear in the ArcEdit command line window. Crosshairs appear in the graphics window.

1) `Vertex.` [Adds vertices.]

2) `Node.` [Select this to add a node. Be sure to begin and end your arc with nodes.]

3) `Curve.` [Calculates a smooth curve between vertices.]

4) `Delete vertex.` [Deletes the last added vertex.]

5) `Delete arc.` [Deletes the last added arc.]

7) `Square on/off.` [Makes a square from two opposite corner points you enter.]

8) `Digitizing Options.` [If digitizing, this selection allows you to customize how new arcs are added. The following text menu appears.]

> 1) `New User ID.` [Select this to manually enter a new ID number for a new arc.]
>
> 2) `New Symbol.` [Use this to specify a symbol for new arcs.]
>
> 3) `Autoincrement OFF.` [Turns off the automatic incrementing of user IDs for new arcs.]
>
> 4) `Autoincrement ON.` [Select this command to turn on the automatic incrementing of unique, sequential IDs for new arcs.]

9) `Quit.` [Select this to return to the Add Lines menu.]

9) `Quit.` [Ends the line-adding session.]

Split (SPLIT)

Split button.

Use the Split button (shown at left) to divide an existing polygon into multiple polygons. First, use the Feature Selection menu to select one polygon. Then click on the Split button. A menu just like that shown in the previous illustration, but without option 8, appears. Use it just as if you were adding a line.

Using the crosshairs, select a point on one edge of the polygon where you want the split to start. Press the <2> key (for node), and then use the crosshairs to create a line through the polygon (pressing the <1> key for each vertex). The line ends at a point on another part of the polygon's edge. Press the <2> key again (for the end node). The arc you add, which will become the boundary between the two new polygons, will split the polygon.

Merge

Merge button.

Use Merge (shown at left) to merge at least two polygons into one polygon. First, use the Feature Selection menu to select as many adjacent polygons as you would like to join into a single polygon. Then select the Merge button. The polygons will automatically collapse into a single polygon containing one label.

NOTE: *Attribute information will be lost during the merge. To control what information the newly created polygon will have, use the New option under the Table Editor button first. For more details on this process, see the ArcEdit Merge and New commands in the on-line help.*

Create Region

Create Region button.

Create Region (shown at left) creates a new region from the currently selected set of polygons. When you select this button, the Create a Region window pops up and you are prompted to type in a subclass name and a region ID for the region.

Table Manager

Table Manager button.

Table Manager (shown at left) activates the "<covername> Polygon Items" menu, which lists the polygon attributes of your current edit coverage. Here you may perform INFO table relate operations (for more information on this, see Chapter 9) and even add new attributes (items). Note that you cannot delete items here. To do this, you must use the Arc command *Dropitem*, as described in Chapter 9. To change values for polygon attributes, use the Table Editor button. Using this tool to add items or relate tables can save you the inconvenience of leaving your editing session and starting up the Arc or INFO modules.

Table Editor

Table Editor button.

Table Editor (shown at left) lets you add an item or change a value for an item for a selected polygon or polygons. Using this tool saves you the inconvenience of leaving ArcEdit and using Arc commands or going into INFO to perform the same operations.

Get from Cover (GET)

Get from Cover button.

Use Get from Cover (shown at left) to copy all polygons from a source coverage into your edit (target) coverage. Note that in using this command the attributes associated with the source cover polygons do not copy over into the target cover, unless the feature attributes are exactly the same for both covers. Selecting this command activates the Get from Coverage window: you simply type in the name (include the path name if it is in another directory) of the source coverage and then click on the OK button. Any messages relating to the "Get" operation will appear in the ArcEdit command line window.

Put to Cover (PUT)

Put to Cover button.

Put to Cover (shown at left) is very useful for taking selected polygons from your edit (source) coverage and putting them into a new or existing second coverage. Note that in using this command the attributes associated with the source cover polygons do not copy over into the target cover, unless the feature attributes are exactly the same for both covers.

Selecting this command activates the Put to Coverage window: you simply type in the name (include the path name if it is in another directory) of the target coverage and then click on the OK button. Any messages relating to the "Put" operation will appear in the ArcEdit command line window.

Edit Environment

Edit Environment button.

Use Edit Environment (shown at left) to set two characteristics of your session's polygon editing environment. These characteristics are saved with the coverage and are active in future editing sessions until they are changed again. The following illustration shows the default settings for editing polygons.

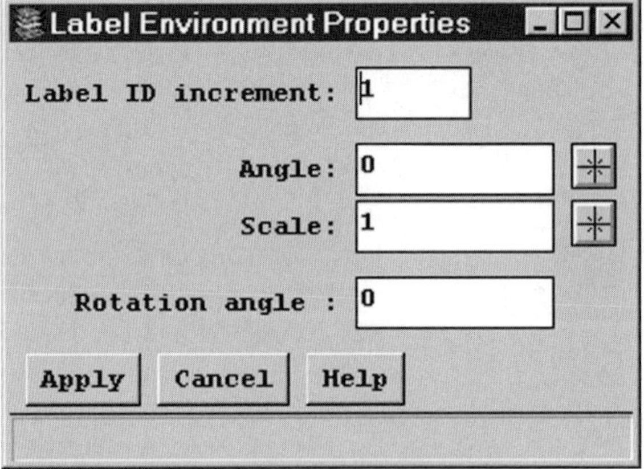

Edit Environment menu for polygons.

When adding polygons, you can choose to have the arcs you add intersect existing arcs, or have the existing arcs dissolved. If the arcs of a polygon you are adding cross existing arcs and you have Dissolve selected, the lengths of existing arcs that lie within the newly added polygon will be deleted. If you select Intersect, the lengths remain and multiple polygons are formed. You can also specify a dangle distance either numerically or interactively. Any arcs that are shorter than the dangle distance will be automatically deleted. This helps correct overshoots when adding arcs and polygons.

Oops (OOPS)

Oops button.

The Oops button (shown at left) erases the results of the last ArcEdit command. You can use it repeatedly to roll back your editing actions. Oops can undo actions back to the last time you saved a coverage. Once you save a coverage, however, Oops cannot undo an edit.

Exercise 6-2: Editing the Blue-eyed Corn Moth Project Area Coverage

You have just gotten word about some changes to the Blue-eyed Corn Moth project area. Beginning next month, the sampling area in Ohio's Clark and Madison counties will be south of U.S. Highway 40, rather than covering the whole of those counties as it does now. Your job is to edit the *proj_cnties* coverage to reflect these changes.

Before beginning, be sure you have completed exercise 6-1, and that the data needed for this exercise has been copied from the companion CD-ROM to your workspace (assumed to be *D:\ai80_exercises\chap_6.*

Start ArcEdit by clicking on Start ➥ Programs ➥ ArcInfo ➥ Workstation ArcInfo ➥ ArcEdit. Arrange the command line window and graphics window to suit your working habits. Now activate the pull-down Edit Tools menu by typing *arctools* at the ArcEdit prompt and then selecting "edit tools" from the menu.

Now specify *proj_cnties* as your edit coverage by clicking on File ➥ Open and selecting "proj_cnties" under "coverages." Select ARC under "Available features" to display the coverage's arcs. The Feature Selection and Editing Arcs & Nodes menus will appear. Your screen should look like that shown in the following illustration.

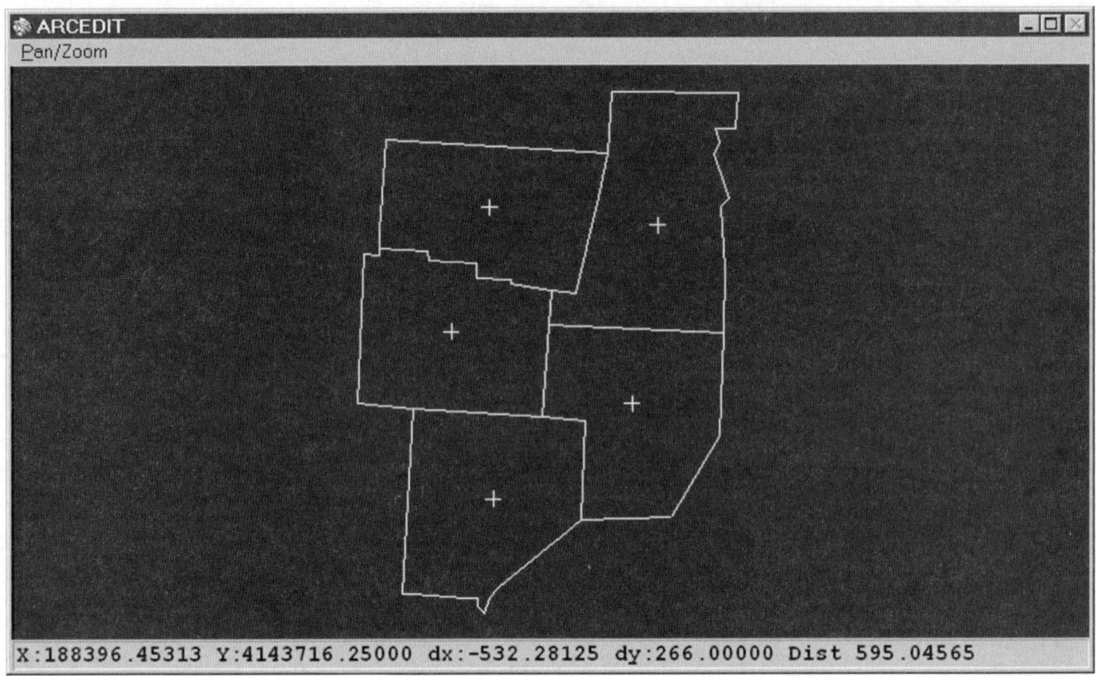

The proj_cnties *coverage.*

Find the two counties (Clark and Madison) you will be editing. Do this by adding labels to your drawing environment; this will allow you to view the county names on your graphics screen. Select Display ➥ Draw env. ➥ General from the Edit Tools menu. A new menu box, shown in the following illustration, appears.

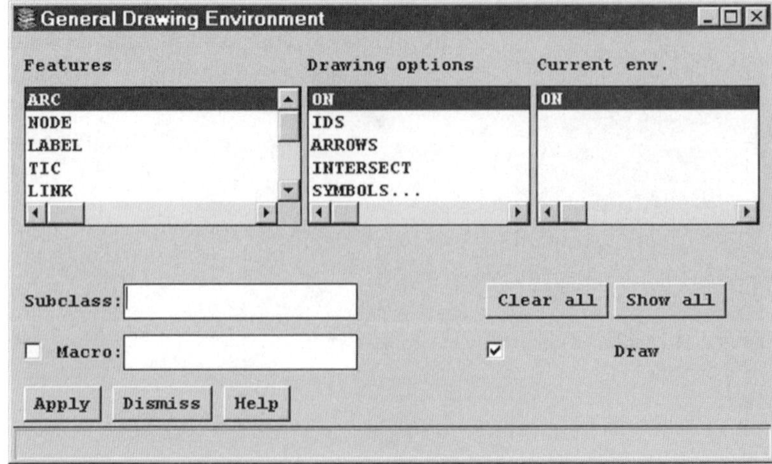

General Drawing Environment menu.

The left-hand window labeled Features contains a list of all available features in the edit coverage. The middle window, "Drawing options," lists the options available for whichever feature is highlighted in the Features window. Finally, the right-hand window, "Current env.," shows which option from the middle window is currently active.

To demonstrate, highlight ARC in the Features window. You will see the options ON, IDS, ARROWS, INTERSECT, and SYMBOLS appear in the "Drawing options" window. In the "Current env." Window, only the word ON appears, telling you that just arcs will appear on the graphics screen each time the Draw command is executed.

Spend a moment exploring the other options. For example, click on ARROWS and you will see it appear under "Current env." Then click on the Apply button. Now view your ArcEdit graphics screen and you will see that small arrowheads, indicating the directionality of the arcs, have appeared. This is shown in the following illustration.

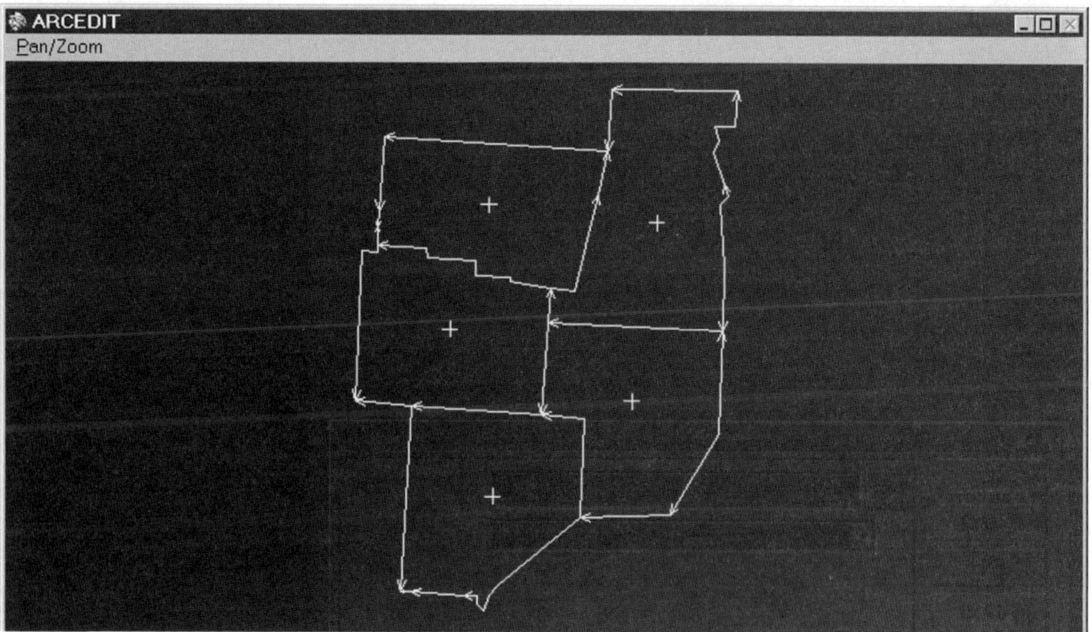

Proj_cnties *coverage shown with arc drawing option ARROWS activated.*

To deactivate the arrows, click on Display ➥ "Draw env: General" again. The "Current env." window lists ARROWS as active. Click on ARROWS in that window to deactivate it. Before leaving this menu, turn the labels on by highlighting LABEL in the Features window and clicking on ON in the "Drawing options" window. Your screen should now show the label points of each county, as shown in the following illustration.

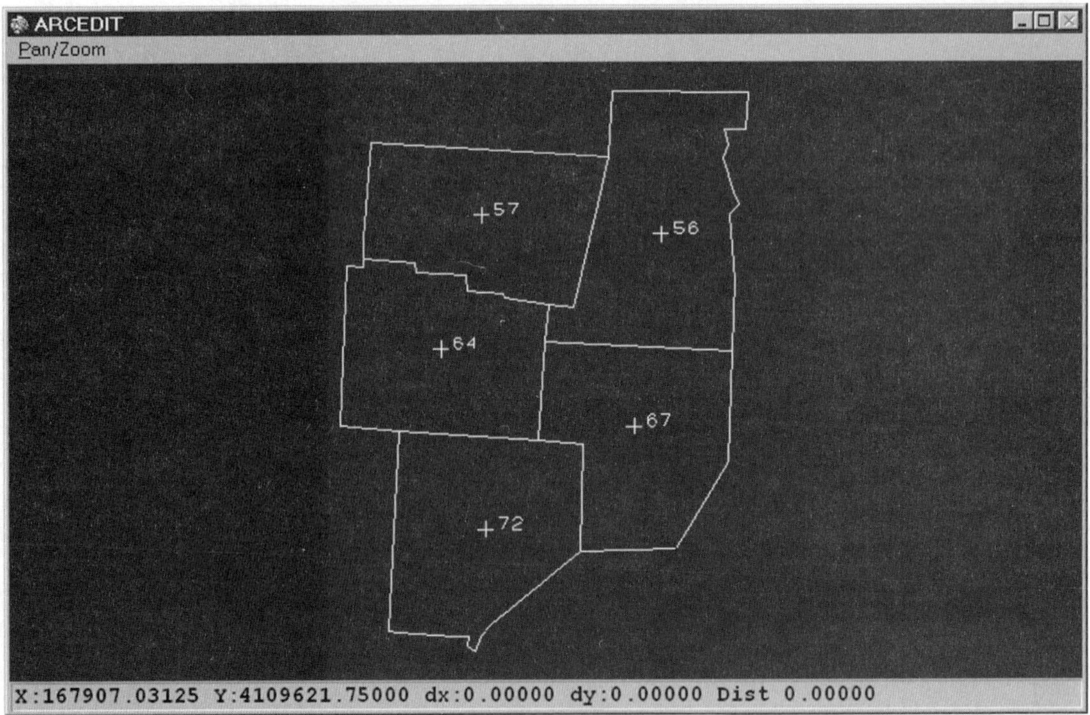

Proj_cnties *coverage with a drawing environment of arcs and labels.*

Display the county names. Do this by selecting Display ➥ Textitem from the Edit Tools menu to reach the Set Textitem Parameters menu, shown in the following illustration. This menu lists which features (arcs, labels, polygons, and so on) are available in a coverage, and what items (also called feature attributes) are associated with each feature.

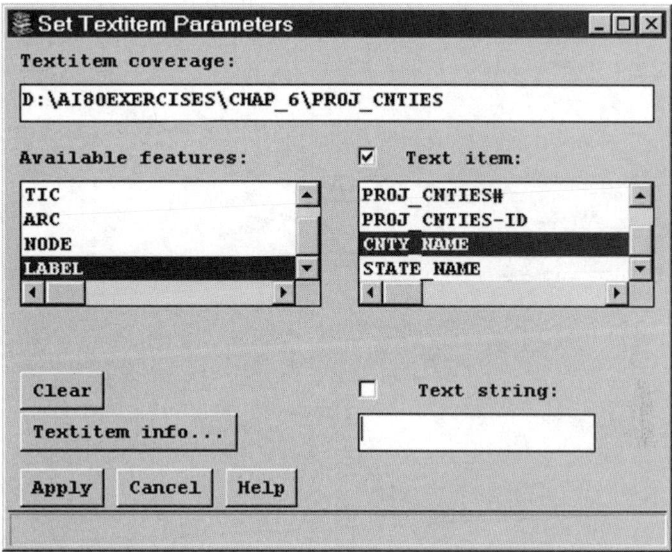

Set Textitem Parameters menu.

To display the county names associated with each label in *proj_cnties*, select LABEL under "Available features" and NAME under "Text item." Click on Apply and you will see county names appear to the right of the county label points. Your screen should look like that shown in the following illustration.

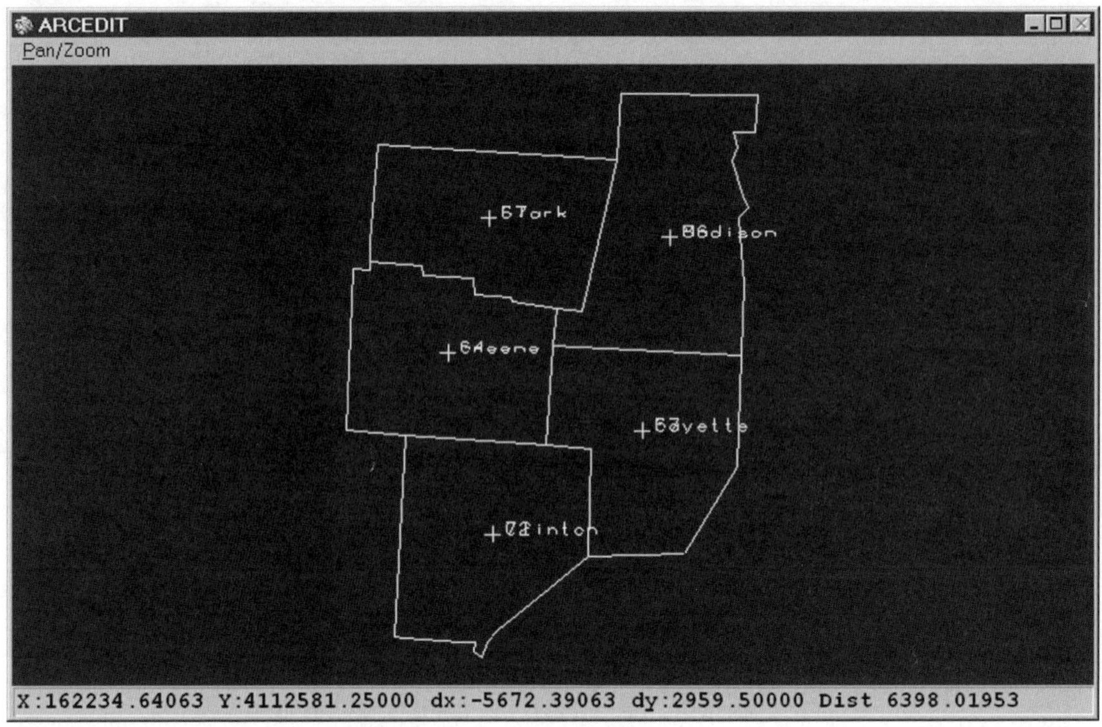

The proj_cnties *coverage with county names displayed.*

Clark and Madison counties are the two northernmost counties. Remember your assignment is to edit the *proj_cnties* coverage to exclude any areas in Clark and Madison counties that are north of U.S. Highway 40. Next, you will locate U.S highway 40 in relation to these two counties.

The *proj_roads* coverage is an arc coverage containing the major highways in your five-county project area. Display it as a background coverage. Do this by selecting Display ➡ "Back env. General" from the Edit Tools menu to activate the Back Object Environment menu, shown in the following illustration. This menu lets you specify which coverages and which features of those coverages you want displayed in the background.

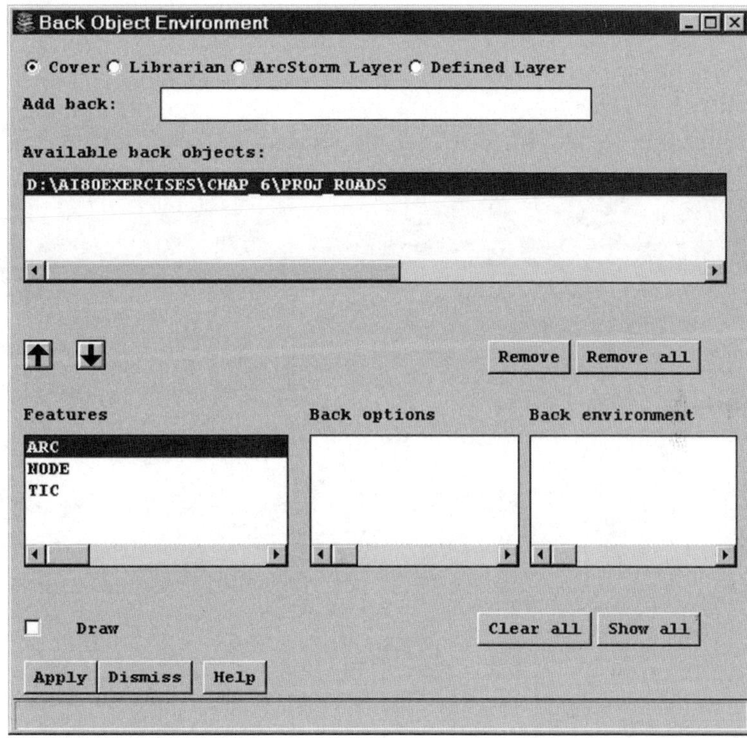

Back Object Environment menu.

You can display various "objects," (such as coverages, Librarian, or ArcStorm layers) as backgrounds for your edit coverage in ArcEdit (both Librarian and ArcStorm are separately purchased modules and their capabilities are not covered in this book). Because the "object" in this case is a coverage, click on Cover. In the "Add back" box, type the coverage name *proj_roads* and press the <Enter> key. (Note that you can also access coverages in different directories by typing in their full path name and name; for example, *D:\ai80exercises\chap_6\proj_roads*.)

Under "Available features" you will see ARC, NODE, and TIC. To display the roads in *proj_roads*, add arcs to your background environment by clicking on ARC. In the "Back options" window, click on ON. It is always helpful to display your background coverages as a different color. Do this by selecting the SYMBOLS option ARC.

A menu pops up from which you can select colors. Select red by clicking on it, and then click on OK. This returns you to the Back Object Environment menu. Finally, check the box labeled Draw and click on Apply. Your graphics screen should look similar to that shown in the following illustration, with your labeled *proj_cnties* coverage in white and the *proj_roads* coverage behind it in red.

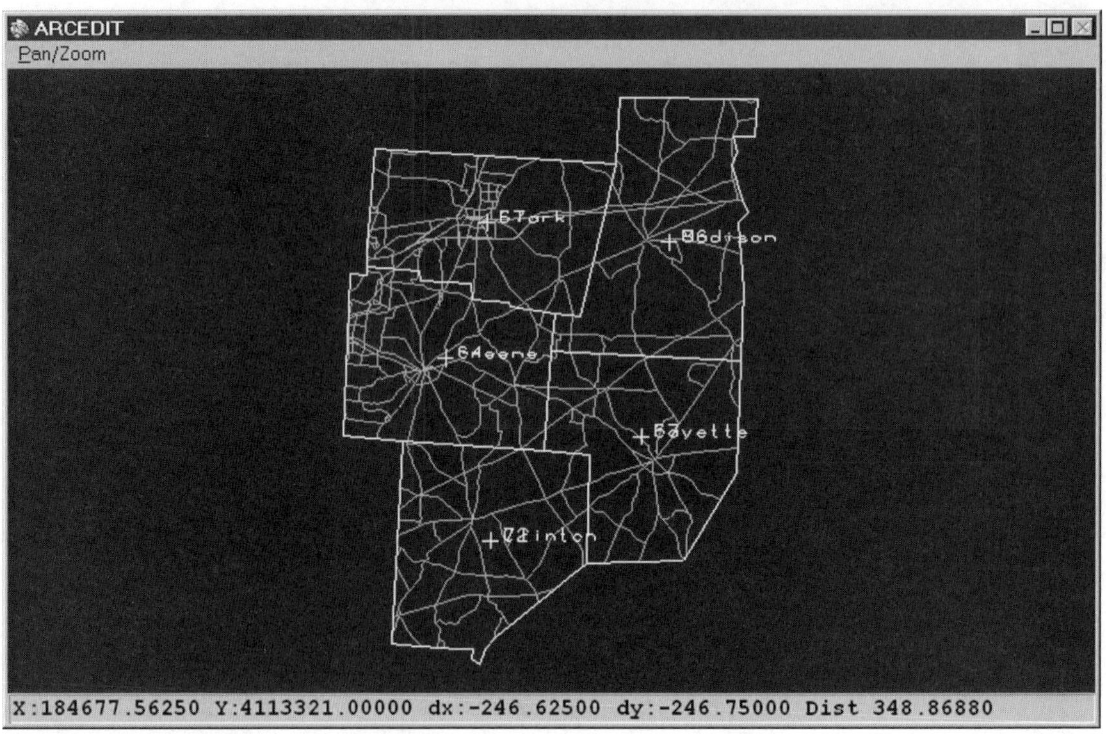

Labeled proj_cnties *coverage with* proj_roads *in the background.*

So far you have identified which counties are Clark and Madison (by displaying the county names), and you have set up the roads coverage as your background. You have also learned how to set additional drawing environment and background environment characteristics. Now you will identify and select the highway (U.S. Highway 40) arcs from the *proj_roads* coverage and add them to your *proj_cnties* coverage. Then you will edit that coverage to correct any errors and remove any county area north of the highway.

To select arcs from *proj_roads*, first you must make *proj_roads* your edit coverage. Do this by selecting File ➥ Open and then selecting *proj_roads*. Add *proj_cnties* as a background coverage by following the steps previously listed. Now you are able to select features from *proj_roads*, and *proj_cnties* is still displayed on your graphics screen. Your edit feature is still arcs, so the menus specific to editing arcs and nodes will pop up.

Second, you must identify and select which arcs in *proj_roads* represent U.S. Highway 40. You will do this by building a selection query. Select Edit ➥ Attribute Selection from the Edit Tools menu to get the Logical Expression menu, shown in the following illustration.

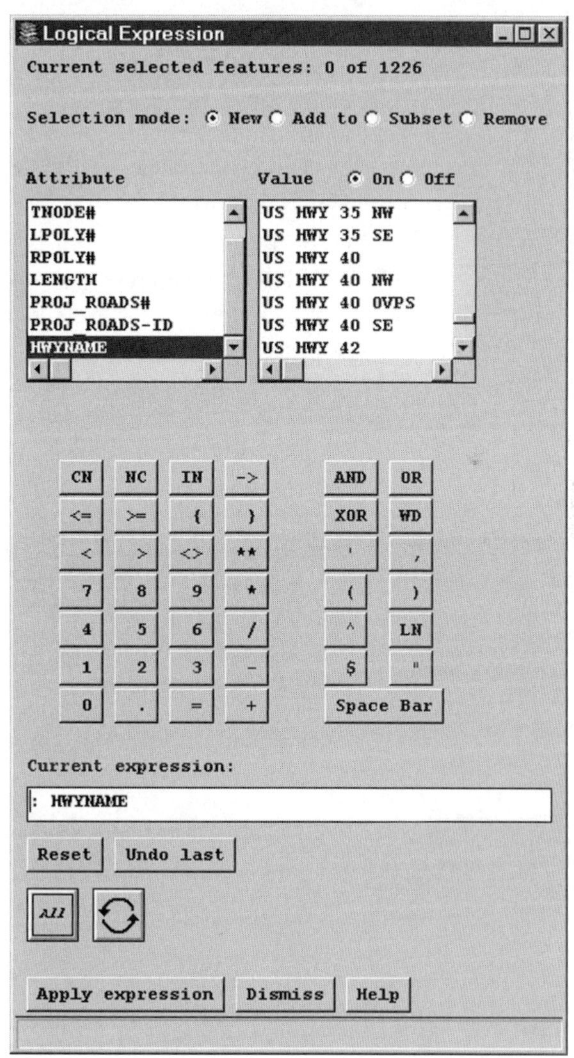

Logical Expression menu.

You will use this menu to build your selection query for U.S. Highway 40. A quick look at it reveals a window that lists the coverage's attributes (Attribute), a window that if turned "On" lists the values of a highlighted attribute (Value), and several operator and number buttons ("=", ">", "1", and so on). As you select components for your query expression, they appear in the "Current expression" box near the bottom of the menu. Thus, you must select the components of your query in a logical order.

Below the buttons is the "Current expression" box, which shows your query as you build it. The Reset button clears the current expression, and the "Undo last" button removes the last item added to the query. Finally, the All button selects all of that feature. The "Switch set" button (with the two half-circle arrows) switches the selected set with the unselected set.

For this exercise, you want to select all of the arcs from the *proj_roads* coverage that represent U.S. Highway 40. To begin, click on New under "Selection mode," because you are creating a new selection set. Note that you could also add to, remove, or form a subset of, a selected set. Under the Attribute window, highlight HWYNAME. This is the item (field) name in the *proj_roads.aat* that contains the values for the road names. Because HWYNAME is a character item, you will use CN (which means "contains") as the operator rather than the "=" sign (the "=" sign is for numerical fields). Click on the CN button. In the Value window, click on On and a list of all values for the item HWYNAME appear, in alphabetical order. Scroll down until you find US HWY 40 and highlight it. The current expression box should look like that shown in the following illustration.

Current expression box showing HWYNAME CN US HWY 40.

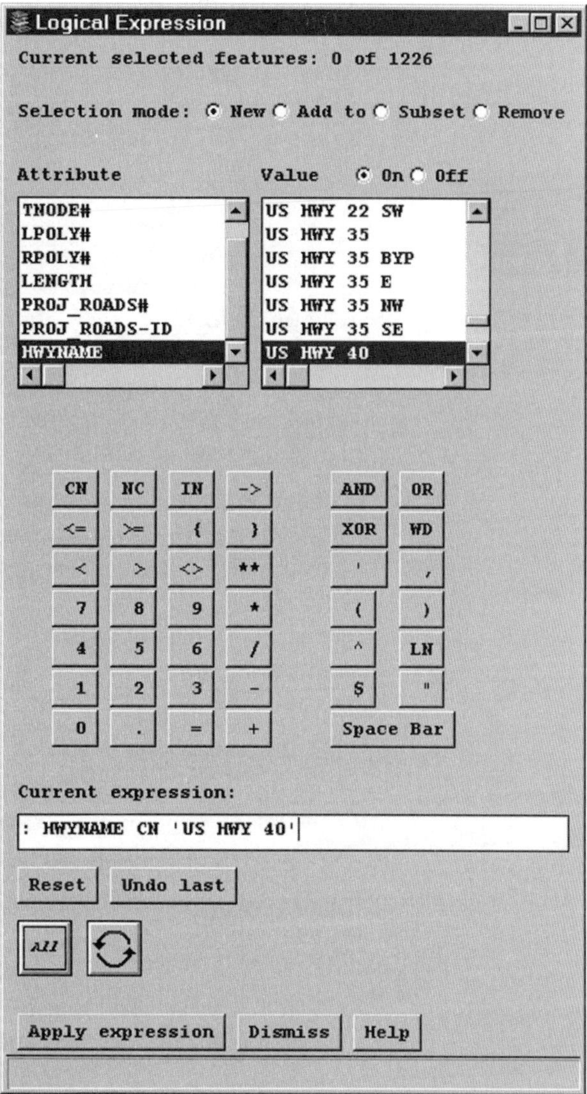

Now click on the "Apply expression" button, and you will see that 50 of 1,226 arcs are chosen. On your graphics screen, these arcs turn yellow.

Now that the arcs representing U.S. Highway 40 are chosen, add them to the *proj_cnties* coverage. From the Edit Arcs & Nodes menu, click on the "Put to cover" button to access the window shown in the following illustration.

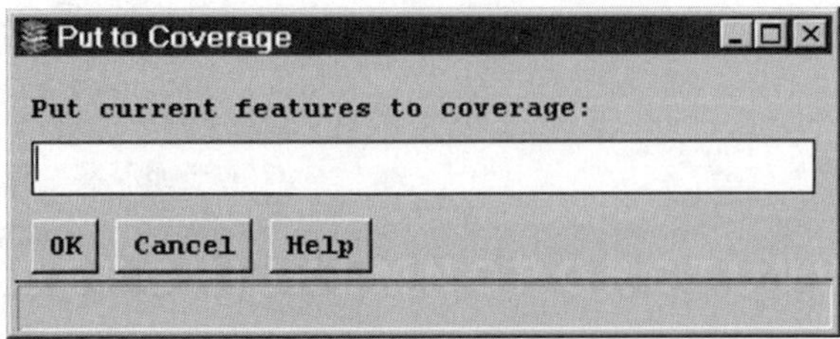

Put to Coverage window.

Type *proj_cnties* into the box and click on OK. A warning pops up telling you that the *proj_cnties* coverage already exists, do you want to append? This warning helps prevent you from accidentally adding arcs to the wrong coverage. Answer "yes" to copy the selected arcs representing U.S. Highway 40 from the *proj_roads* coverage to the *proj_cnties* coverage.

In the command line screen you will see messages, as follows, corresponding to the menu choices. ArcEdit tells you that the arc attribute items of the arcs in *proj_cnties* and *proj_roads* are not the same. In other words, you will not suddenly find the item HWYNAME in your *proj_cnties.aat*. Because the items in the two coverages' *.aat* table are different, ArcEdit copies only the arcs, not the attributes.

```
The arc attribute items between coverages, do not match

Only the User-ID and any matching items will be copied over

Copying the arcs into d:\ai80exercise_6\proj_cnties...

Please wait... d:\ai80exercise_6\proj_cnties is an open cover,

use SAVE to keep changes
```

It is important to note that *proj_cnties* is now an open cover. This means that it has had changes made to it and those changes have not been saved. Save those changes now by selecting File ➥ Coverages: Open from the Edit Tools menu. Select *proj_cnties* from the coverages window and click on the OK button. Select File ➥ Coverage ➥ Save As from the Edit Tools menu and type in the new name *proj_cnty_rev* (rev for "revised") in the "Save as" box. Click on the Save button. ArcEdit automatically makes this newly saved coverage your edit coverage.

Next, you will interactively remove a couple of dangling nodes. Dangling nodes are small arcs that extend past an intersection with another arc, but are often of very short length. They are usually removed because they contribute nothing to a coverage. In this case, they are small arcs not needed in the coverage.

First, display the dangling nodes by selecting Display ➥ Draw env: General from the Edit Tools menu. Click on NODE in the Feature Attributes box and on ERRORS in the drawing options box. This makes node errors part of the current drawing environment. Now assign a color to the dangling nodes by selecting DANGLE and SYMBOL from the drawing options box. You can choose how colors nodes, dangles, and pseudonodes will appear on your graphics screen. However, for this exercise, do not worry about assigning colors to nodes or pseudonodes, just highlight red for the dangle color.

Your graphics screen should now show a number of small white diamonds and two small red squares along the newly added U.S. Highway 40. Using the Pan/Zoom options on your graphics window, select Create and use the crosshairs to zoom in on a box around the two dangling nodes. You will see that they are actually small arcs. Delete them with the DEL option from the Edit Arcs & Nodes menu.

Because you have now added and edited some arcs to a polygon cover, you must restore topology to it. Do this by selecting Clean under Tools ➥ Topology from the Edit Tools menu. The Clean option, unlike the Build option, will intersect added arcs and create nodes at those intersections.

Finally, you will end your editing session by deleting the areas in both Madison and Clark counties that now fall north of U.S. Highway 40. Because you have just restored topology, those areas are now represented by polygons. These two areas are partitioned with a dividing line in the following illustration.

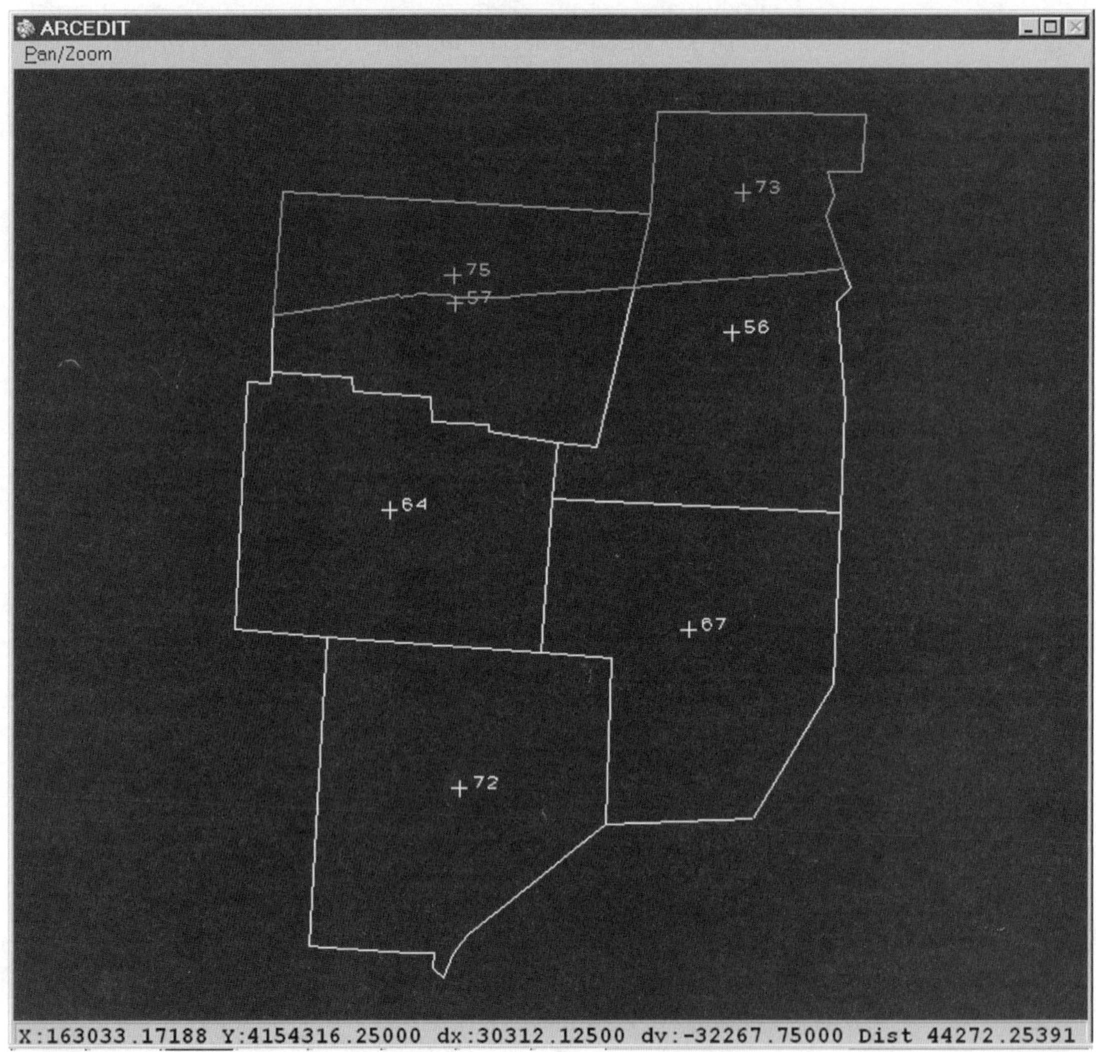

Shading shows areas in Madison and Clark counties that are north of U.S. Highway 40.

Change your edit feature to polygon by selecting Edit ➥ "Change edit feature" from the Edit Tools menu. Highlight POLYGONS and click on OK. The Edit Polygons menu now appears. Use the Feature Selection menu to select the two polygons representing the area in Madison and Clark counties that is north of Highway 40. You will see their outlines turn yellow, indicating that they are selected. Select the DEL button from the Edit Polygons menu. Both polygons are now deleted from the *proj_cnty_rev* coverage. Because ArcEdit knows that you are editing polygons, it automatically restores topology after the deletion.

✓ **TIP:** *When selecting polygons for deletion, it may seem as if you will "ruin" other polygons that share sides with the selected ones. ArcEdit "knows" that those polygons may share sides, and unselected polygons remain intact.*

Save your changes once again using File ➥ Save, and exit this editing session by selecting ArcTools ➥ Quit. This returns you to the Arc Tools menu. Click on Quit again to exit the menu system entirely. Type *quit* or *q* at the ArcEdit prompt, to close the ArcEdit module.

Congratulations! You have completed an advanced exercise that demonstrates some time-saving editing techniques. Using the "Put" command, you added arcs to a coverage. You removed some dangling nodes. By rebuilding topology after these edits, you were able to select and delete entire polygons rather than performing the same function by selecting and deleting arcs and labels. You will be able to incorporate these techniques to save time in your own editing sessions.

Editing Labels

Edit Labels menu.

This section provides descriptions of all editing commands used to edit labels in the Edit Labels menu (shown at left) of Arc Tools. Equivalent command-line names are provided in parentheses after the ArcEdit command names. These can be used outside Arc Tools.

Add (ADD)

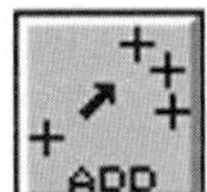

Add button.

Use the Add button (shown at bottom left) to add new labels. When selected, crosshairs appear in the graphics window and a text menu with the following add options appears in the ArcEdit command line window.

1) `Add label.` [In the graphics window, position the crosshairs at the point where you want the label added. Then either press the <L> key on the keyboard or click the left mouse button.]

5) `Delete last label.` [Removes the last label you added. It functions as a mini "Oops" command.]

8) `Digitizing options.` [If digitizing label locations, you can use the options that follow to control your editing session.]

    1) `New User-ID.` [Lets you specify a number manually for the label's user ID.]

    2) `New Symbol.` [Select this to change the symbol used to display labels.]

    3) `Autoincrement off.` [Turns off the automatic numbering of user IDs for new labels. All new labels will have the same user ID.]

        1) `Autoincrement resume.` [Starts the automatic numbering of new labels with unique sequential numbers.]

        2) `New angle.` [Changes the angle of the symbol used to display new labels.]

        3) `New scale.` [Use this to change the scale of the label as it is displayed after being created.]

    9) `Quit.` [Quits the Digitizing options menu and returns to the Add menu.]

9) `Quit.` [Stops the current Add session.]

Delete (DELETE)

Delete button.

Use the Delete button (shown at left) to interactively select and delete labels. The following text menu appears in the ArcEdit dialog (command line) window, and crosshairs appear in the graphics window.

`1) Select.` [Selects a label.]

`2) Next.` [Selects the nearest label to the one last selected.]

`3) Who.` [Makes all selected labels blink, helping to identify their locations visually.]

`9) Quit.` [Stops the current Delete session and deletes all selected labels.]

Copy (COPY)

Copy button.

Copy (shown at left) lets you select labels and then copy them to new locations. The following text menu appears in the ArcEdit dialog window, and crosshairs appear in the graphics window.

`1) Select.` [Selects a label.]

`2) Next.` [Selects the nearest label to the one last selected.]

`3) Who.` [Makes all selected labels blink, helping to identify their locations visually.]

`9) Quit.` [Stops the current selection process.]

Upon pressing the <9> key the crosshairs change to a small hand icon. Use it to drag the cursor from the current position to the copy location. The labels will be copied en masse the distance and angle away from their original location.

Move (MOVE)

Move button.

Use the Move button (shown at left) to move a label to a new location. Upon selecting this command, crosshairs appear in the graphics window and the following menu appears in the ArcEdit dialog window. Use the crosshairs to select the labels to be moved, and then move them to a new location.

1) **Select.** [Selects a label.]

2) **Next.** [Selects the nearest label to the one last selected.]

3) **Who.** [Makes all selected labels blink, helping to identify their locations visually.]

9) **Quit.** [Stops the current selection session.]

After pressing the <9> key the crosshairs will change to resemble a small hand. Drag the cursor from the selected label's current location to its new location. The label will then be moved to the new location.

Scale (SCALE)

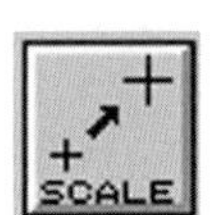

Scale button.

Select the Scale button (shown at left) to change the scale of labels you select. You will see the following text menu appear in the ArcEdit dialog window.

1) **Select.** [Selects a label.]

2) **Next.** [Selects the nearest label to the one last selected.]

3) **Who.** [Makes all selected labels blink, helping to identify their locations visually.]

9) **Quit.** [Stops the current Scale session and scales the labels.]

Once you press the <9> key, the labels are scaled via the value specified in the Edit Environment menu, which is accessed via a button described later in this section.

Rotate (ROTATE)

Rotate button.

The Rotate button (shown at left) changes the rotation of labels you select. Labels will be rotated via the value you specify in the Edit Environment menu, described later in this section. You will see the following text menu appear in the ArcEdit dialog window.

1) **Select.** [Selects a label.]

2) **Next.** [Selects the nearest label to the one last selected.]

3) **`Who.`** [Makes all selected labels blink, helping to identify their locations visually.]

9) **`Quit.`** [Stops the current Rotate session and rotates the labels.]

Once you press the <9> key, the labels are rotated via the value specified in the Edit Environment menu, which is accessed via the Edit Environment button, described later in this section.

Table Manager

Table Manager button.

Table Manager (shown at left) activates the "<covername> Labels Items" menu, which lists the label attributes of your current edit coverage. Here you may perform INFO table relate operations (for more information on this, see Chapter 9) and even add new attributes (items). Note that you cannot delete items here; you must use the Arc command Dropitem, as described in Chapter 9. To change values for label attributes, use the Table Editor button. Using this tool to add items or relate tables can save you the inconvenience of leaving your editing session and starting up the Arc or INFO modules.

Table Editor

Table Editor button.

Table Editor (shown at left) lets you add an item or change a value for an item for a selected label or labels. Using this tool saves you the inconvenience of leaving ArcEdit and using Arc commands or going into INFO to perform the same operations.

Get from Cover (GET)

Get from Cover button.

Use Get from Cover (shown at left) to copy all labels from a source coverage into your edit (target) coverage. Note that in using this command the attributes associated with the source-cover labels do not copy over into the target cover, unless the feature attributes are exactly the same for both covers. Selecting this command activates the "Get from Coverage" window, where you simply type in the name (include the path name if it is in another directory) of the source coverage and then click on the OK button. Any messages relating to the "Get" operation will appear in the ArcEdit command line window.

Put to Cover (PUT)

Put to Cover button.

Put to Cover (shown at left) is very useful for taking selected labels from your edit (source) coverage and putting them into a new or existing second coverage. Note that in using this command the attributes associated with the source cover labels do not copy over into the target cover, unless the feature attributes are exactly the same for both covers.

Selecting this command activates the Put to Coverage window, where you simply type in the name (include the path name if it is in another directory) of the target coverage and then click on the OK button. Any messages relating to the "Put" operation will appear in the ArcEdit command line window.

Edit Environment

Edit Environment button.

Use Edit Environment (shown at left) to set the label edit environment for your session. The following illustration shows the default settings for editing labels.

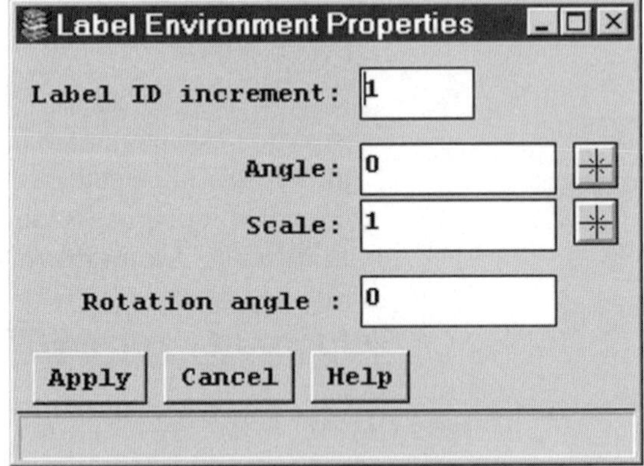

Edit Environment menu for labels.

By default, the Label-ID (represented by the <covername>-ID item) increments by 1 each time you add a label. In some GIS applications it may be useful to change this increment, but a discussion of those applications is beyond the scope of this book. If you need to change the increment value, select the Edit Environment button. The Label Environment window appears. Here you can change the increment value, and all new labels will automatically increment by this number.

Here you can also specify the rotation angle and scale of symbols, as well as the symbol angle, in case you need to have a symbol appear partially rotated. The default rotation angle is 0°. At 0°, the label symbol could be an "8", at 90° it becomes an "∞." The symbol angle can be typed in or specified interactively. Symbol scale, the default of which is 1, can also be changed either interactively or by typing in a value.

Oops (OOPS)

OOPS

Oops button.

The Oops button (shown at left) erases the last action performed in ArcEdit. For each editing session, ArcEdit keeps track of every command executed since the last Save command. When you undo a command's action with the Oops command, you will see a message in the ArcEdit dialog window telling you that the action has been undone, and that you are back at the previous transaction (command). You can use Oops repeatedly to roll back your editing actions. Remember, Oops can undo actions back to the last time you saved a coverage. Once you save a coverage, Oops cannot undo an edit.

Exercise 6-3: Combining Coverages

Remember that the Blue-eyed Corn Moth project sets out a grid of traps each year, to discover which areas have high moth populations. Each county has its own trapper, a person assigned to place and check the traps. This exercise consists of two parts. In Part I, you will add an item and values to the labels in *proj_cnties*. In Part II, you will use an Arc command to combine two coverages to create a third. The third coverage will contain the traps within a county, and each trap will have a county name and trapper name associated with it.

Part I: Adding an Item and Values to Labels

You have just received a list of this year's trappers and the county in which each will work. You must now edit the trap site coverage (named *grid_traps*) to assign each site the correct county name and trapper name. *Grid_traps* is a point coverage containing a grid of traps that cover the five-county project area. The traps are spaced 5 km apart.

First, view the *grid_traps* coverage in ArcEdit. Start ArcEdit by clicking on Start ➥ Programs ➥ ArcInfo ➥ Workstation ArcInfo ➥

ArcEdit. Arrange the command line window and graphics window to suit your working habits. Now activate the pull-down Edit Tools menu by typing *arctools* at the ArcEdit prompt and then selecting "edit tools" from the menu.

Specify *grid_traps* as your edit coverage by clicking on File ➥ Open and selecting *proj_cnties* under "coverages." Select LABEL under "Available features" to display the coverage's label points. In this case they represent trap sites. The Feature Selection and Editing Labels menus will appear. Your screen should look like that shown in the following illustration.

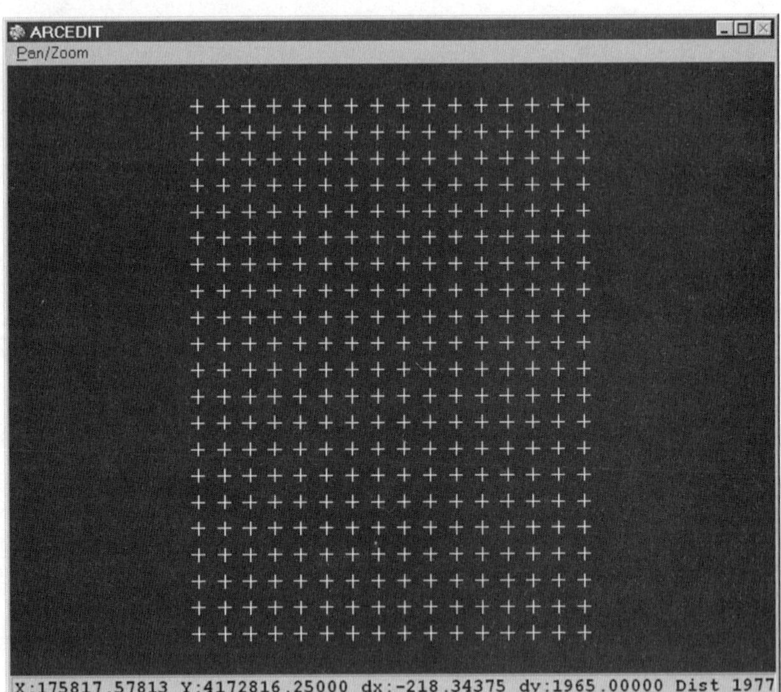

Grid_traps *as the edit coverage.*

To orient this coverage with your project area, bring up *proj_cnties* as a background coverage by selecting Display ➥ Back env. General. Type in the name *proj_cnties* and press the <Enter> key. Turn ARCS on and, using the SYMBOLS menu, specify that the background arcs be shown in red. If you need to refresh your memory on the steps for doing this, see the previous exercises. Your screen should now look like that shown in the following illustration.

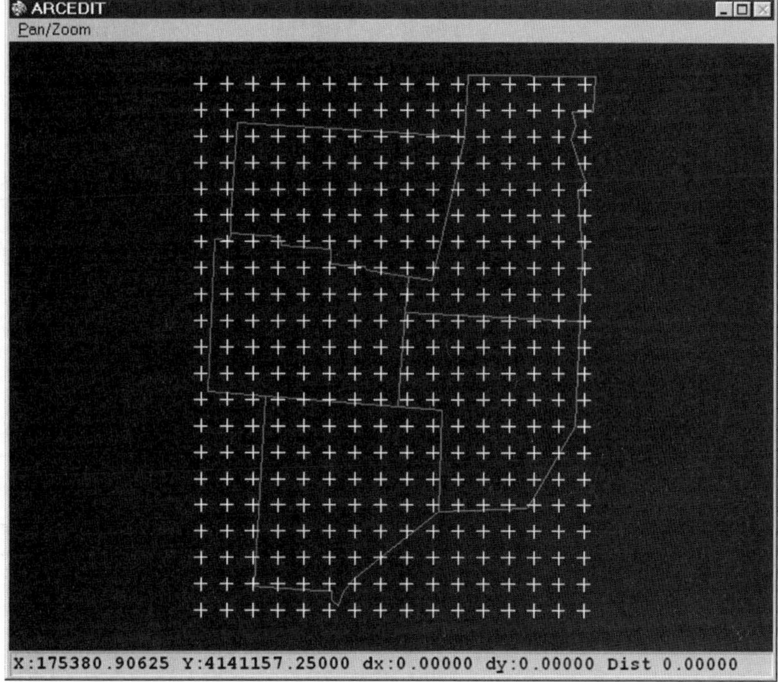

Revised Grid_traps *as the edit coverage.*

You will notice that the trap coverage is an evenly spaced grid that covers the entire five-county project area (the trap coverage was created using Arc's Generate command and the map extents of *proj_cnties*). The map extent of a coverage is a rectangle (based on the smallest and largest coordinates of the coverage) that encompasses all of the features of a coverage. Because of its regular rectangular shape, many of the trap sites fall outside the actual five-county project area. You will remove these "extra" traps from the *grid_traps* coverage.

You now have three goals for this exercise: (1) assign the correct county name to each trap, (2) assign the correct trapper name to each trap, and (3) remove any traps outside the Blue-eyed Corn Moth Project area. Completing these three tasks will give you an accurate and competent project coverage.

Using Delete from the Edit Labels menu, you could select and delete the trap sites that fall outside the project area. However, what about the trap sites that fall on the line? Are they inside or outside the project area? In addition, guessing which traps lie within which counties could be equally difficult, as some trap sites fall on county borders. You would have to make selections to assign the correct trapper name (based on the county name) to each trap. This would be extremely tedious. This exercise demonstrates an easier method, which makes ArcInfo work for you.

First you will use the Table Manager menu to create a new item in the *proj_cnties* coverage to hold the trapper's name. Do this by changing your edit coverage to *proj_cnties*, and your edit feature to polygon (remember, counties are represented by polygons in the *proj_cnties* coverage). Click on the Table Manager button to activate the menu shown in the following illustration.

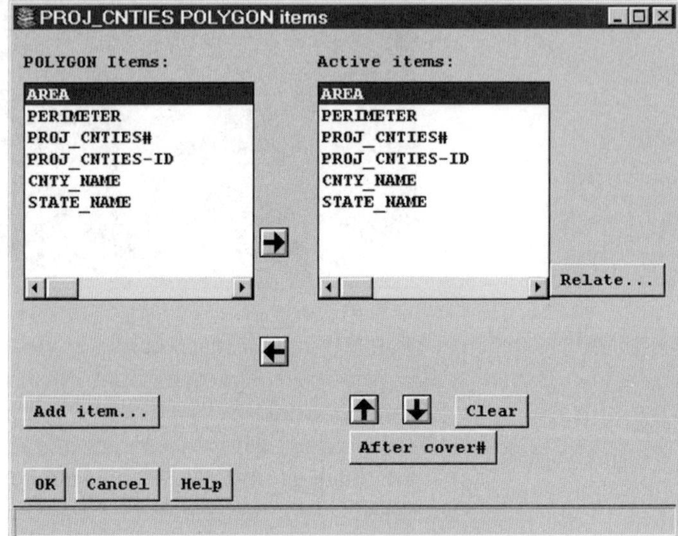

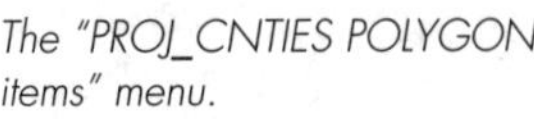
The "PROJ_CNTIES POLYGON items" menu.

Notice that this menu lists the feature attributes associated with the current edit feature. Because you will be adding the new item (or attribute) to all of the polygons (counties), the item will be part of the .PAT (polygon attribute table). Now, use the Feature Selection menu to select all of the polygons in *proj_cnties*. You will see "5 of 6" appear on the Feature Selection menu (no, one polygon is not missing; the background polygon, although counted in the total, is never included in the selection).

Now select Additem from the Table Manager menu. The Add Items window, shown in the following illustration, appears. Here you will specify the new item's name and characteristics.

➻ **NOTE:** *For in-depth information on item characteristics, see Chapter 9.*

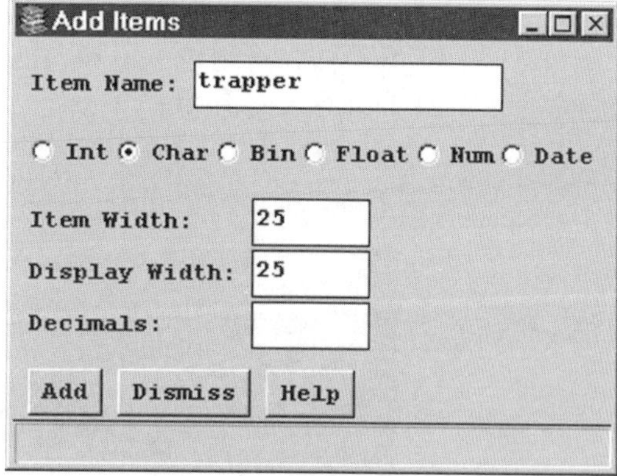

Add Items window showing the new item's characteristics.

For this exercise, type in TRAPPER as the new item name. It is a character (rather than a numerical) field; therefore, next select the Char option. By default, the "width" and "output width" show 25 (that is, this field can have names of up to 25 characters). As these lengths are suitable for this exercise, click on the Add button to add this new item to *proj_cnties.pat*. The message "item TRAPPER added" appears at the bottom of the Add Items menu. Dismiss this menu to return to the "PROJ_CNTIES POLYGON items" menu, shown in the following illustration. Scroll down and you will see the new TRAPPER item.

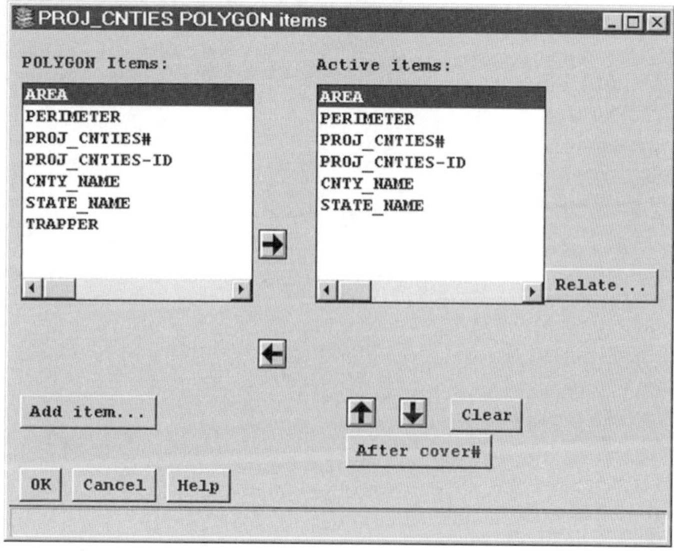

"PROJ_CNTIES POLYGON items" menu with the new item added.

The new item is now added, but as yet contains no values. Next, you will add values by assigning a trapper name to each county. For this part of the exercise it would be helpful to have the county names appear on screen. Remember that you did this in exercise 6-2 by selecting Display ➥ Textitem from the Edit Tools menu. Select POLYGON from Available Features and NAME from "Text items." Click on Apply and the county names should appear on your graphics screen as shown in the following illustration.

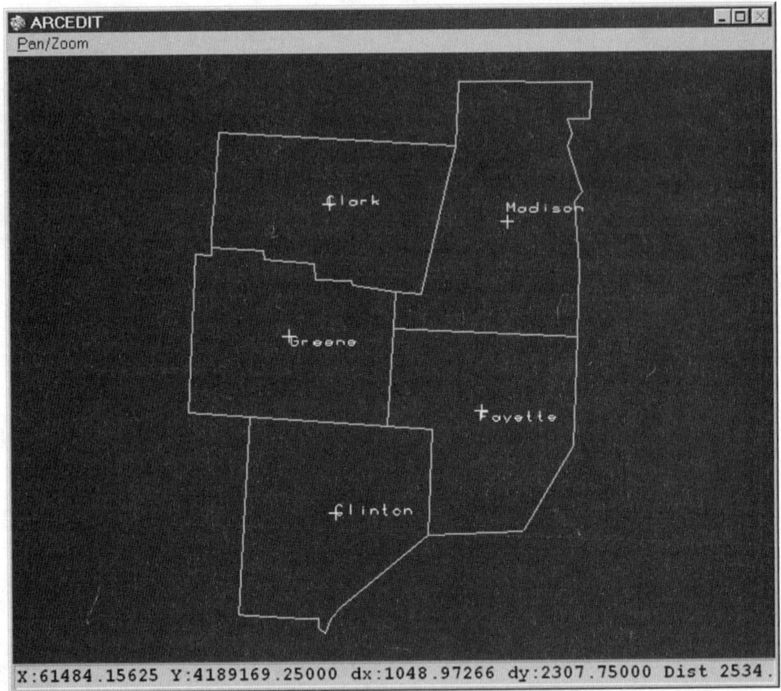

Proj_cnties *coverage with county names displayed.*

Next, using table 6-1, which follows, you will type in a trapper name for each county.

Table 6-1: Associated Feature Data

County Name	*Trapper*
Clark	Barbara Stewart
Madison	Connor Murphy
Greene	Ethan DeSilvey
Fayette	Katie Flynn
Clinton	Ed Windsor

To do this, first select a county's polygon label. Use the Feature Selection menu to select Clark County. The selected polygon and its label will turn yellow (or whichever color you have specified as your selection color; see exercise 6-1). Now click on the Table Editor button, found on the Edit Polygons menu, and then click on the Edit button. This brings up the Edit Attributes menu, shown in the following illustration.

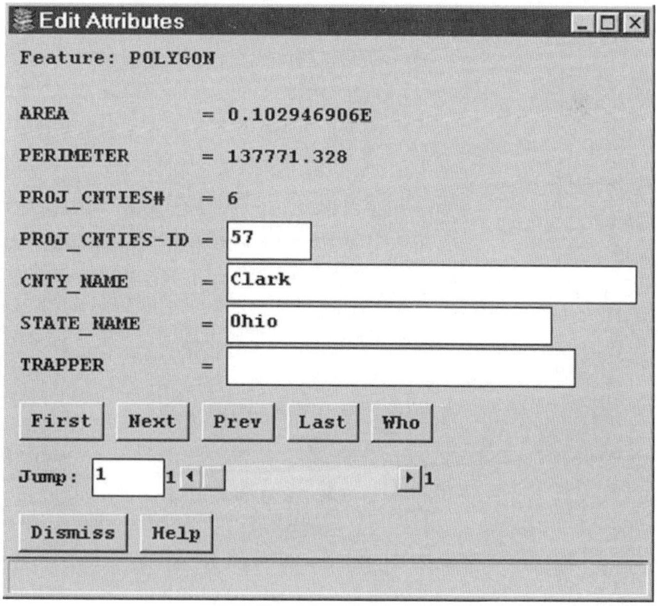

Edit Attributes menu.

This menu lists the items and item values for the selected polygon, which in this case is Clark County. You will notice that the item TRAPPER has no value. Move the cursor to this box, type the trapper's name (Barbara Stewart), and click on the <Enter> button.

Now select the other four counties (Madison, Greene, Fayette, and Clinton). Again, use the Table Editor button to bring up the Table Editor menu and the Edit Attributes menu. Type in the trapper name for the next county shown and click on the Next button. You can use the First, Next, Prev, Last, and Who buttons to move through your selected set. ArcEdit highlights, on the graphics screen, the polygon associated with the record you are editing.

Near the bottom of this menu you see a Jump box and a slider bar. The Jump box tells which record number of the selected set is currently displayed. Note that the records are ordered by their "cover-name>-ID" number; they are not necessarily in the same order as selected. You can also type a number into the Jump box to move through records. The slider bar has a "1" to its left and the total number of selected records to its right. You can also use the slider bar, shown in the following illustration, to move through selected records.

When you are done adding trapper names to the county labels in *proj_cnties*, dismiss the Edit Attributes and the Table Editor menus. At this point you have added the new item TRAPPER to the *proj_cnties* coverage and have assigned trapper names, corresponding to the county names, to it. In Part II of this exercise you will use the Arc command Intersect to combine *proj_cnties* with *grid_traps* to create a new coverage.

The new coverage will contain only the traps within the project area (only those traps that fall within the five counties), and those traps will be assigned their proper county name and trapper name. In other words, by intersecting *proj_cnties* and *grid_traps*, the resulting coverage will contain only those traps within the five-county area, and those traps will be assigned the same attributes as their corresponding counties.

Part II: Some Arc Commands for Editing Coverages

The Arc commands mentioned here are generally used for analysis; that is, a user would combine the coverages to look for some sort of spatial patterns. These commands can also serve as a convenient method of transferring a lot of information from a polygon coverage to another polygon, line, or point coverage. These commands fall under the title "Commands for Analysis and Manipulation" in Arc's on-line help. The family of overlay commands includes Clip, Erase, Identity, Intersect, and Union.

All of these commands require that at least one of the coverages has polygon topology. They allow you to transfer information associated with the polygon labels of one coverage to the features in another coverage. Because they are each discussed in the Analysis Tools section of the ArcToolbox chapter (Chapter 3), this section of the ArcEdit chapter serves to introduce you to them by name, and to walk you through a demonstration one of them. For Part II , you will use the Intersect command to assign the polygon features of *proj_cnties* to *grid_traps.*

The Intersect command combines two coverages to create a third coverage. The third coverage contains all features in the area common to both the first and second coverage. This command also "assigns" polygon feature attributes from the first coverage to features in the second coverage.

To get started, first be sure you completed and saved your work from Part I. Next, start Arc by clicking on Start ➡ Programs ➡ ArcInfo ➡ Arc, and navigating to your exercise workspace. If you like, use what you have learned in the previous exercises to view the coverages in ArcEdit before proceeding. At this point the two coverages should look like those shown in the following illustration.

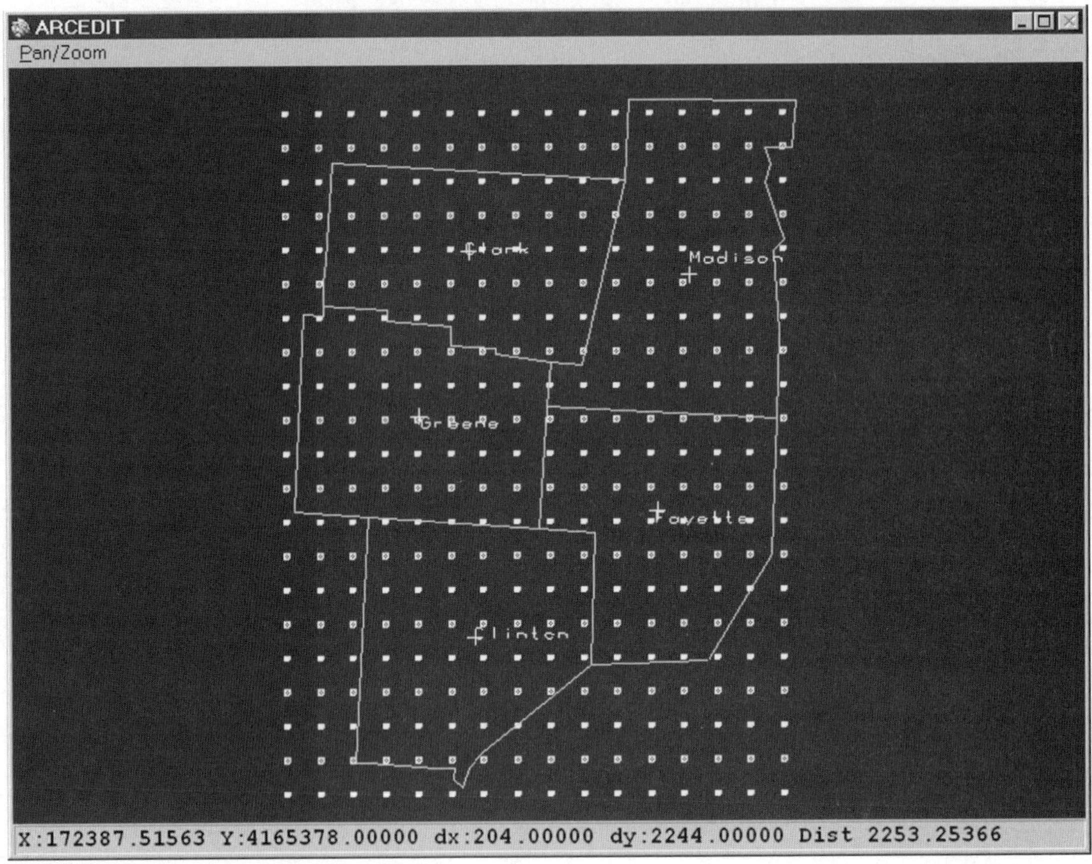

Proj_cnties *and* grid_traps *coverages in ArcEdit.*

Either start Arc or leave ArcEdit to return to the Arc prompt. Type in *w* (an abbreviation for "workspace") to find out which workspace you are in. If you are not in the same workspace as your Chapter 6 exercises, navigate by using the Workspace command once again: *w d:\ai80exercises\chap_6* (this exercise assumes that this is the directory name). Now type *lc* (the abbreviation for the "listcoverages" command) and you will see a list of available coverages in your workspace. That list should include the coverages shown in the following illustration.

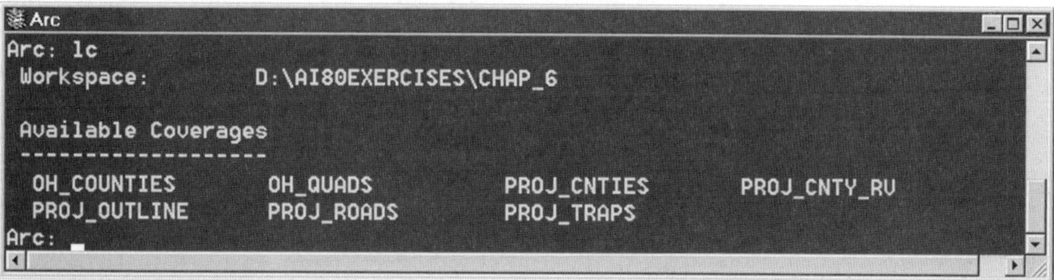

Using the "listcoverages" command ("lc") to show coverages in a workspace.

Before actually using the Intersect command, check out its usage and parameters (described in the list that follows) by typing *usage intersect* at the Arc prompt, as shown in the following illustration.

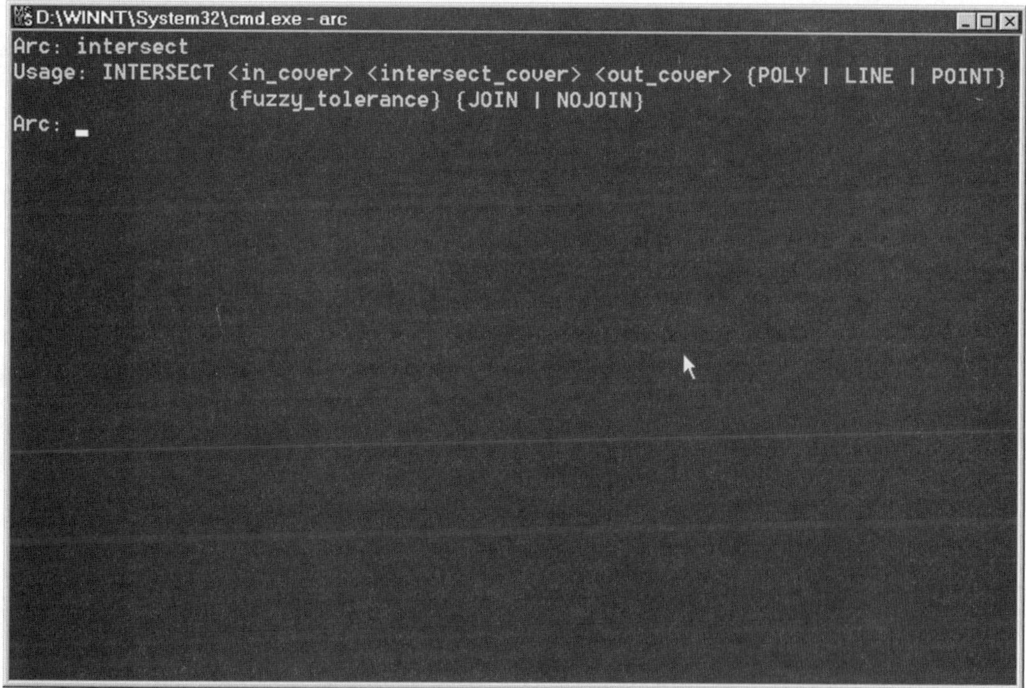

Viewing the parameters of the Intersect command.

- The <in cover> parameter is the point, line, or polygon coverage to which (in this example) you want to add attributes.
- The <intersect cover> parameter is the polygon coverage that contains the attributes. It must be a polygon coverage.
- The <out cover> parameter is the name you assign to the resulting coverage. It cannot be the same name as an existing coverage.
- The {POLY | LINE | POINT} parameter describes the feature of the <in cover> parameter to which you want the <intersect cover> parameter's attributes added. The default is polygon.
- The {fuzzy tolerance} parameter specifies the minimum distance between coordinates in the <out cover> parameter. For this example, the default values are fine. For more information on how the default values are determined, see the on-line help.
- The {JOIN | NOJOIN} parameter specifies whether all attributes in the <in cover> and <intersect cover> parameters will be joined together in the <out cover> parameter. The default option is JOIN, which is what you want for this exercise.

Now that you are familiar with the parameters, go ahead and execute the Intersect command by typing in the following: *intersect grid_traps proj_cnties proj_traps point* (see the following illustration). Notice that the name of the resulting coverage will be *proj_traps*. Several messages will appear as the command runs.

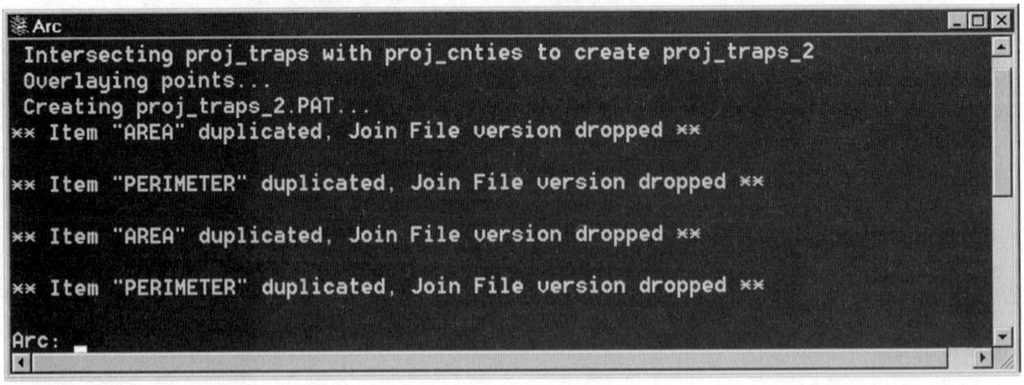

```
Arc
 Intersecting proj_traps with proj_cnties to create proj_traps_2
 Overlaying points...
 Creating proj_traps_2.PAT...
** Item "AREA" duplicated, Join File version dropped **

** Item "PERIMETER" duplicated, Join File version dropped **

** Item "AREA" duplicated, Join File version dropped **

** Item "PERIMETER" duplicated, Join File version dropped **

Arc:
```

Typing in the Intersect command.

The items AREA and PERIMETER appear in both *proj_cnties* and *grid_traps*. Because ArcInfo does not allow a coverage to contain two items containing the same name, the information in the item from the <in cover> parameter (in this case, *grid_traps*) is dropped.

Now type in *lc* again and you should see the new coverage, *proj_traps*, listed. Restore its point topology by typing *build proj_traps point.*

✓ **TIP:** *Watch out! Do not ever "build" a point coverage as a polygon coverage. Remember, "polygon" is the default of the build command. If you had typed* build proj_traps, *the labels would remain, but each label would now contain no usable attribute information.*

Now start up ArcEdit to view the new *proj_traps* as the edit coverage. Select LABELS as your edit feature. Bring up *proj_cnties* as a background coverage, with the county names appearing. Your ArcEdit graphics screen should now look like that shown in the following illustration.

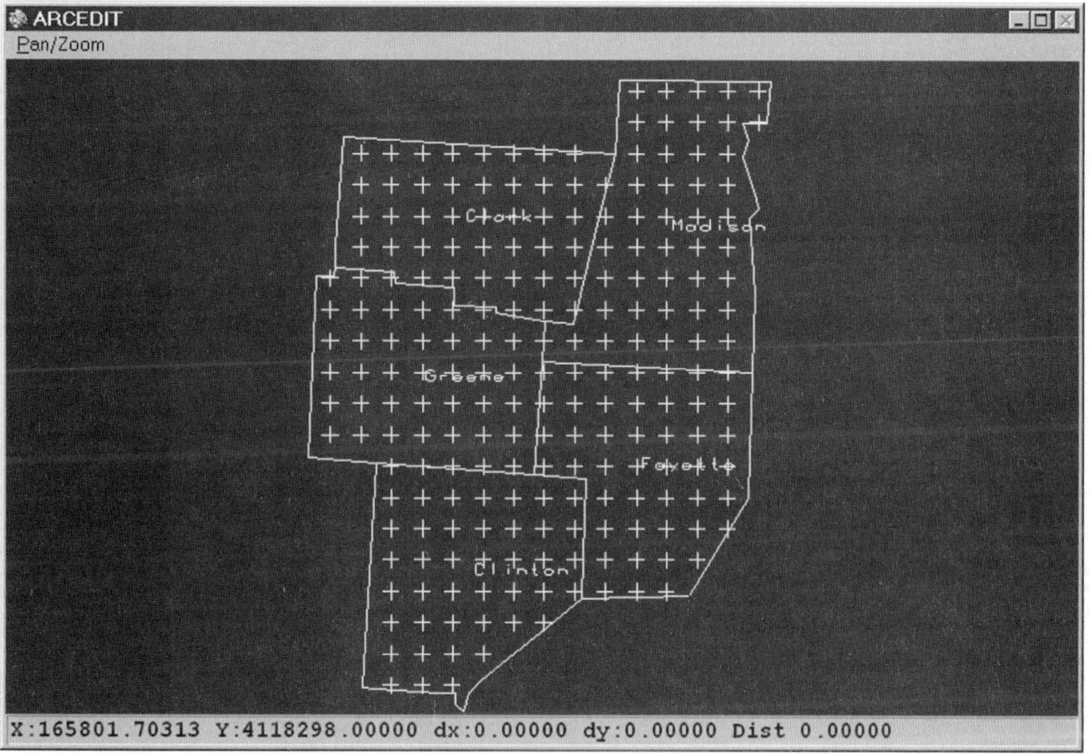

Proj_traps *with* proj_cnties *in the background.*

Use the Table Manager button to view the attributes of *proj_traps*. You will see that the county name and trapper name items are present. Go ahead and select a few labels, and use the Table Editor ➥ List button to list their attributes. You should see that they contain values for the county name and trapper items.

Congratulations! You have just completed an advanced editing exercise. You have learned how to add items and values to a coverage using the Table Manager and Table Editor buttons in ArcEdit. You have also used an Arc command to quickly transfer information from one coverage to another. This exercise also took you through more of the pull-down menus and gave you additional practice in setting edit and background environments. The next section contains the menu references for editing annotation, a coverage's text items.

Annotation

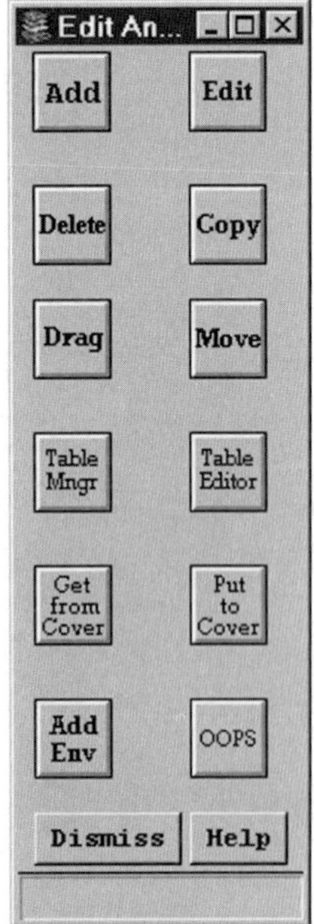

Annotation Feature menu.

Annotation allows you place text anywhere in a coverage. You have complete control of where it is placed, whether it is placed at an angle, as well as the font used for it. Annotation is stored as a separate feature class, meaning that you make it your active edit feature before adding, changing, or deleting it. Common uses for annotation include street names and tax map numbers on property maps, county names (as you haven seen), and so on. Because streets and parcels can be irregularly shaped, annotation offers a convenient means of labeling these features in a way that is visually pleasing.

Use this section as a reference to the annotation editing commands available in Edit Tools. (The Annotation Feature menu is shown in the illustration at left.) Again, browse through the buttons and commands available, and perhaps follow along on your computer. Then complete exercise 6-4 to learn about editing annotation.

Add (ADD)

Add button.

Adding annotation is straightforward. Selecting the Add button (shown at left) brings up the Add Annotation menu, shown in the following illustration.

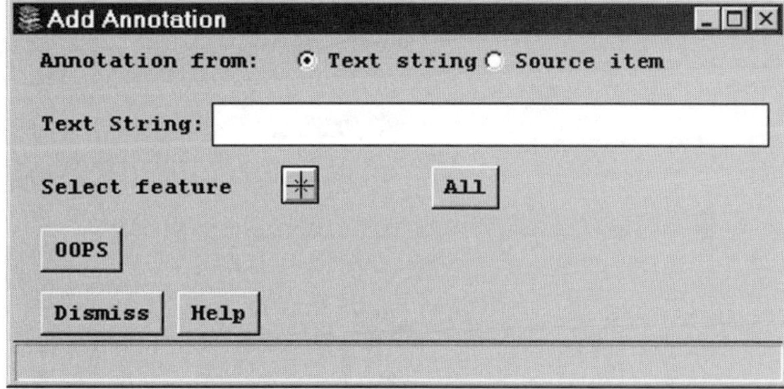

Add Annotation menu.

You have two means of adding text, either by typing some text interactively in the Text String box or by adding it from an item in the coverage's feature attribute table. To add text interactively, select the Text String button and type your text in the box provided. Then, use the "Select feature" button to enter text anywhere on the map. Use the crosshairs to position the beginning point of your text string. Move the crosshairs to where you want the text to end and press the <9> key. The text string will now be drawn.

To add annotation from an item in the feature attribute table (for instance, a parcel number from a parcel coverage), use the Add Environment menu to select the feature attribute coverage from which annotation will be retrieved. Then, in the Add Annotation menu, select the Source Item button. This changes the menu so that you are then prompted to enter the name of the item to use as annotation.

After adding annotation either from an item or from a string, the Annotation Edit menu appears. This menu is described in the material that follows.

Edit

Edit

Edit button.

Selecting the Edit button (shown at left) starts the Annotation Edit window, shown in the following illustration.

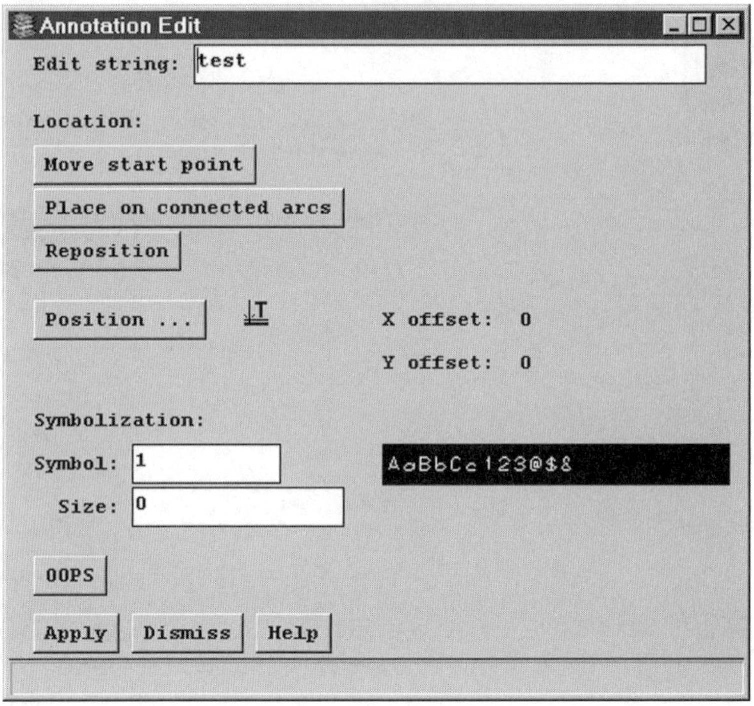

Annotation Edit window.

This window displays the text you either just entered using the Add Anno command or the annotation you select using the Edit Feature menu. Using this window, you can change the starting point for the annotation, place it on an arc so that it now follows the arc's curve, and change the symbol and size of the lettering. Once you have completed your edits, click on the Apply button. There is also an Oops button in case your edits did not turn out as you wanted.

Delete (DELETE)

Delete button.

The Delete button (shown at left) deletes annotation that has already been selected using the Feature Selection menu, or prompts you to select annotation using the crosshairs. Pressing the <9> key then deletes the selected annotation.

Copy (COPY)

Copy button.

The Copy button (shown at left) copies selected annotation from one location to another. Upon pressing the Copy button, you are prompted to use the crosshairs to identify the point to move from. Select the lower left corner of your text. You are then prompted to position the crosshairs where you want the text copied to. Do this and press the left mouse key. The annotation will then be copied to the new location.

Drag

Drag button.

Drag (shown at left) is a useful tool for moving annotation about the screen. Upon selecting this command, the cursor becomes a small hand. Position the hand on the annotation to move and drag it to a new location. The annotation is now moved to that position.

Move (MOVE)

Move button.

The Move button (shown at left) is similar to the Drag command, except that Move makes you select a beginning point and a new point to move to. It is just like the Copy command except that it moves instead of copies the annotation.

Table Manager

Table Manager button.

Select Table Manager (shown at left) to edit annotation subclasses you may have. Subclasses are layers of annotation you can turn on or off when displaying coverages. A good example would be street names. You can have one subclass that shows major road names and one that shows neighborhood street names. If you only need to see major road names, display only the annotation for this subclass when drawing.

Table Editor

Table Editor button.

Use Table Editor (shown at left) to change values in any annotation subclasses you may have.

Get from Cover (GET)

Get from Cover button.

Get from Cover (shown at left) prompts you to specify a coverage whose annotation will be copied into the current edit coverage.

Put to Cover (PUT)

Put to Cover button.

Put to Cover (shown at left) takes all currently selected annotation and puts it into another existing coverage. You are prompted to supply the name of the coverage.

Add Environment

Add Environment button.

Add Environment (shown at left) brings up the Annotation Add Properties menu for annotation, shown in the following illustration.

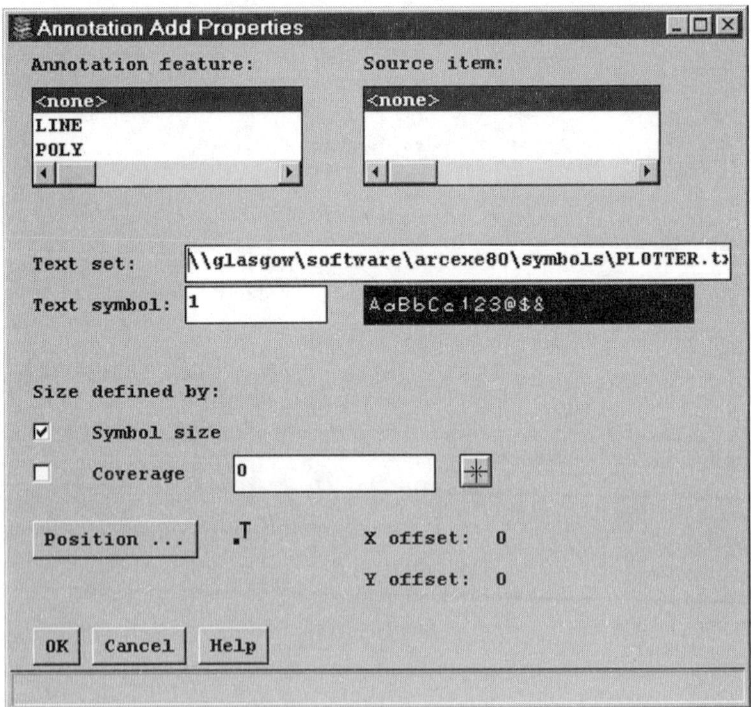

Annotation Add Properties menu.

Use the Annotation Add Properties menu to supply the color, symbol, and size for annotation. Use it also to specify a feature class in the edit coverage that will be used to supply annotation.

Oops button.

Oops (OOPS)

The Oops button (shown at left) erases the last action undertaken in ArcEdit. You can use it repeatedly to roll back your editing actions. Oops can undo actions back to the last time you saved a coverage. Once you save a coverage during an ArcEdit session, Oops cannot undo an edit.

Exercise 6-4: Editing Annotation

This exercise acquaints you with some of the common procedures used to add and edit annotation. Working with the polygon coverage *proj_cnty_rev*, you will learn how to place annotation from an item in the coverage's polygon attribute table, and you will learn how to place some text interactively.

You will also become acquainted with how to change the characteristics of text. Before beginning you should have completed the previous exercises in this chapter. Make sure you are in the directory to which you have copied the Chapter 6 exercises. This exercise assumes the directory is named *D:\ai80exercises\chap_6*.

Step 1: Starting ArcEdit and Setting Up the Work Environment

Your task now is to create a graphic showing which trappers are assigned to which counties in the Blue-eyed Corn Moth project. The first step is to start ArcEdit. Now, start ArcTools and select Edit Tools.

Next, select the *proj_cnty_rev* coverage and set up an editing environment. Select ANNOTATION as your edit feature.

The screen should now display the five counties in your project area. It should look like that shown in the following illustration.

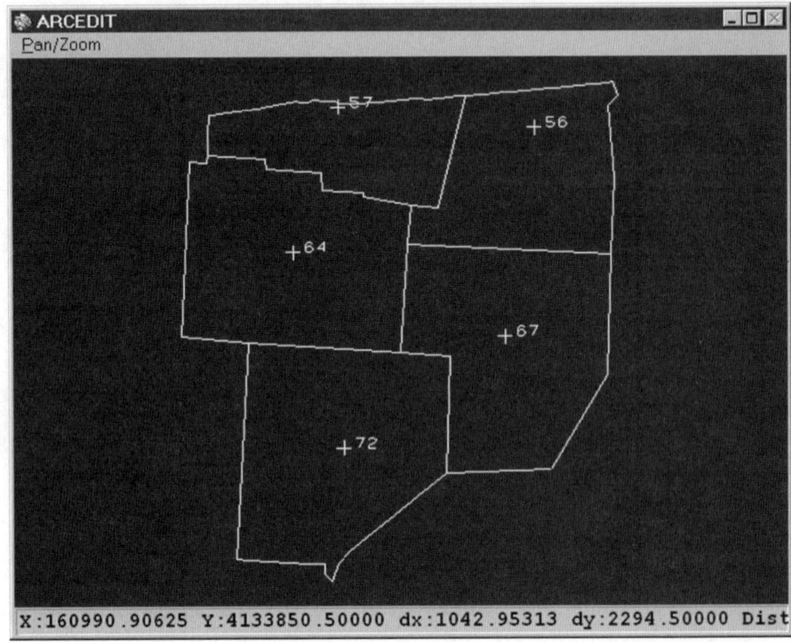

Five-county Blue-eyed Corn Moth project area.

Notice that just the county outlines appear. The next steps will add trapper names as annotation.

Step 2: Adding Annotation from a Coverage Item

The trapper names are stored in an item named TRAPPER in the *proj_cnty_rev's* polygon attribute table. Use this item as the source to create the annotation for the coverage.

To do this, click on the Add Environment button in the Edit Annotation menu. The Annotation Add Properties menu appears, which is shown in the following illustration.

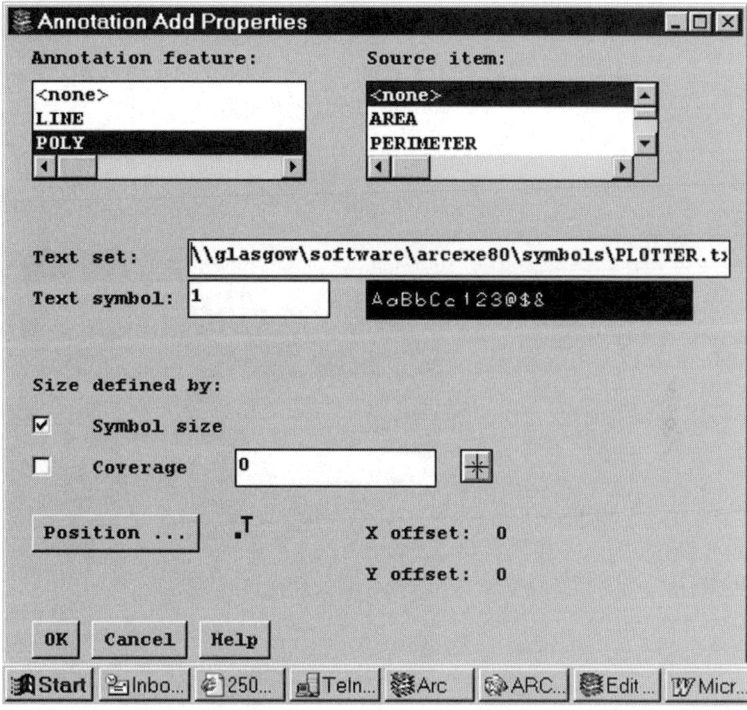

Annotation Add Properties menu.

In this menu, select POLY as the Annotation feature, select TRAPPER as the source item, and click on the OK button. This tells ArcEdit that the item used to create annotation comes from the coverage's polygon attribute table and that the item is named TRAPPER.

Next, create the annotation by selecting the Add button in the Edit Anno menu. The menu shown in the following illustration appears.

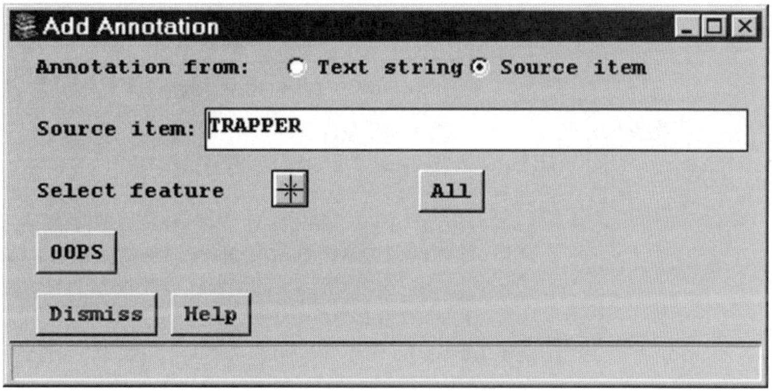

Add Annotation
Annotation from: Text string Source item
Source item: TRAPPER
Select feature All
OOPS
Dismiss Help

Add Annotation dialog menu.

Notice that all of the information is filled out for you. The dialog box identifies TRAPPER as the item where annotation will be supplied for you. Click on the All button to add annotation to all counties. Annotation will automatically appear centered on the label point for each county (you can see this for yourself if you make "labels" part of your draw environment). Examine the screen shown in the following illustration to see the annotation.

Annotation added to five counties.

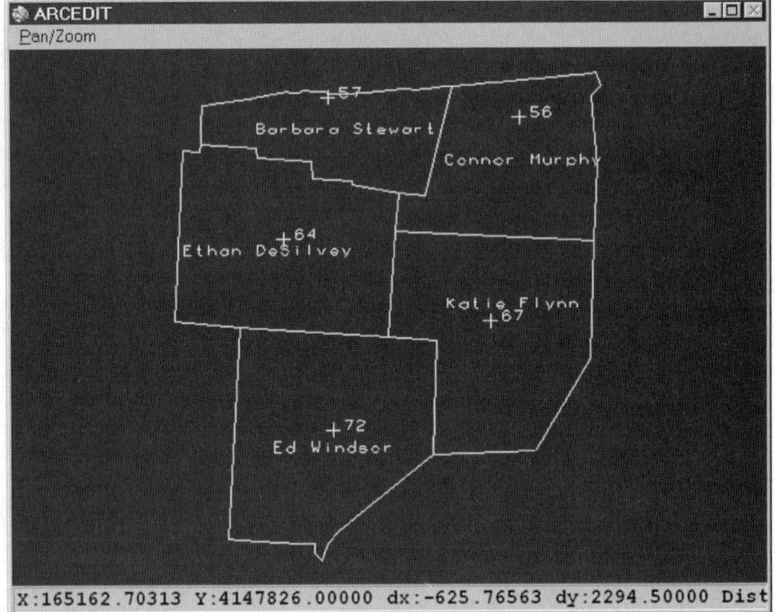

Adding annotation from an existing item is simple. Use this to create annotation for maps and screen graphics. Sometimes, however, the information you need to place on the map is not available in an item. Follow the next step to see how to add annotation interactively.

Step 3: Adding Annotation Interactively

In this step, you will practice placing the state name on the map, as well as moving it around and deleting it.

First, use the Add button to add new annotation. In the Source area of the Add Annotation menu, select the Text String option. Then, type the words *North Carolina* in the Text area. Your screen should look like that shown in the following illustration.

Entering North Carolina in the Add Annotation menu.

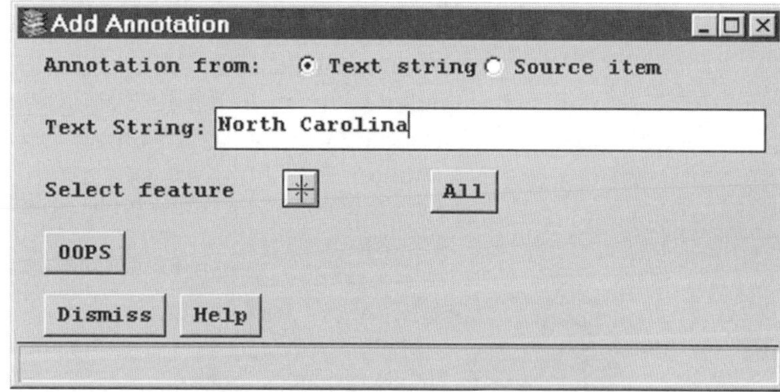

Now, press the * button to add text interactively. You are prompted to use the crosshairs in the graphics window to select a polygon in which to place the label. Select any polygon and press the <9> key. The text now appears on the screen, as shown in the following illustration.

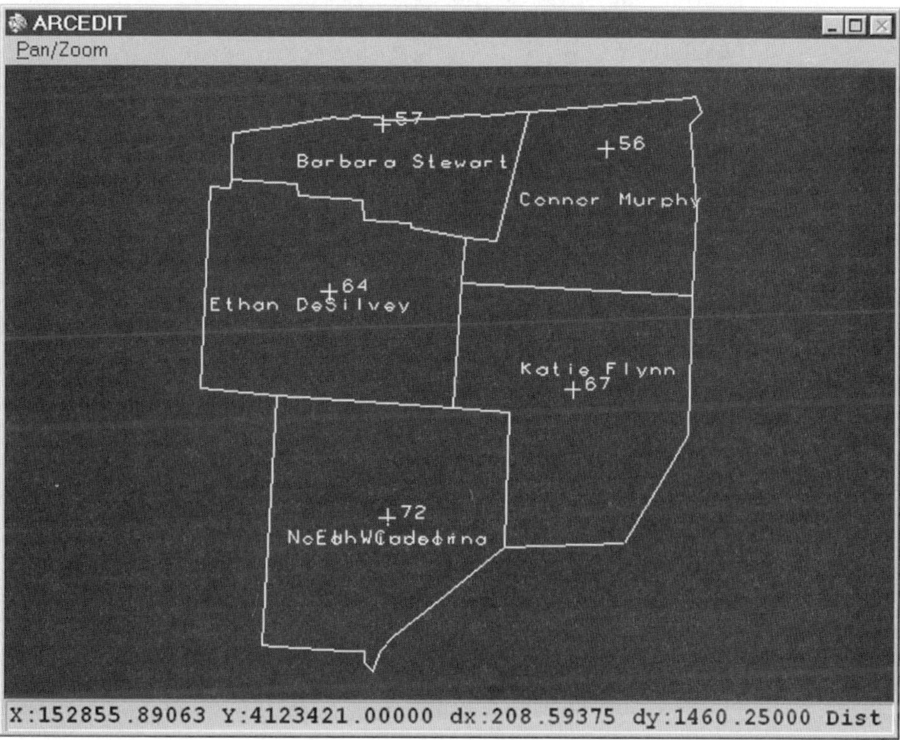

North Carolina added as annotation.

Unfortunately, the words *North Carolina* overwrite the trapper's name. Dismiss the Annotation Edit menu and click on the Move button in the Edit Anno menu. The crosshairs again appear on the graphics screen. In the ArcEdit dialog window, a prompt asks you to select the point to move from. Select a point on the words *North Carolina.* You are then prompted to select a point to move to. Select a point away from the trapper's name. The text now moves, as shown in the following illustration.

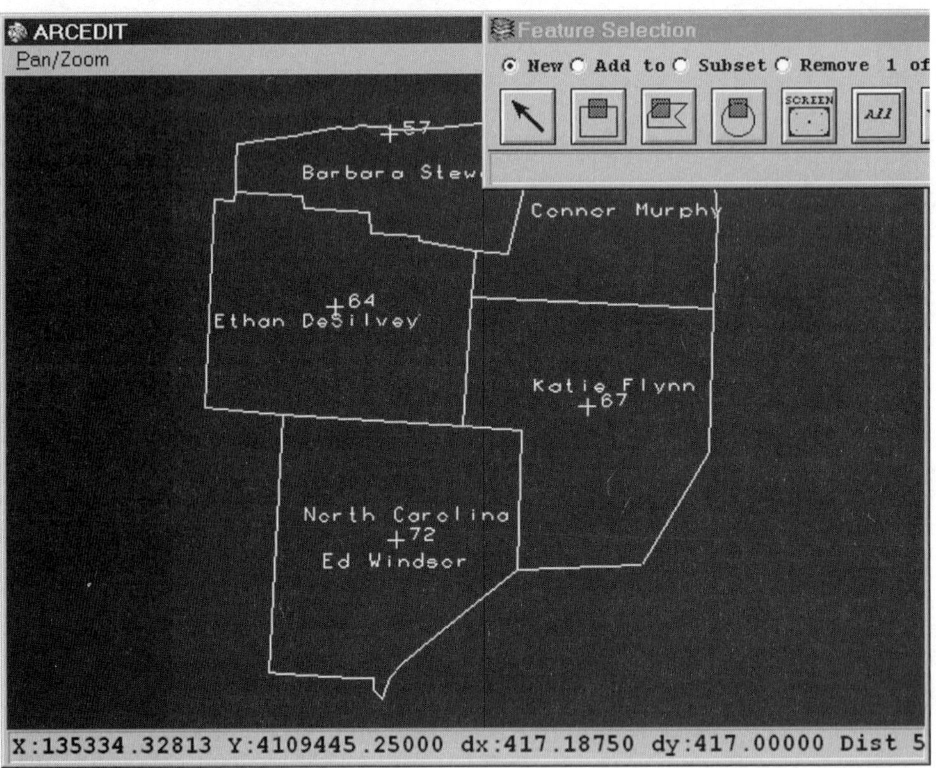

North Carolina annotation moved to a new location.

Now that you have it where it looks good, you realize that this truly has been a hectic day. You labeled the five-county area as North Carolina when it ought to be Ohio! Use the Select button in the Edit Feature menu to select the North Carolina text. Then use the Delete button from the Edit Annotation menu to delete it. The screen will look like that shown in the following illustration.

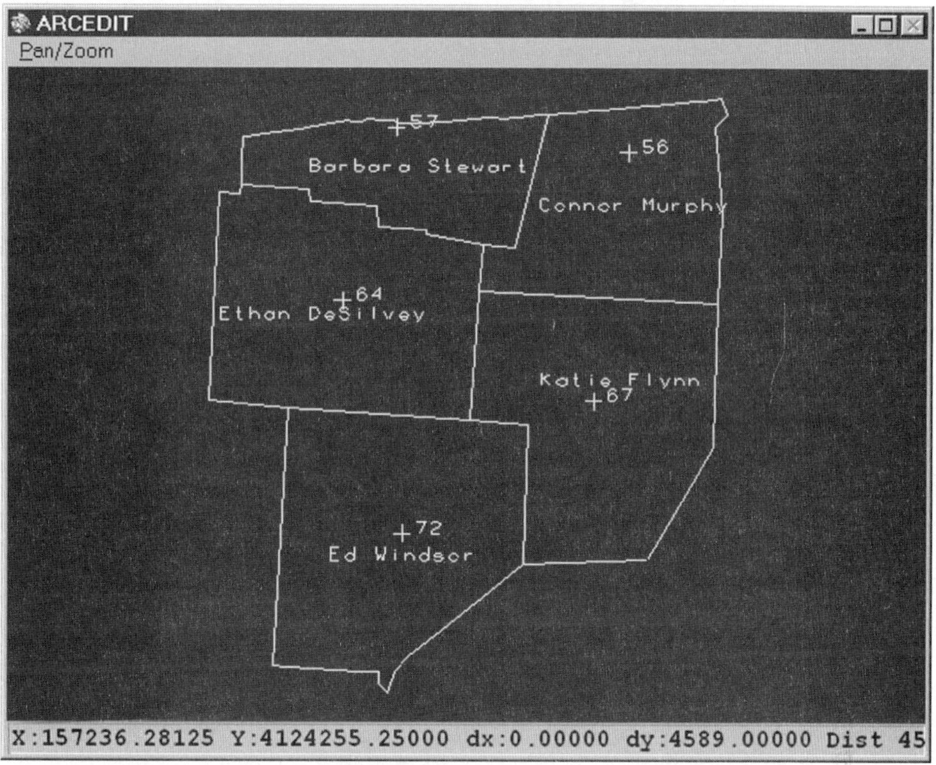

Proj_cnty_rev *after deleting North Carolina annotation.*

Now repeat the previous steps, except this time add the correct state name, Ohio. Your screen should look similar to that shown in the following illustration.

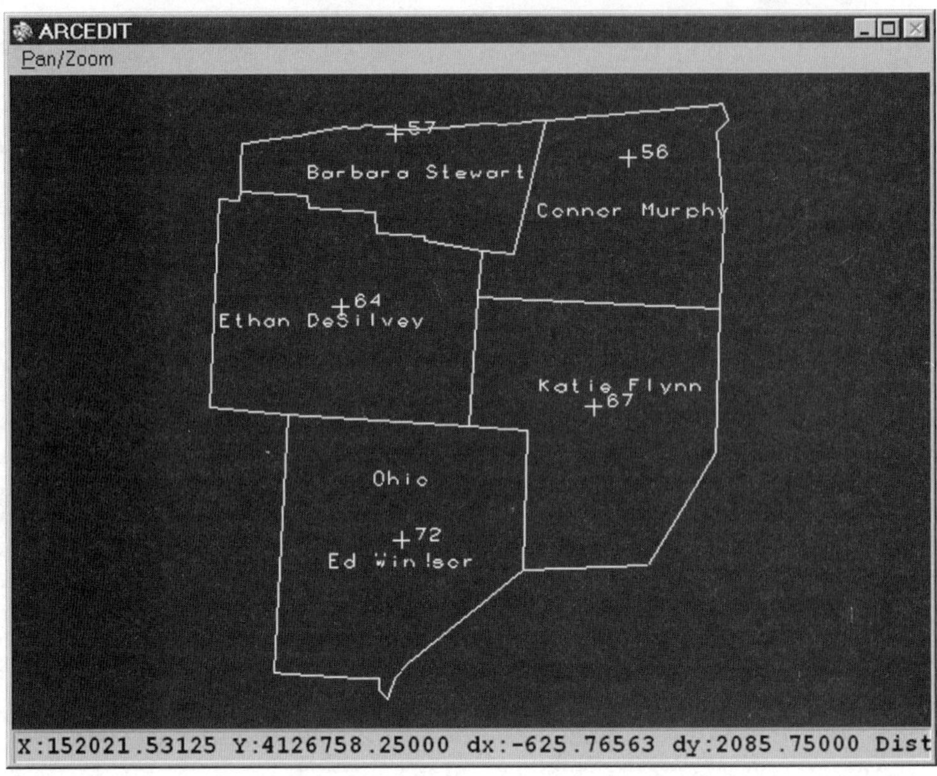

Proj_cnty_rev *correctly labeled as Ohio.*

By using the Move tool to reposition annotation that either runs over other text or appears in a poor location, you can create effective and visually pleasing maps. As a final note about adding annotation, know that it is possible to add text that follows a curve, such as a road or a stream. This can be done with the Annoadd command. For more information on the Annoadd command, see this command name, the "Basics of adding annotation," and "Adding annotation with the ANNOADD menu" in ArcInfo's on-line help.

Summary

Through much discussion and four exercises, this chapter has introduced you to editing with ArcEdit, one of the traditional modules of ArcInfo. With ArcEdit, you can use either a series of pull-down menus or command line entry for editing. ArcEdit requires that you set up a work environment, and that you edit only one feature type at a time, in one coverage at a time.

You have learned how to set up a work environment and how to perform basic and advanced editing techniques on arcs, labels, polygons, and annotation. ArcEdit is probably the largest and most powerful of the traditional modules. Unlike Arc of ArcPlot, its functionality is an irreplaceable component of the software.

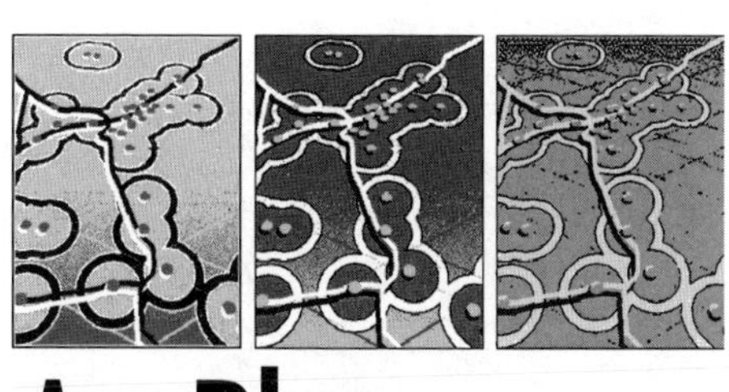

Chapter 7

ArcPlot

An Introduction

ArcPlot is the traditional ArcInfo module used for designing map graphics. The new Version 8.0 ArcMap module has replaced much of its functionality. The new module incorporates a drag-and-drop interface that will have veteran ArcInfo users rejoicing. New ArcInfo users should rely on ArcMap as their primary map composition tool. It allows you to easily select coverages, features, colors, and symbols—even North arrows—interactively.

That being said, an ArcInfo user might ask, "If the ArcMap module is so great, why did ESRI still include ArcPlot? The simple answer is probably so that any Arc Macro Language (AML) programs your organization has, which use ArcPlot commands, can still run successfully.

This chapter introduces you to the basics of ArcPlot, but does not go into extensive detail. You are encouraged to use ArcMap as your primary map composition tool. If you will be creating AML programs that use ArcPlot commands and need a comprehensive lesson in ArcPlot's functionality, see previous editions of this book. In addition, refer to the on-line help section "Map display and query using ArcPlot" under "Cartography." This chapter covers the following topics.

- An introduction to ArcPlot
- ArcPlot's pull-down menus
- Creating views and querying spatial data in ArcPlot
- Creating a map composition in ArcPlot
- Printing paper maps and Internet-ready maps

ArcPlot, one of the traditional ArcInfo modules, is one of ArcInfo's data viewing and map creation modules. It can provide cartographic tools for you to create map compositions. These tools include data selection and display, labeling or symbolizing features, cartographic additions such as North arrows and scale bars, importing graphics, and so on.

You can create and display map compositions on the ArcPlot display canvas, as well as create plotfiles ready for hardcopy plotting or printing. You have full control of map layout and design, including access to a comprehensive collection of cartographic symbols, North arrows, and fonts.

ArcPlot's interactive query functions also allow you to use the displayed maps to retrieve data directly from your database. You can also reselect features from a coverage and create new coverages.

ArcPlot includes some simple pull-down menus; however, the interface, because it was not originally designed for Microsoft Windows NT, is not as sophisticated as Version 8.0's new modules ArcCatalog, ArcMap, and ArcToolbox. Because it is one of the original modules, you can also use command line for editing commands, and AML programs can contain and execute ArcPlot commands (AML is introduced in Chapter 10).

Because of ArcInfo's command line history, many of the longer commands ("listcoverages," for example) have abbreviations ("listcoverages" is abbreviated as "lc"). Where appropriate, a command's abbreviation is also given.

The First section of this chapter discusses the pull-down menus. Then exercise 7-1 gets you started using ArcPlot and its pull-down menu system. It shows you how to create a view (a group of coverages) and how to display and query data on the screen. Exercise 7-2 takes you through creating a map composition complete with cartographic elements. The last section discusses how to create files for printing maps or displaying them on the World Wide Web.

ArcPlot contains dozens of commands. Again, it is not the purpose of this chapter to describe every command in detail. Some common commands are demonstrated to give you a feel for ArcPlot's capabilities. Remember that detailed information on commands can be found in the on-line help documents.

Exploring ArcPlot's Pull-down Menus

To start ArcPlot, simply select Start ➥ Programs ➥ ArcInfo ➥ ArcInfo Workstation ➥ ArcPlot. A blank, black background window pops up containing the copyright information and the ArcPlot command line prompt. This window is used to execute editing commands.

A few moments later another blank, black background window appears. This is your display canvas, where the graphics will appear. You can resize and move these windows to suit your working habits and needs. One configuration, as shown in the following illustration, is to have the command window be long and narrow and at the bottom of the screen, with the canvas window a large rectangle above it. This is how the windows will be shown in the exercises for this chapter.

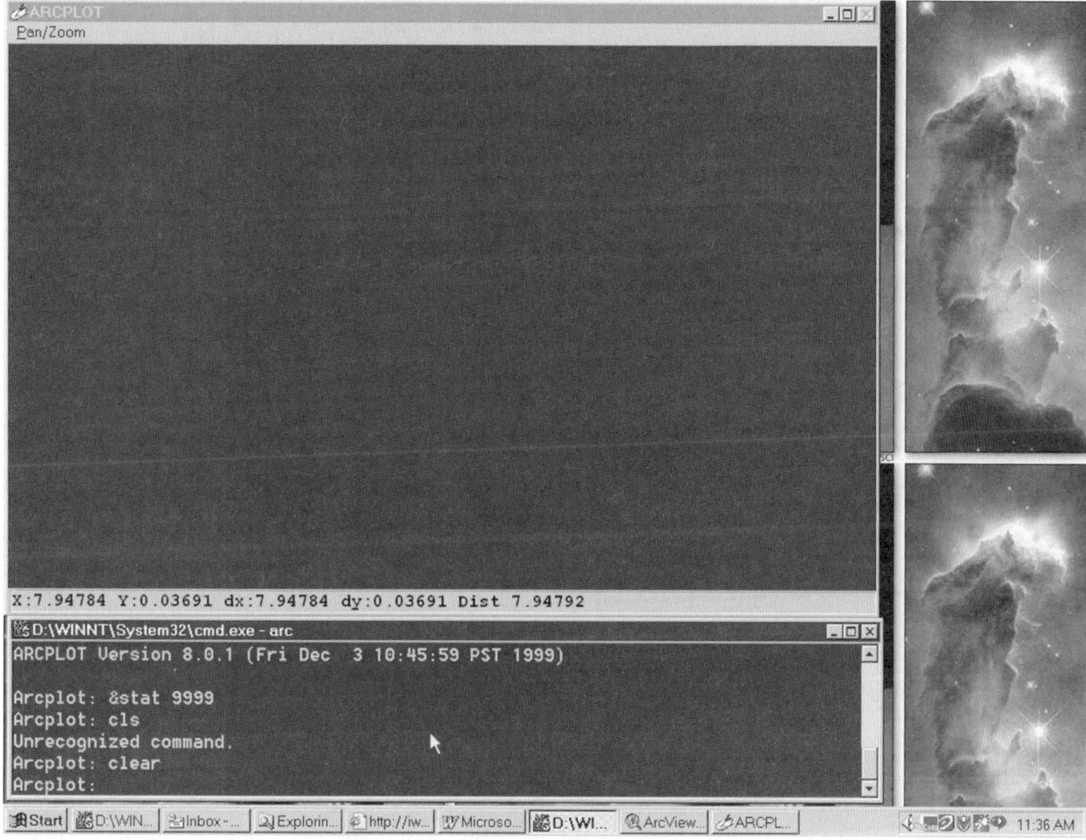

One way of arranging your ArcPlot command and canvas screens.

Now launch ArcPlot 's pull-down menu system by first typing *arctools* at the ArcPlot prompt. the ArcTools main menu, shown in the following illustration, pops up in the upper center of your screen. It lists several menu choices. Highlight Map Tools and click on the OK button.

✓ **TIP:** *Sometimes ArcInfo Workstation menus are so long they will not entirely fit on the screen. You can fix this problem by changing your screen's resolution. Simply right click on the Windows Desktop (your Windows background) and select Properties. Click on the Settings tab and use the Desktop Area slider bar to change to a higher resolution (from 800 X 600 to 1024 X 768, for example).*

ArcTools main menu.

A narrow menu bar named Map Tools, shown in the following illustration, appears in the upper left of your screen. Again, you may move these menus around to suit your needs. Simply click in the blue bar area and drag them to their new location.

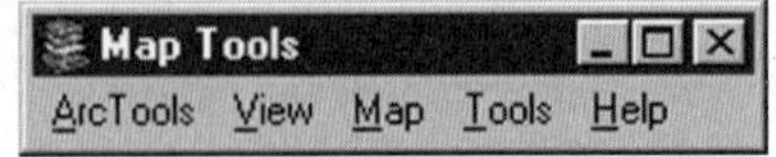

Map Tools menu bar.

The Map Tools menu contains five pull-down menus. These menus contain various commands for managing your ArcPlot session. The following are descriptions of the functions of these menus.

ArcTools: Contains commands for changing workspaces, managing coverages, viewing data from other sources, and quitting Map Tools.

View: Use this menu to create a view. A view is simply a set of coverages you refer to by a single name. Reopen a view in ArcPlot to gain access to the same set of coverages you used in an earlier session. You can display a view on the display canvas and even perform analyses on it while in ArcPlot.

Map: Contains commands you would use to create a map composition, complete with title, North arrow, and other common map elements. You can also use this menu to save your ArcPlot commands as a time-saving macro.

NOTE: *For more information on macros, see the sidebar in Chapter 6.*

Tools: Contains many commands for performing spatial analysis functions in ArcPlot. These functions are generally commands you would normally execute at the Arc prompt. This menu also contains a "Units calculator" you can use to convert various units (for example, to convert inches to meters).

Help: Provides access to help on specific Map Tools operations, as well as access to on-line ArcInfo documentation. Use this to explore in detail the commands available in Map Tools.

This gives you a brief introduction to the pull-down menus available in ArcPlot. In the following exercise you will use the menus to query data on screen and to create a view.

Creating Views and Querying Spatial Data in ArcPlot

This section introduces you to creating views in Map Tools. Views are used to group coverages and other spatial data sets, such as grids and images, so that they may be viewed together and accessed under a single name.

Your GIS project may frequently use the same few coverages each time you create a map composition. Creating a view has the benefit of saving you composition time. This way, you do not have to remember the coverages' names and path names the next time you work with them. Once you create a view, it can then be used in the Map Composer for inclusion on a paper map.

Exercise 7-1: Using Views and Querying Spatial Data

Before beginning, be sure to load the Chapter 7 exercise data from the companion CD-ROM to your workspace (assumed to be *D:\ai80exercises\chap_7*). You should also be familiar with the basic concepts of GIS and ArcInfo Workstation (if you are not, see chapters 1 and 5). The purpose of this exercise is to create a view of the United Kingdom and the Republic of Ireland. You will then use this view to query major cities to determine their populations, and you will change the symbols that represent the cities.

Step 1: Getting Started

Start an ArcPlot session by selecting Start ➡ Programs ➡ ArcInfo ➡ ArcInfo Workstation ➡ ArcPlot. Now type *arctools* at the ArcPlot prompt and from there launch Map Tools. A message will display in the ArcPlot dialog box, stating "Warning, the mapextent is not defined." Ignore this warning, because it always appears when starting ArcPlot. Once you add a coverage, your map extent will default to that of the coverage. You can also change it to the map extent of another coverage.

First, view the names of your coverages by selecting View ➡ Open Workspace (to get the Select menu) and navigating to your workspace. Table 7-1, which follows, lists the coverages you should see.

Table 7-1: Coverages for Exercise 7-1

Coverage	*Content*
uk_eire_city	Point coverage of cities in the UK and Ireland
uk_eire_bdrs	Polygon coverage of country boundaries
uk_eire_river	Line coverage of rivers in the UK and Ireland
uk_leire_akes	Polygon coverage of water bodies in the UK and Ireland

✓ **TIP:** *You will find that using the " ↑ " button (the up arrow) in the Select menu gets you no further than the D drive of your machine. For accessing data from another drive, simply type that drive letter into the directory box and press the <Enter> key.*

Step 2: Creating a New View

Now begin creating your view. To do so, click on View ➡ New to get the Theme Manager window. A theme is a coverage, image, or a grid, all of which are GIS data layers. Your view will consist of themes. The Theme Manager window is shown in the following illustration.

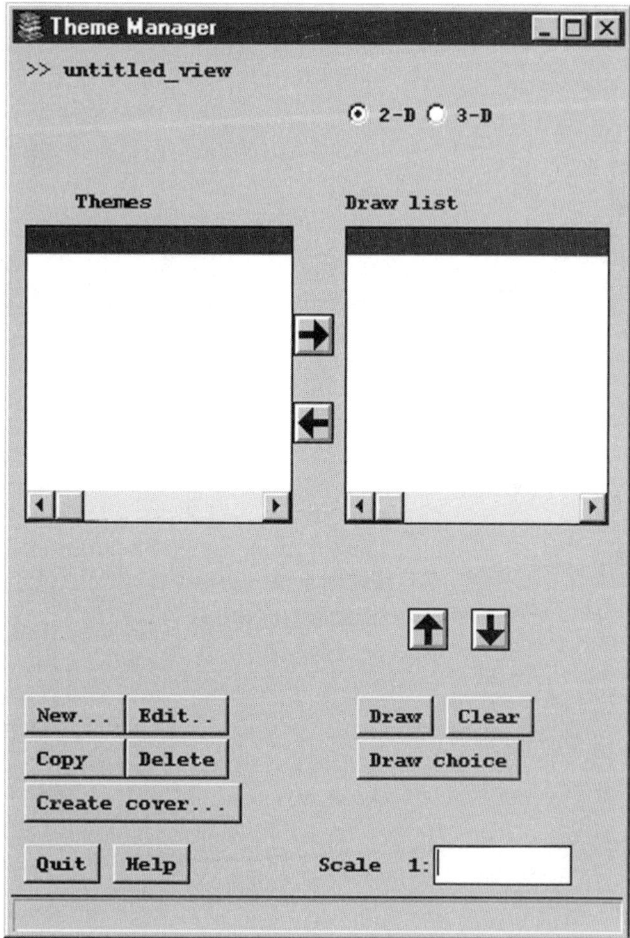

Theme Manager window.

The Theme Manager window is used to add new themes, as well as to modify existing ones. Because you are creating a new view, there are presently no themes listed in the Themes box.

Click on the New button in the Theme Manager window. The Add New Theme window appears, which is shown in the following illustration.

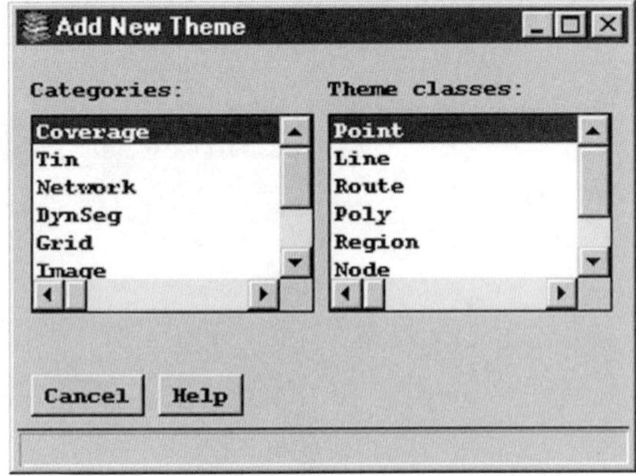

Add New Theme window.

There are two scrollable windows present in the Add Theme window. The left-hand window, named Categories, lists types of spatial data sets (coverages, grids, networks, and so on). The right-hand window, "Theme classes," lists the feature available for the highlighted category. For example, highlighting Coverage in the left-hand window lists features such as Point, Line, and Poly (polygon) in the right-hand window.

To add a theme, first tell ArcPlot its category and feature class. You will now add the *uk_eire_bdrs* theme by selecting Coverage in the Categories window and Poly in the "Theme classes" window. Double click on the word *Poly* and the POLY Theme Properties menu appears, as shown in the following illustration.

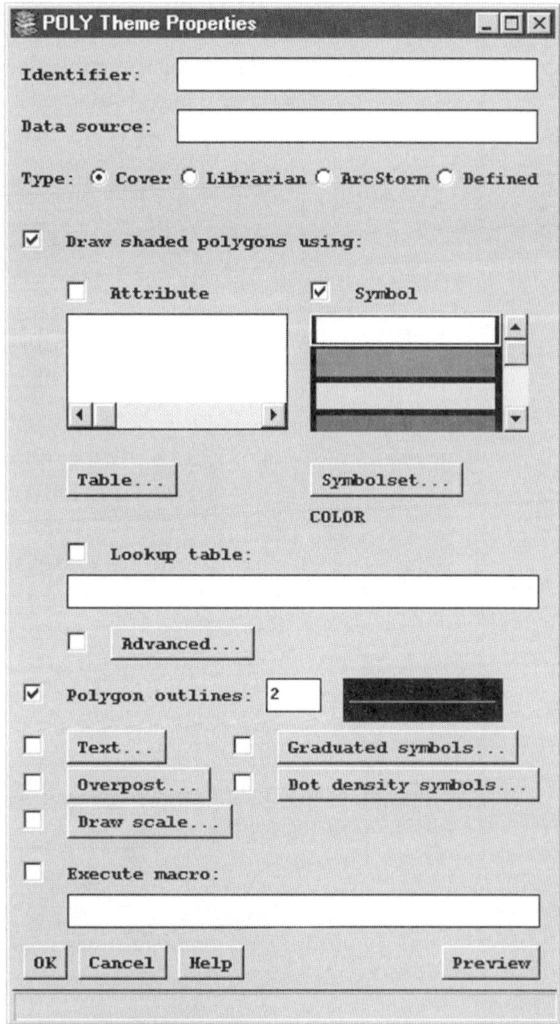

POLY Theme Properties menu.

Using the mouse, right click in the "Data source" text box. This displays a list of polygon coverages present in the current working directory (although other coverages may be present). Highlight the coverage named *uk_eire_bdrs* and click on OK.

Now specify the drawing attributes for *uk_erie_bdrs*, as shown in the following illustration. Here you can choose the color used to draw polygon outlines, the fill colors, and so on, as well as view the items and data in the polygon attribute table (*uk_eire_bdrs.pat*).

Notice also that the coverage is currently displayed using red lines for the outlines and white to shade the polygons. Change the polygon shading color now by scrolling down in the Symbol window until you see light blue. Select it.

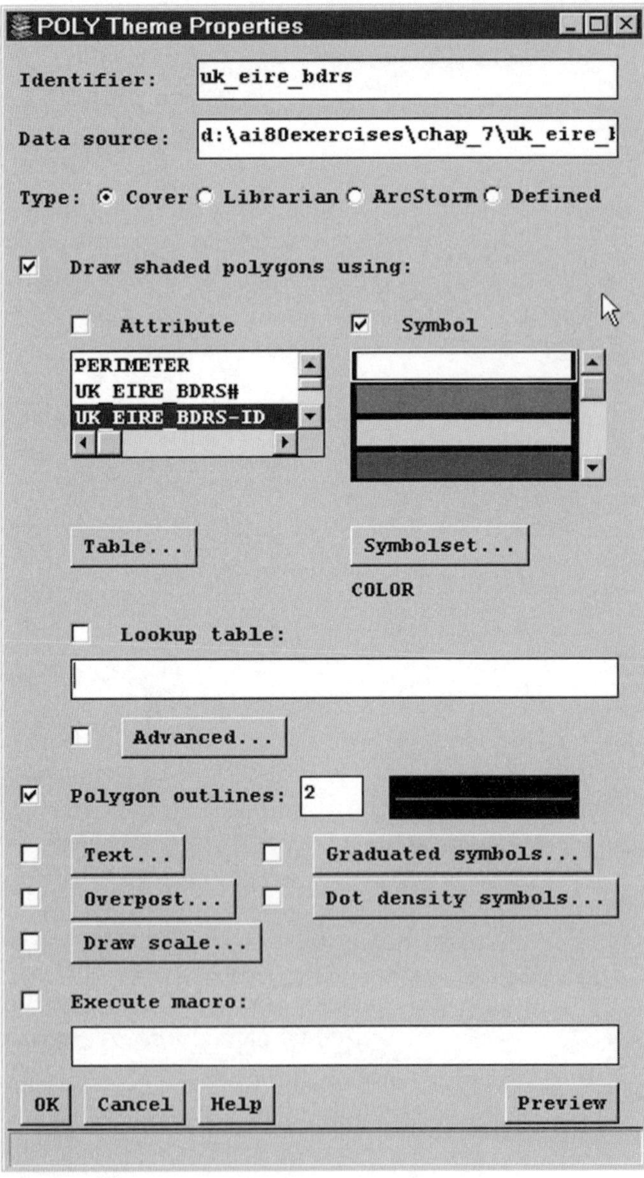

POLY Theme Properties menu with uk_eire_bdrs selected.

The POLY Theme Properties menu now shows a listing of the coverage's items in the Attribute scrollable window. The coverage name now appears in the top-most text box, named Identifier. Here you can type in a longer, more descriptive, name for your theme. Do so now by clicking on the Identifier text box and typing in *Country Boundaries.* This is simply a descriptive explanation, in your view, of what the coverage portrays. It does not change the coverage's name.

✓ **TIP:** *Use the Identifier feature to give themes in your view more descriptive names. This can help remind you (and others in your organization) of what is in that theme.*

Now click on the OK button and the coverage will be added to the Theme Manager window. Nothing has been drawn in the ArcPlot canvas window yet. To do this, go back to the Theme Manager window, shown in the following illustration.

Theme Manager window showing Country Boundaries.

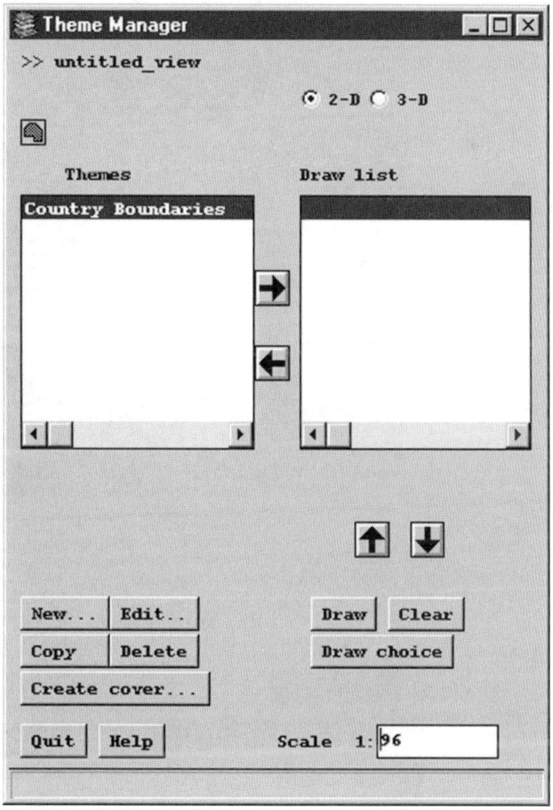

Notice that the Theme Manager now shows Country Boundaries as the theme name of the *uk_eire_bdrs* coverage. To display this theme, press the "➥" button (the right arrow) located between the Themes and "Draw list" windows. This moves the Country Boundaries theme to the Draw list. Now click on the Draw button, and the theme will be displayed on ArcPlot's canvas. Your screen should look like that shown in the following illustration.

Country Boundaries theme displayed on screen.

The Draw list enables you to have many themes in your view (although currently you have only one), without all of them being displayed all of the time. Use this to help organize your GIS projects.

✓ **TIP:** *You can store many themes in a view and access them as needed for map compositions.*

Finish adding themes to your view by following the previous steps to add the other three coverages. Remember that one coverage is a point coverage and one is a line coverage. Use table 7-2, which follows, to give each theme its own identifier and color.

Table 7-2: Theme Identifiers and Color Codes for Exercise 7-1

Coverage	Identifier	Color
uk_eire_city	Cities	Red
uk_eire_river	Rivers	Blue
uk_eire_lakes	Lakes	Blue

When you are finished, the Theme Manager window should look like that shown in the following illustration.

Theme Manager window showing all four themes.

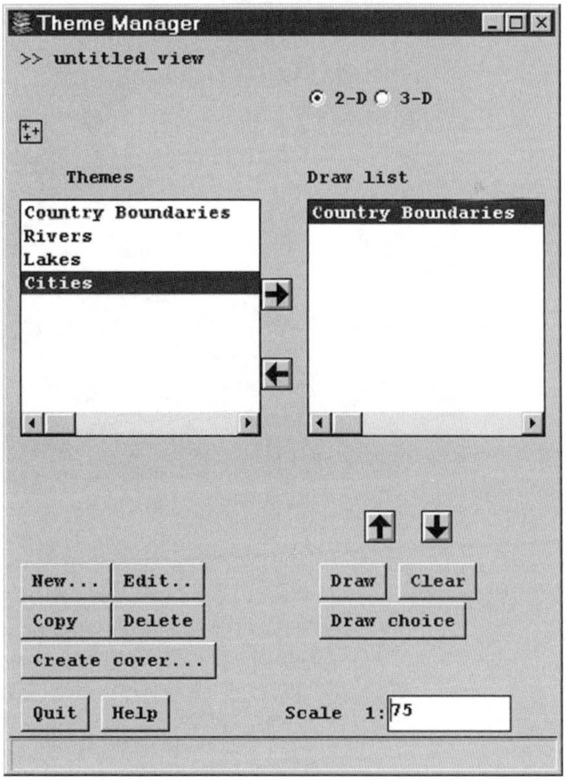

Now use the "➥" arrow to move all three new themes to the Draw list. Notice that you can use the "←" button to remove themes from the Draw List. Now click on the Draw button and the screen should now display all four themes, as shown in the following illustration.

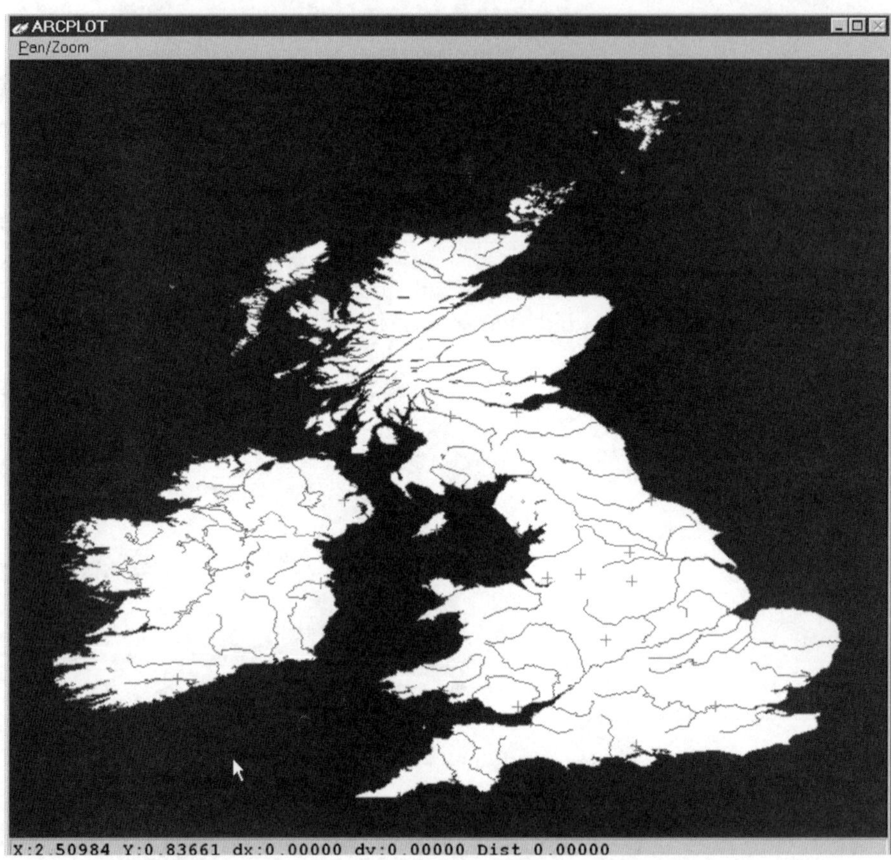

The four themes depicted on the ArcPlot canvas.

Step 3: Saving the View

Now that you have the view set up, the next step is to save it. Go to the Map Tools menu and select the View ➥ Save menu option. A new window appears, prompting you to enter a name for the view. Type *uk_eire_view* and click on the OK button. All views must end with "_view" to be recognized as such by ArcPlot. The view is now saved and may be reopened at any time by going to the View ➥ Open menu selection.

☛ **WARNING:** *Name your views with "_view" extensions, or ArcPlot will not recognize them as views.*

✓ **TIP:** *Although the new ArcInfo modules may, ArcInfo Workstation does not recognize blank spaces in file names, so be sure never to use them!*

Step 4: Querying the Cities Theme

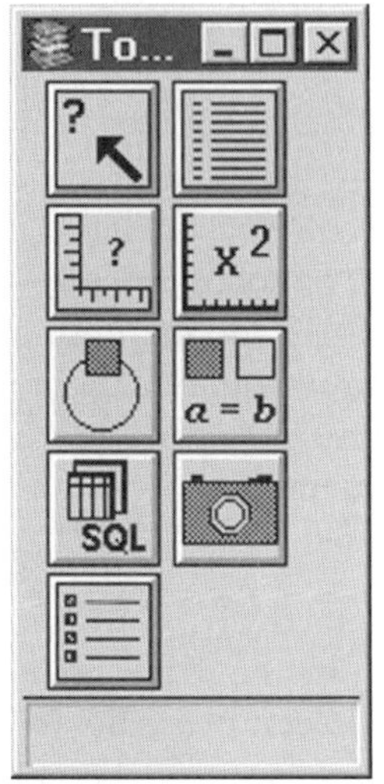

Tools menu.

Now that you have the view constructed, this step demonstrates how to identify its features interactively on the screen, and how to list their attributes. You can point and click on a particular feature in a theme, and information about it will be displayed. For this step you will query the cities coverage to discover the populations of some of the United Kingdom's larger cities. To begin, click on the Map Tool menu's View ➥ Tools option to get the Tools menu, shown in the illustration at left.

This menu contains querying and snapshot tools. This exercise will only use two of the buttons, but it is at this menu that you can select other useful tools. One of these is the Snapshot tool (it looks like a small camera), which allows you to print the view or to save the view as an ArcInfo plot file, an encapsulated postscript (EPS) file, or a CGM file. To find out what the other tools do, right click with the mouse on the button and the button's name appears at the bottom of the menu. You can then reference that button in the on-line help.

✓ **TIP:** *Save views as an EPS or CGM file for easy transfer to another graphics program such as Adobe PhotoShop, or for inclusion in a word processing program.*

Click first on the Cities theme in the Theme Manager menu. This makes its features selectable. Now click on the Identify button (the arrow and question mark) in the Tools menu. Crosshairs appear in the ArcPlot canvas window, and selection options appear in the ArcPlot dialog window. Using the crosshairs, select a city on the screen. Press the <1> key or left click with the mouse to display information about the city. An example is shown in the following illustration. You can see that the population of Edinburgh, Scotland, is in the category "between 250,000 and 500,000."

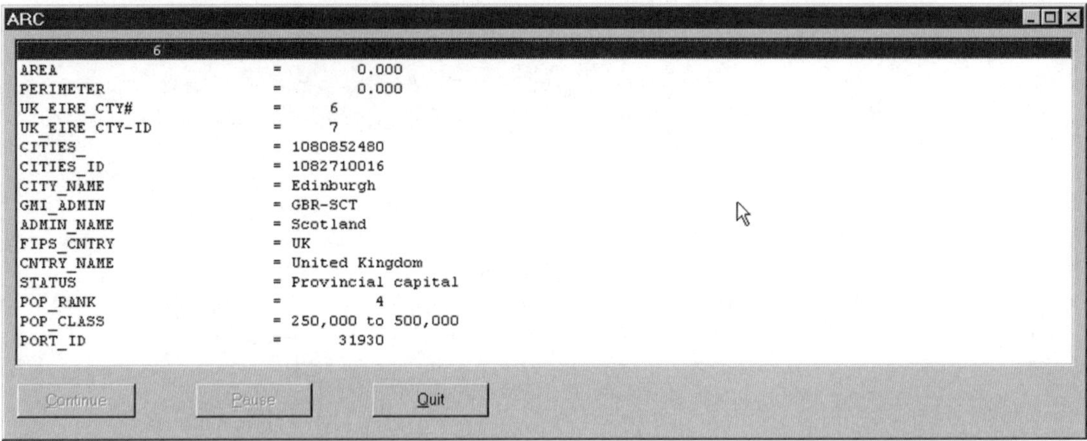

Descriptive data for Edinburgh.

To quit the query process, press the <9> key. Try querying the other themes by first selecting them in the Theme Manager and then using the Identify tool. Selecting interactively is one way to query a theme for information about its features. You can also construct more complex queries using the Logical Expression menu.

Step 5: Selecting a City Using the Logical Expression Menu

Just as you used the mouse to select a city interactively, you may also create a query using the Logical Expression menu. This allows you to construct sentence-like queries using Boolean elements (=, >, and so on). The Logical Expression menu is similar to the Query Builder found in ArcMap.

For this step you want to view population information about London, but pretend you are not sure which city it is on the map. Press the Logical Expression button (the button with two squares and "a = b" on it) on the Tools menu. The Logical Expression menu appears, shown in the following illustration.

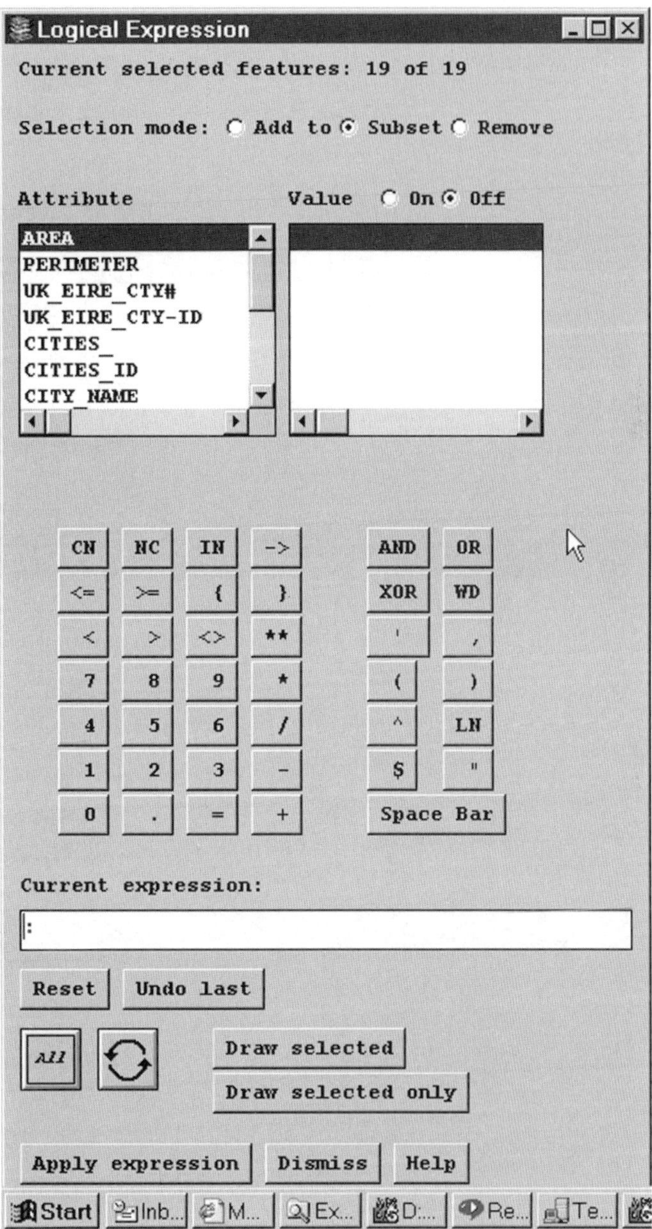

Logical Expression menu.

Now begin building your query expression. First, select CITY_NAME in the scrollable Attributes window. Next, select the "=" key from the menu's keypad. Finally, set the Values window to On (to list the available values for CITY_NAME) and select London from the scrollable list of cities. The expression you have just built, which follows, appears in the "Current expression" box.

```
: CITYNAME = 'London'
```

Now click on the "Apply expression" button to launch this query. At the very top of the Logical Expression menu you will see that it now says "Current selected features: 1 of 22." Dismiss the Logical Expression menu and select the List button (which looks like a sheet of lined paper) from the Tools menu. This displays a window that contains the attribute information for the currently selected set of features (in this case, there is only one item). The attributes are shown in the following illustration. When you are done examining its attributes, dismiss this window.

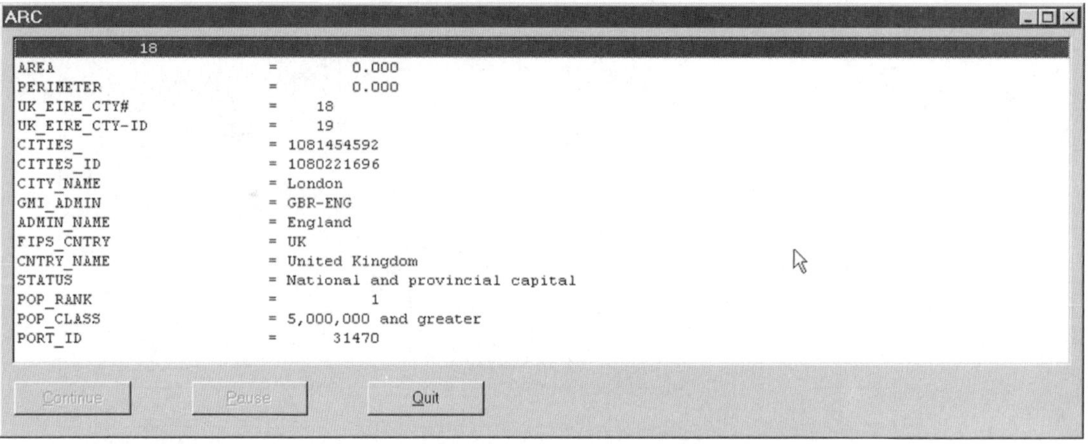

Listed attributes for the selection "London."

You now have one point selected from the coverage of 22 points. All of your future operations will be performed on the currently selected set. In other words, anything you do now (delete, for example) would be performed on "London" only. Reselect all of the cities by selecting the All button. This selects all features in a coverage. It is a good idea to remember to do this reselection because any further operation conducted on the Cities point coverage would affect only the selected set of points.

Step 6: Changing Symbols

The cities are currently visible on ArcPlot's display canvas as small squares. You will now change these squares to a different, larger symbol (a red "+") that is easier to see.

Using the Theme Manager window, highlight the Cities theme in the Themes box. Now select the Edit button to access the POINT Theme Properties menu, shown in the following illustration.

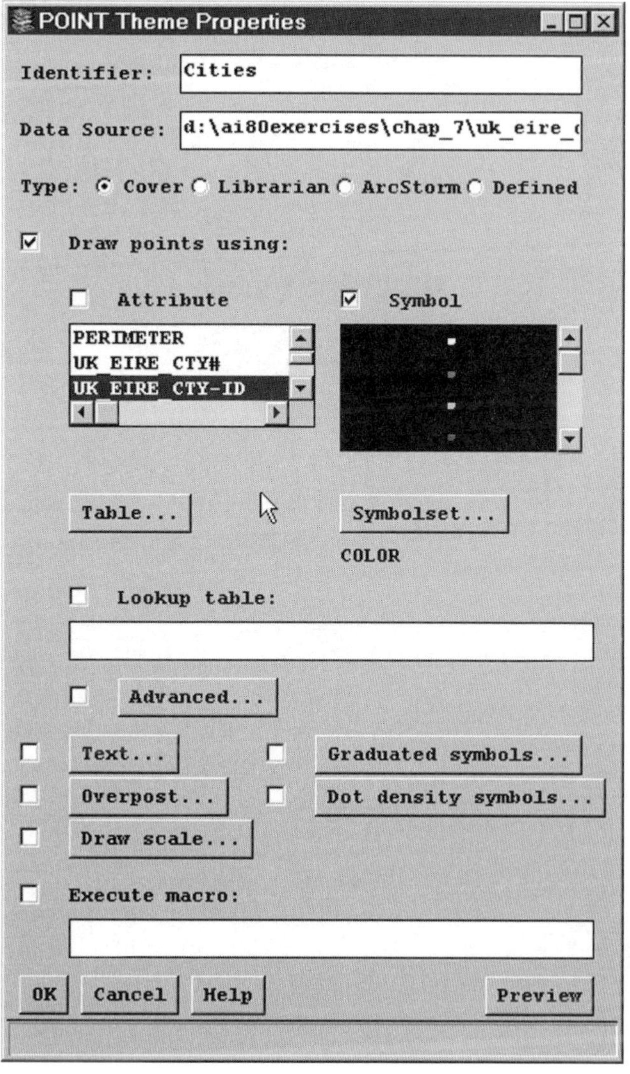

POINT Theme Properties menu.

Now click on the Symbolset button. The Select a Symbolset File menu appears, which is shown in the following illustration.

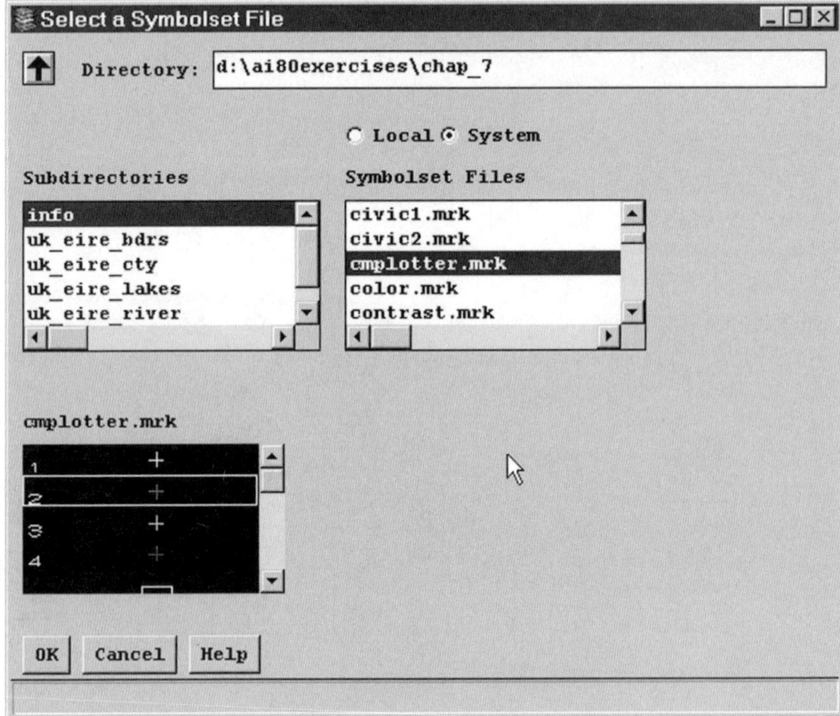

Select a Symbolset File menu.

Use this menu to select from the many symbol sets provided with ArcInfo. For this exercise, scroll down in the Symbolset Files window to select the *cmplotter.mrk* symbol set. Its name and symbols appear in the scrollable box above the OK, Help, and Cancel buttons. Now click on the OK button, and you are now back in the Point Theme Property menu. Next, select the red "+" from the scrollable "Symbol" list of available markers in *cmplotter.mrk*.

Click on the OK button when done, and you will be returned to the Theme Manager menu. Here, click on the Draw button to display the newly chosen symbol for the Cities theme. Your screen should look like that shown in the following illustration.

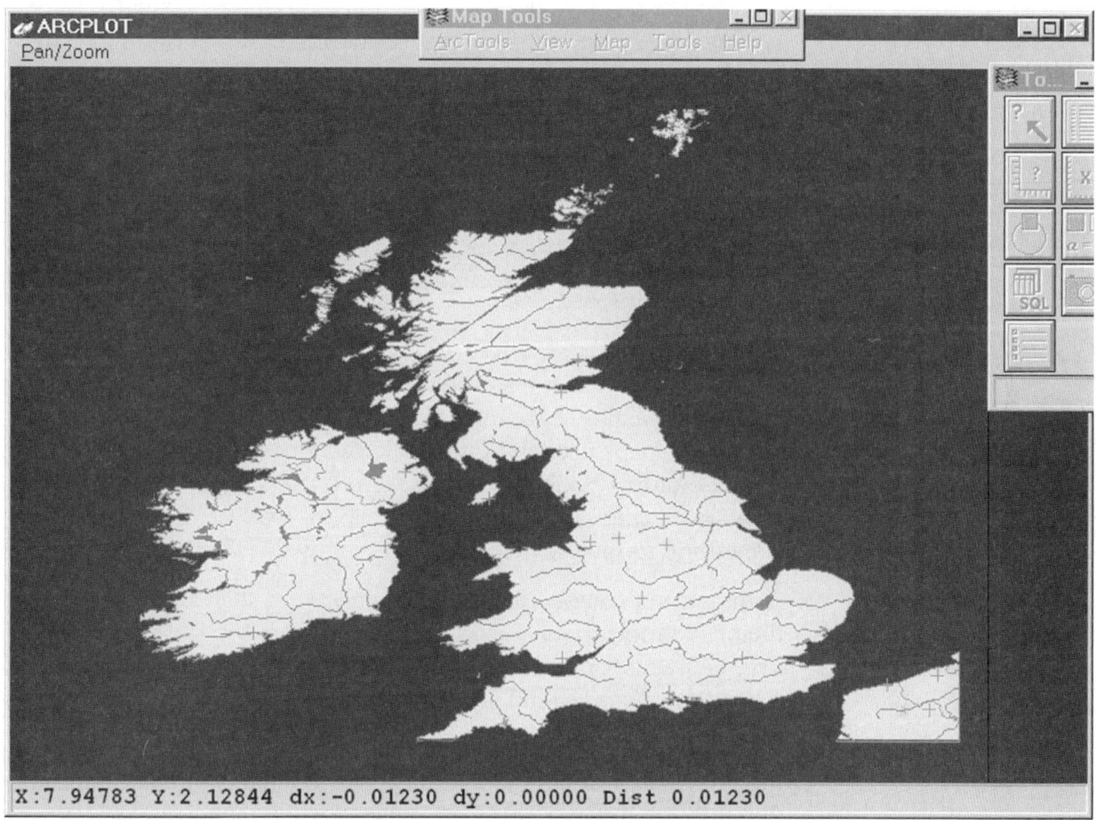

Updated symbols for Cities theme shown in ArcPlot.

You are done! Congratulate yourself for having worked through this exercise. Now, quit ArcPlot by first dismissing the Map Tools menu (do this by clicking on the X in its blue top bar) and then selecting Quit from the ArcTools menu. At the ArcPlot prompt, type *quit* or *q*. This exits you from the ArcPlot module.

In this exercise you have learned how to create a view, how to query themes in a view, and how to change symbols in themes. Remember that these functions of ArcPlot are also available in the ArcMap module, which has an easier-to-use interface. Also remember that the commands and pull-down menus of ArcPlot can be incorporated into AML programs (introduced in Chapter 10) to better automate map composition functions.

The next section introduces the Map Composer tools, which are used to create map compositions for printing, plotting, posting on an Internet site, and so on. Map compositions include views, and map elements such as North arrows and scale bars.

Creating a Map Composition in ArcPlot

Creating a map is an effort worthy of study at the university level. No map can ever represent reality; rather, maps represent our perceptions of reality. Care must be taken in how maps represent data, especially maps used in policy-making arenas. Although it is not the purpose of this book to instruct ArcInfo users on good mapmaking techniques, the "Cartographic Principles" section in Chapter 3 is something to review before proceeding with exercise 7-2.

A map composition differs from a view in that a view contains themes (various data layers), whereas a map composition contains map elements, one of which might be a view. Map elements include a North arrow, a legend, a scale bar, a data source, and so on.

Exercise 7-2 introduces you to ArcPlot's map composition tools. In this exercise you will make a map of the United Kingdom and the Republic of Ireland, complete with map elements. Before beginning you should have completed exercise 7-1. You will use the same data sets as in exercise 7-1, assumed to be stored in your workspace *D:\ai80exercises\chap_7.*

Exercise 7-2: Creating a Map Composition in ArcPlot

Using the view and themes from exercise 7-1, you will create a map composition of the United Kingdom and the Republic of Ireland that shows rivers, water bodies, and cities. This exercise steps you through Map Tools and shows you how to create a professional-looking map, complete with a title and North arrow.

Step 1: Opening a Map Composition

Start an ArcPlot session by selecting Start ➡ Programs ➡ ArcInfo ➡ ArcInfo Workstation ➡ ArcPlot. Now type *arctools* at the ArcPlot prompt and from there launch Map Tools. From the Map Tools menu, select the Map ➥ New option. The Select a Template Layout window, shown in the following illustration, will open.

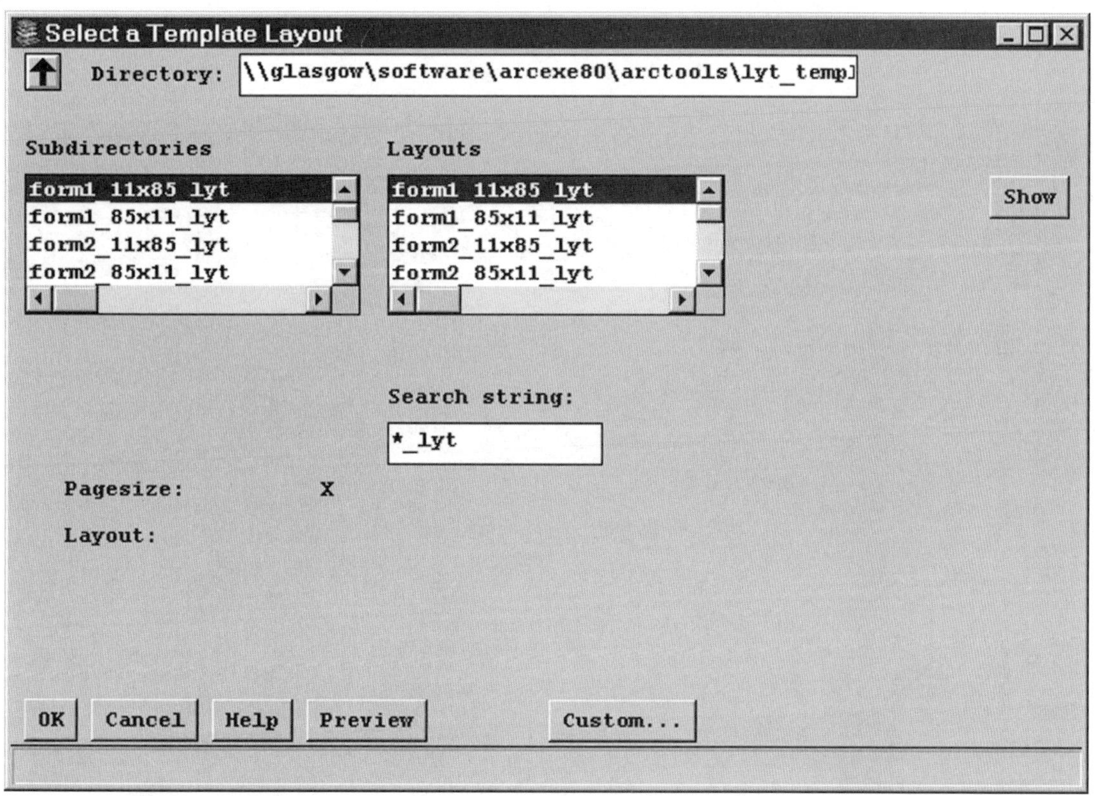

Select a Template Layout window.

First, examine the Layouts scrollable window. Listed are a number of prepared map templates that fit on an 8.5 x 11-inch sheet of paper. Each of these map templates has its own way of laying out the page. Some come complete with North arrows and scales. Highlight one and you will see the layout previewed at the bottom of the window above the Cancel button. Spend some time now exploring the various templates.

✓ ***TIP:*** *Select the Custom button if you want to make a map larger than 8.5 x 11 inches. You can type the dimensions for the map but will not have any cartographic items, such as legends and scales, added for you beforehand. You can also create your own templates. See the on-line help section under "Templates."*

For this exercise, highlight the first template in the list, *(form1_11x85_lyt)* which is 8.5 x 11-inch (landscape) with no North arrow or other accessories. Click on OK. Two white-line boxes appear on the ArcPlot display canvas, shown in the following illustration.

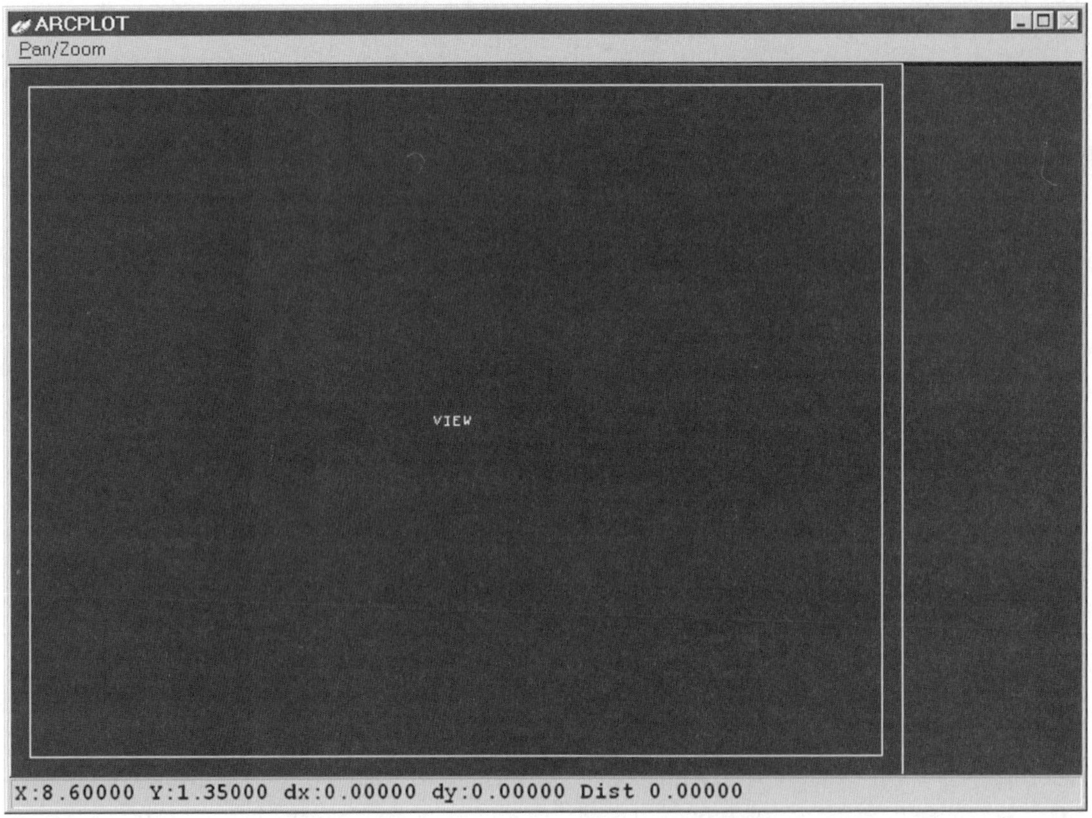

ArcPlot canvas.

The ArcPlot canvas displays two white rectangles, with the word *view* placed in the inner rectangle. The outer white rectangle shows the outer limits of the map page, where items can be placed for inclusion on the map. The inner white rectangle shows the outer dimensions of the view, the place where coverages will be drawn. The Map Object Manager and Add New Object menus also appear, which are shown in the following illustrations.

NOTE: *If the Add New Object menu does not appear, simply select New from the Map Object Manager menu to launch it.*

Map Object Manager and Add New Object menus.

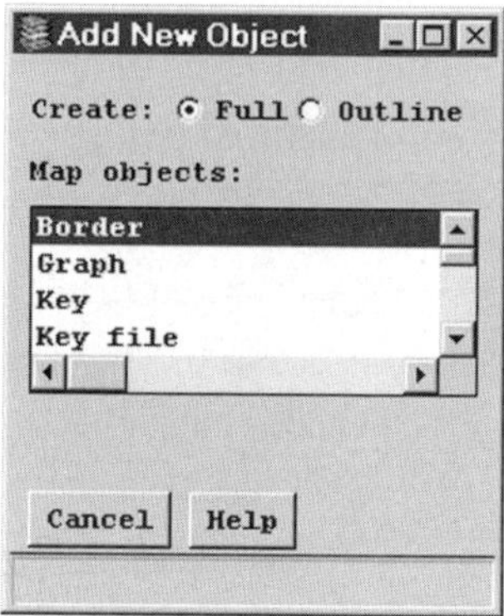

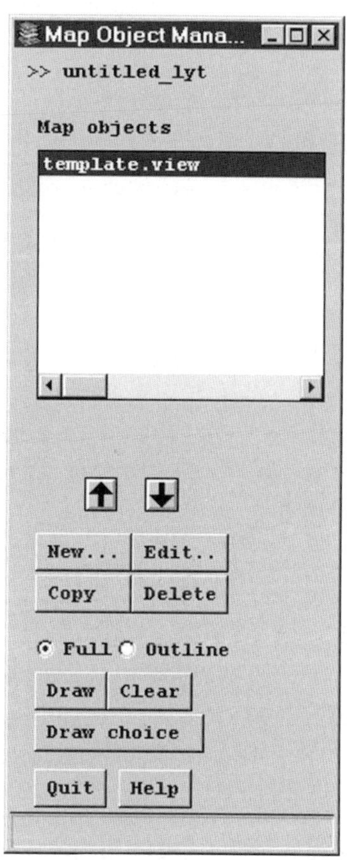

The Add New Object menu is the "control panel" for the map elements you add to the map. The Map Object Manager menu lets you edit added elements. Remember, these elements can be views, freehand lines, titles, scale bars, and so on. The Map Object Manager manages each element individually, allowing for easier map composition and layout.

Step 2: Adding an Element to a Map

Use the Add New Object menu to add elements to your new map composition. In the scrollable "Map objects" window, scroll down and click on "view." The View Properties menu appears, which is shown in the following illustration.

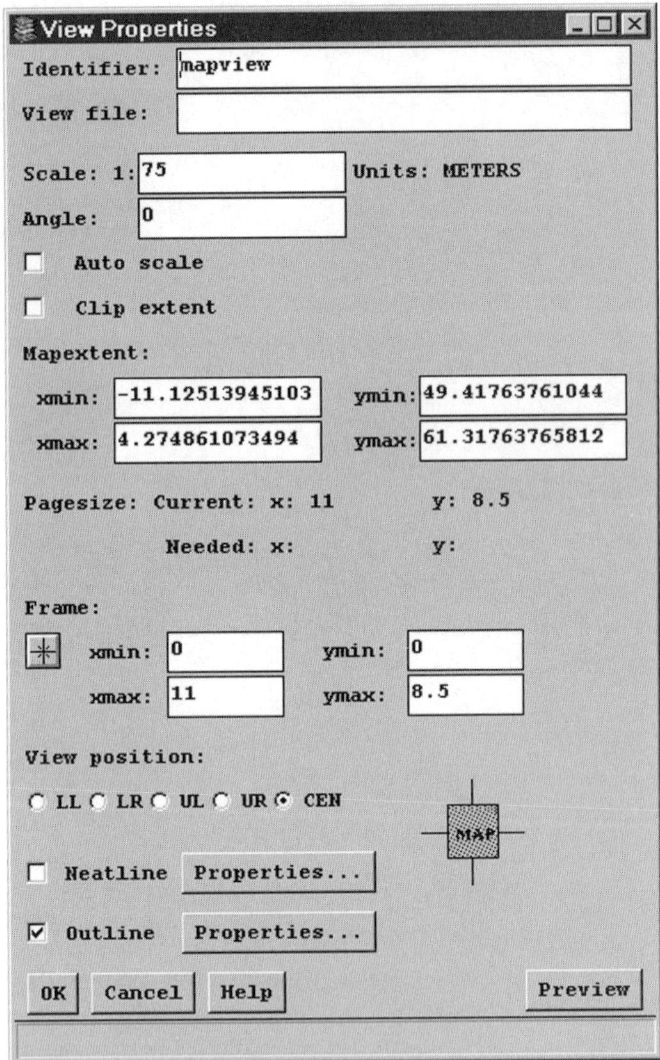

View Properties window.

Right clicking in the "View file" text box will show the views available in the current workspace, and allow you to navigate to other workspaces. For this exercise, select the view you created in exercise 7-1, *uk_eire_view*. Click on the OK button and the view is added to the View file box. Now click on OK in the View Properties menu, and you are returned to the Map Object Manager menu. Here you see *uk_eire_view* listed in the "Map objects" box. Now click on the Draw button to display this view on the ArcPlot canvas. Your screen should look like that shown in the following illustration.

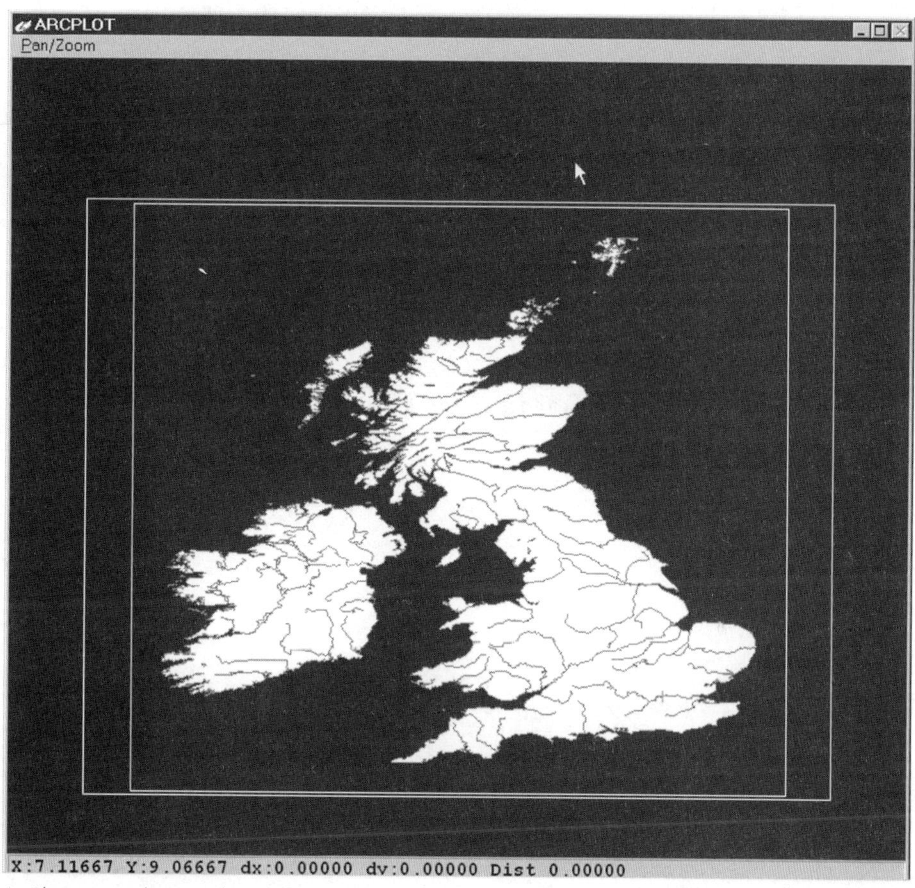

ArcPlot canvas showing uk_eire_view.

Step 3: Adding Another Element and Changing Its Size

Now add a title to your map composition. Select "text string" from the Map objects scrollable box on the Add New Object menu. This brings up the Text String Properties menu. Select the Text String element in its scrollable window, shown in the following illustration.

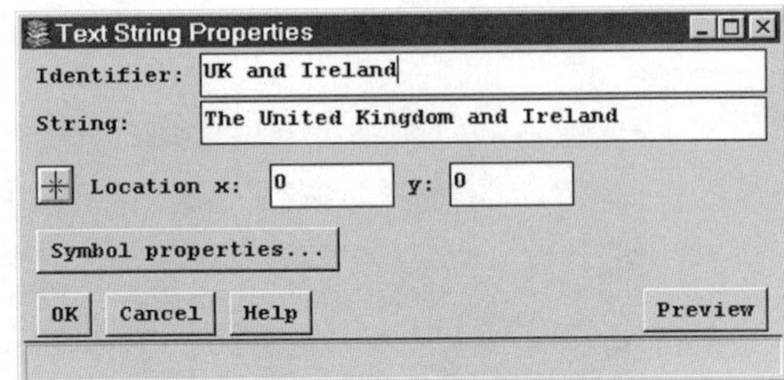

Text String Properties menu.

As in the Text String Properties box previously shown, type *UK and Ireland* in the Identifier box. This identifier is a short descriptive label that appears in the Map Object Manager menu. Next, type *The United Kingdom and Ireland* in the String box. This is the longer string of text that will be used as the title for the map composition. Finally, press the "*" button to interactively position the text, using the crosshairs on the display canvas. Go ahead and position the text above the word *view*, as shown in the following illustration.

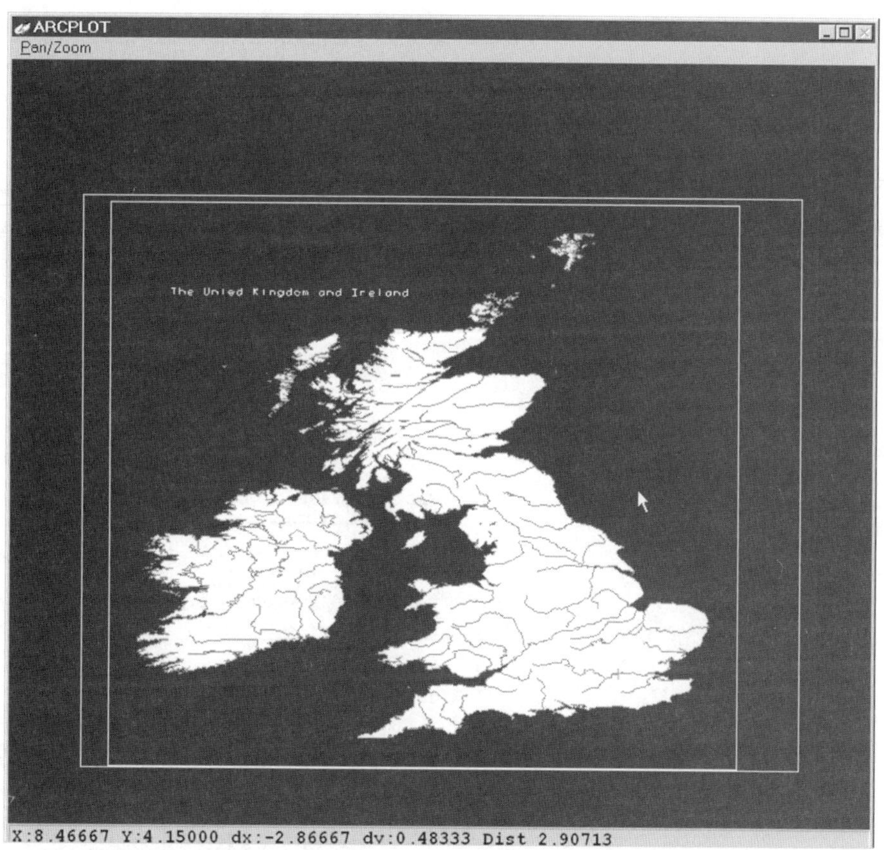

Placing the title on the ArcPlot canvas.

You can see that this title is too small. Enlarge the font by clicking on the Symbol Properties button in the Text String menu. The Text Symbol Properties menu pops up, which is shown in the following illustration.

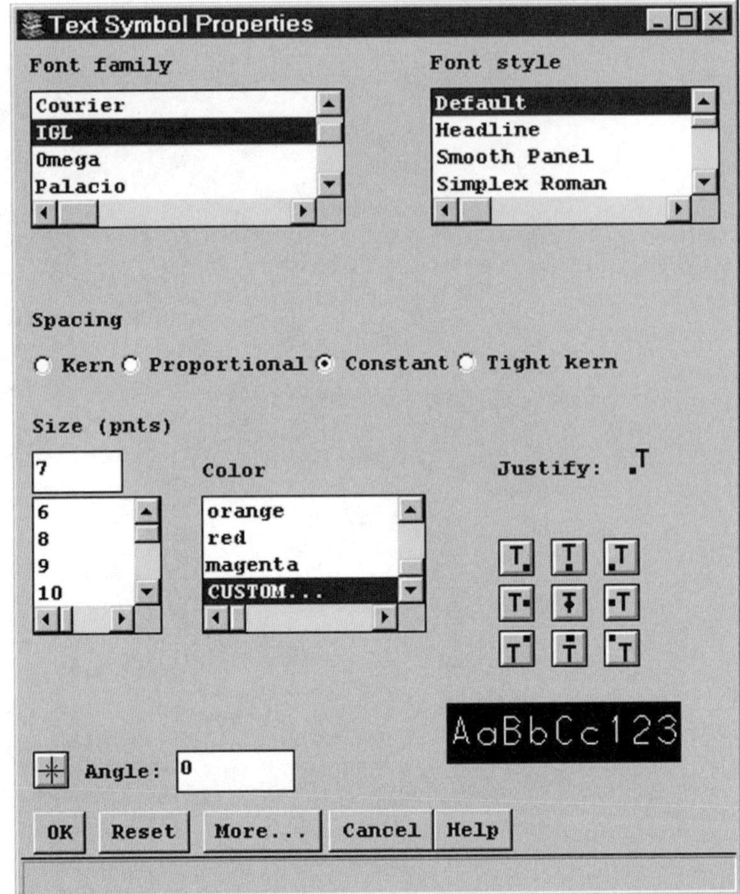

Text Symbol Properties menu.

The Text Symbol Properties menu has many options for controlling the appearance of text. There are scrollable lists of available fonts, font colors, and point sizes. Point size controls the size of text. Because the default font size of 7 points is too small, try doubling it to 14. Click on the OK button. The result is shown in the illustration that follows.

✓ **TIP:** *There are 72 points in one inch. Therefore, a font that is 12 points will be displayed 12/72 of an inch tall.*

ArcPlot canvas with larger title.

Although the font now appears readable, it may not quite be centered. You can adjust its position on your screen, if necessary, using the "*" button in the first Text String Properties box. The result is shown in the following illustration.

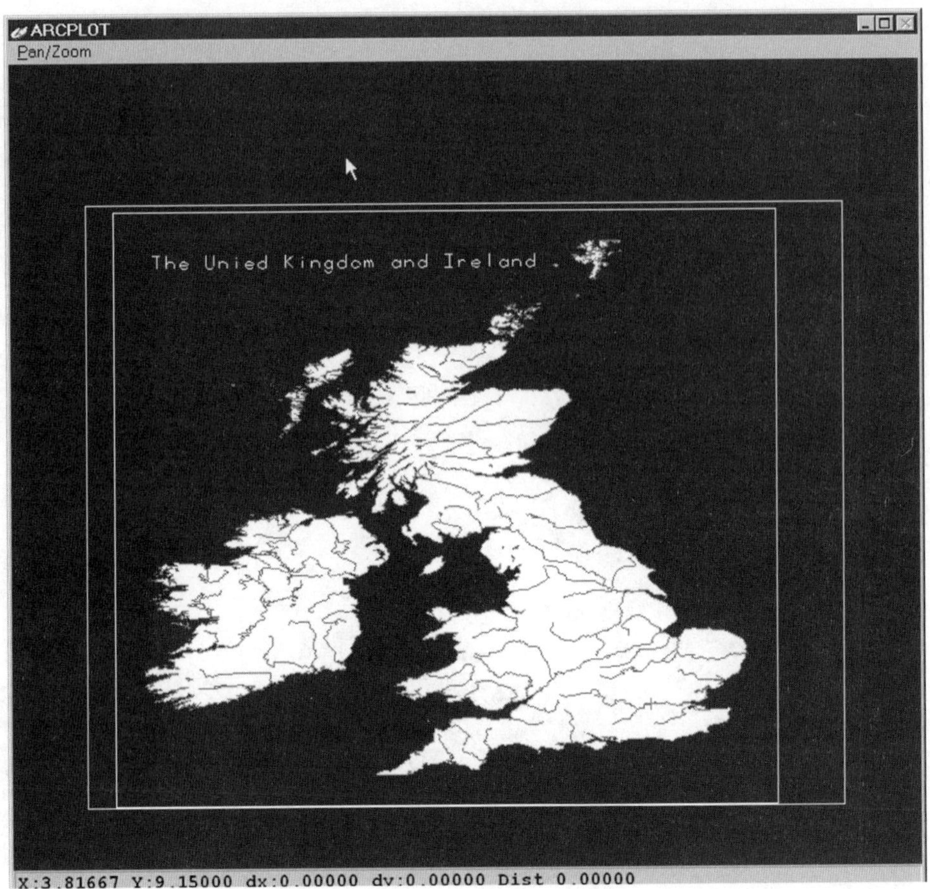

ArcPlot canvas with centered title.

✓ **TIP:** *You sometimes see the text both in its new location as well as in the old location after performing a move or a resizing of text. Click on the Draw button in the Map Object Manager menu to refresh the display canvas.*

Step 4: Adding a North Arrow

Next, add a North arrow. In the Add New Object menu, select the "North arrow" option in the scrollable window. The North Arrow Properties menu appears, which is shown in the following illustration.

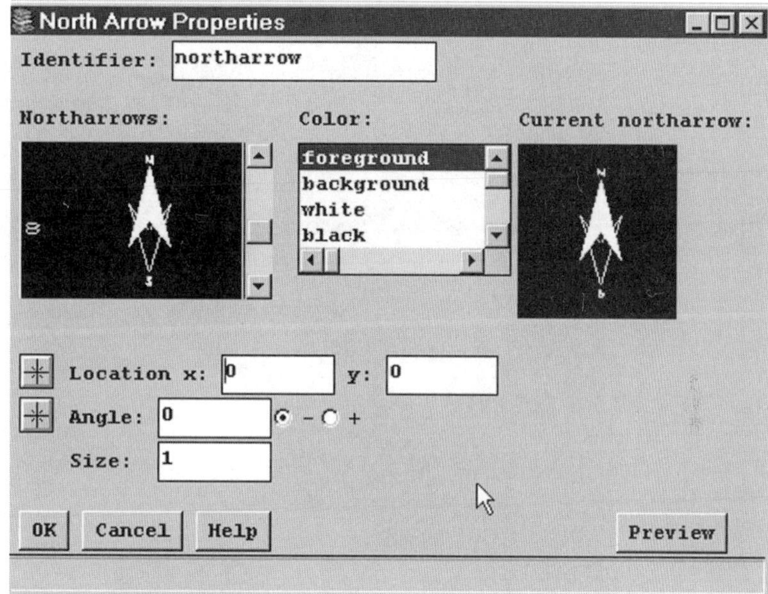

North Arrow Properties menu.

Use the scrollable Northarrows window to select a style you find pleasing. For illustrating this exercise, the fourth arrow in the list was chosen. Press the "*" button to interactively position the arrow to the right of the view in the ArcPlot canvas. This exercise's completed map composition should look similar to that shown in the following illustration.

Completed map of the United Kingdom and Republic of Ireland.

For purposes of brevity, this exercise showed you how to add only two of the many map elements available. You may continue this exercise on your own and add other elements such as a border, a pie chart, a scale bar, and so on. You may also want to redo this exercise from the beginning using a different map template.

The next section discusses how to print paper maps and create Internet-ready maps from your map compositions.

Printing Paper Maps and Internet-ready Maps

Once you have created a map composition and are ready to print it, select the Map ➥ Create Metafile pull-down menu, shown in the following illustration.

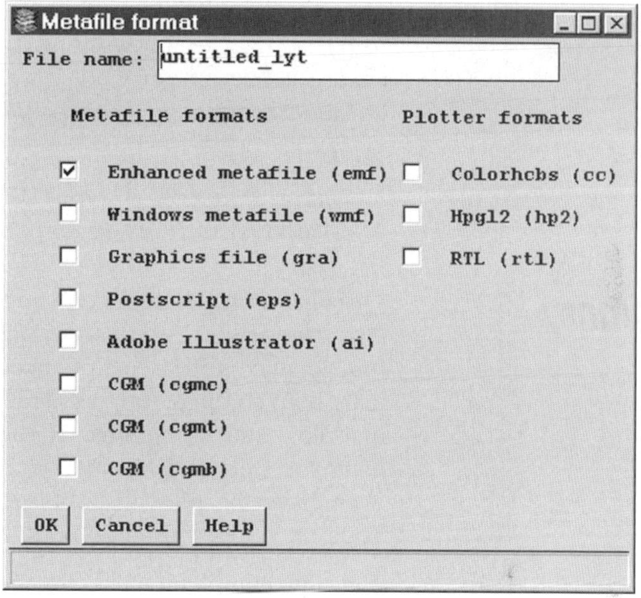

"Metafile format" menu.

Metafile format

File name: untitled_lyt

Metafile formats	Plotter formats
☑ Enhanced metafile (emf)	☐ Colorhcbs (cc)
☐ Windows metafile (wmf)	☐ Hpgl2 (hp2)
☐ Graphics file (gra)	☐ RTL (rtl)
☐ Postscript (eps)	
☐ Adobe Illustrator (ai)	
☐ CGM (cgmc)	
☐ CGM (cgmt)	
☐ CGM (cgmb)	

OK Cancel Help

Here you would type in a name for your plot file and select the graphics file format you want to use. The most common choice is an ArcInfo graphics file. You may, however, choose EPS, CGM, Adobe Illustrator, HPGL2, or RTL formats. Select the CGM, EPS, or Adobe Illustrator formats if you intend to use the graphics file in another software program.

To print the file you just created, select Map ➥ Plotting from the Map Tools menu. A window appears asking you to enter the name of the plot file. Right click in the text window and a list of plot files in the current directory will display. Once you specify a plot file, click on the Submit button and you will be asked which printer you want to use. Select the printer and the file will be sent.

✓ **TIP:** *The map colors may not always look the same on the screen as when printed. If your printer or plotter has a limited range of colors, design your map in ArcPlot to use only these colors. If you have a black-and-white printer, avoid using color in ArcPlot maps. Instead, use shades of gray.*

If you want to put the map you created on the Internet, you should save it in a file format compatible with Internet graphics formats such as JPEG and TIFF. Use the ArcPress module to convert your ArcInfo graphics file to a JPEG or TIFF. Once converted, include the new graphics file in your web site and it will be viewable to anyone using an Internet browser. Because ArcPress is a separately purchased product from ESRI, however, a detailed discussion of it is outside the purview of this book.

NOTE: *For more information, see the "Printing and exporting maps with ArcInfo" and "Printing and exporting maps with ArcPress" sections under "Cartography" in the on-line help.*

Summary

This chapter has introduced you to some of the functionality of ArcPlot. The new Version 8.0 ArcMap module has replaced most of its functionality, and users are encouraged to use ArcMap as their primary map composition tool. ArcPlot is one of the traditional ArcInfo modules, and its commands can be executed by command line entry and by pull-down menus. The pull-down menus do not, however, have the level of sophistication of the newer ArcMap, ArcToolbox, and ArcCatalog modules.

Through completing exercise 7-1, you learned how to create a view, query features in that view, and how to change features in a view. Exercise 7-2 showed you the basics of creating a map composition. The last section discussed creating paper and Internet-ready maps.

Although ArcPlot has been largely replaced by ArcMap, its commands and menus can still be incorporated in AML programs (discussed in Chapter 10, including an exercise that uses ArcPlot commands). These programs can help make your organization's use of ArcPlot more automatic and efficient. Most users will find uses for both modules; for example, ArcMap for map composition and ArcPlot for working with AML programs.

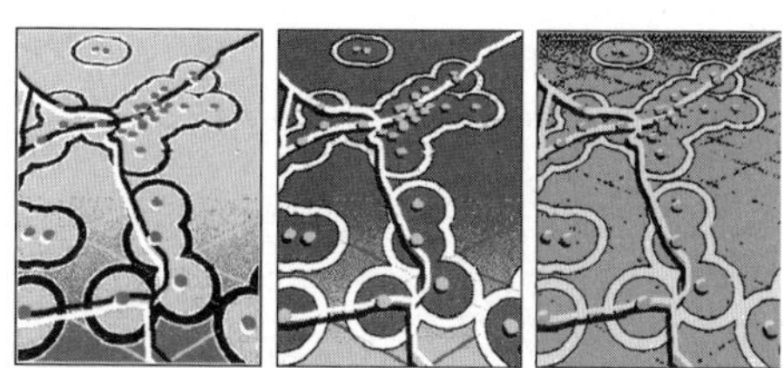

Chapter 8

Arc

An Introduction

Arc is the traditional ArcInfo module used for conducting analyses (such as spatial overlay operations) and for data management tasks (such as building topology, transforming coordinates, and organizing coverages and workspaces). As one of the traditional modules, Arc offers both command line entry and pull-down menus for executing commands.

The new Version 8.0 ArcToolbox and ArcCatalog modules replace much of the functionality of Arc, however, and offer a more sophisticated user interface. The new modules incorporate a drag-and-drop interface that will delight veteran ArcInfo users (no more remembering syntax). New users will find that the ArcToolbox and ArcCatalog interfaces make learning and using the commands much easier.

Both new and veteran ArcInfo users are encouraged to use ArcToolbox and ArcCatalog as their primary analysis and management tools. If your organization has an investment in AML programs, Arc will remain crucial to your work. Note that the pull-down menus can be incorporated into your own custom AML programs.

This chapter introduces you to some basics of Arc, but does not go into extensive detail. All commands presented here are already discussed at length in the reference section of Chapter 4 (ArcToolbox). Look to that reference section for command syntax information. If you will be creating AML programs that use Arc commands, and need a comprehensive lesson in Arc's functionality, see the previous edition of this book. Also refer to the on-line help section titled "A functional list of ARC commands" found in the "ARC" section of

"Command references for ArcInfo prompts." This chapter covers the following.

- An introduction to Arc
- Exploring Arc's pull-down menus
- Exercise 8-1: Stepping Through a Spatial Analysis

Unlkie ArcEdit and ArcPlot, the Arc module has no graphic display capabilities. To graphically view the results of a spatial analysis, for instance, users will have to use ArcMap, ArcEdit, or ArcPlot. It would be best to work through either Chapter 3, Chapter 6, or Chapter 7 before doing exercise 8-1. This way, you could display your results with ease. For those of you in a hurry, however, some ArcPlot commands are included in exercise 8-1 so that you may view the results of its simple spatial analysis.

Exploring Arc's Pull-down Menus

This section takes you through the options available under Arc's pull-down menus. Not all of Arc's commands are included in these menus, and you may occasionally have to "exit" the menu system to type in a command at the Arc prompt. Follow along with the menu descriptions that follow by starting Arc and launching its "Command Tools" menus.

To start Arc, select Start ➥ Programs ➥ ArcInfo ➥ ArcInfo Workstation ➥ Arc. A blank, black background window pops up containing the copyright information and the Arc command line prompt. This dialog (or command line) window is used to execute Arc commands. To access the pull-down menus, type *arctools* at the Arc prompt. The ArcTools menu appears in the upper center of your screen. Select Command Tools, shown in the following illustration, and click on OK.

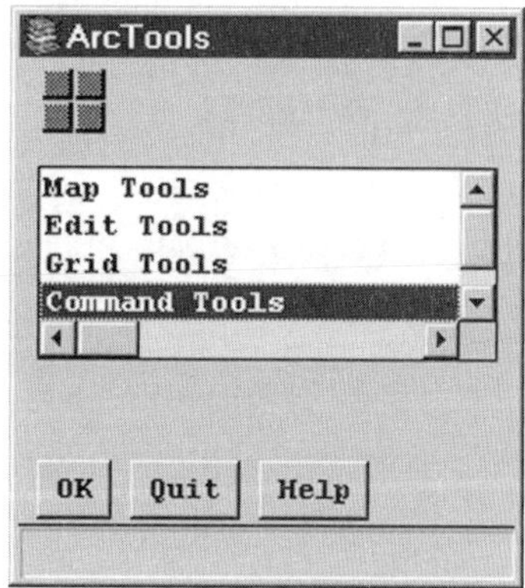

Command Tools selection under ArcTools.

Now two Command Tools menus appear, which are shown in the following illustration. The menu on the left contains three options: ArcTools, Tools, and Help. This menu is referred to as the primary Command Tools menu because the secondary menu (the Command Tools menu on the right) can be launched from this primary menu.

Arc Command Tools menus.

The options on the primary menu contain commands for data management. The secondary menu (on the right) contains the options Edit, Analysis, Conversion, and Help. These options are used for creating and maintaining topology, and performing spatial analyses and data conversion.

✓ **TIP:** *Sometimes ArcInfo Workstation menus are so long they will not fit on the screen. You can fix the problem by changing your screen's resolution. Simply right click on the Windows Desktop (your Windows background) and select* Properties. *Click on the Settings tab and use the Desktop Area slider bar to change to a higher resolution (from 800 X 600 to 1024 X 768, for example).*

Command Tools Data Management Menu

First, examine the options available at the primary Command Tools menu. Begin by clicking on the ArcTools option, shown in the following illustration.

ArcTools option of the primary Command Tools menu.

The ArcTools option contains six selections for executing commands and managing your data. The selections are discussed in the sections that follow.

Commands

Use the Commands selection to temporarily "exit" the menu system and execute an Arc command interactively. Upon selecting Commands, you will see the Enter Command window with an Arc prompt and text box. You would then type your command in at the text box and press the <Enter> key. Try it by typing *workspace* in the text box and pressing the <Enter> key. In the dialog window, you will see a message telling you what workspace you are in (probably "Current location: *D:\ai80exercises\chap_8*").

Click on the More button and the menu will expand. Here you can type in several commands to be executed sequentially. Check the "echo" box and the commands will print to the dialog window as they are executed; that is, they "echo" back to the screen. (For more information on this useful capability, see the AML command "&echo" in the on-line help.) The TTY button gives input to the keyboard; typing in *&return* returns you to the pull-down menus.

Workspace

Use the Workspace selection to change workspaces, or to create or delete workspaces. Selecting Workspace brings up the Workspace menu. From there you can use the "up" ("↑") arrow button to navigate through your hard drive directories. To access zip and disk drives, type their name in at the text box and press the <Enter> key.

Manage

Click on Manage to get a list of other menus, each of which contains options for managing a type of data (coverages, grids, and so on).

Click on Coverages to access the Coverage Manager menu, shown in the following illustration.

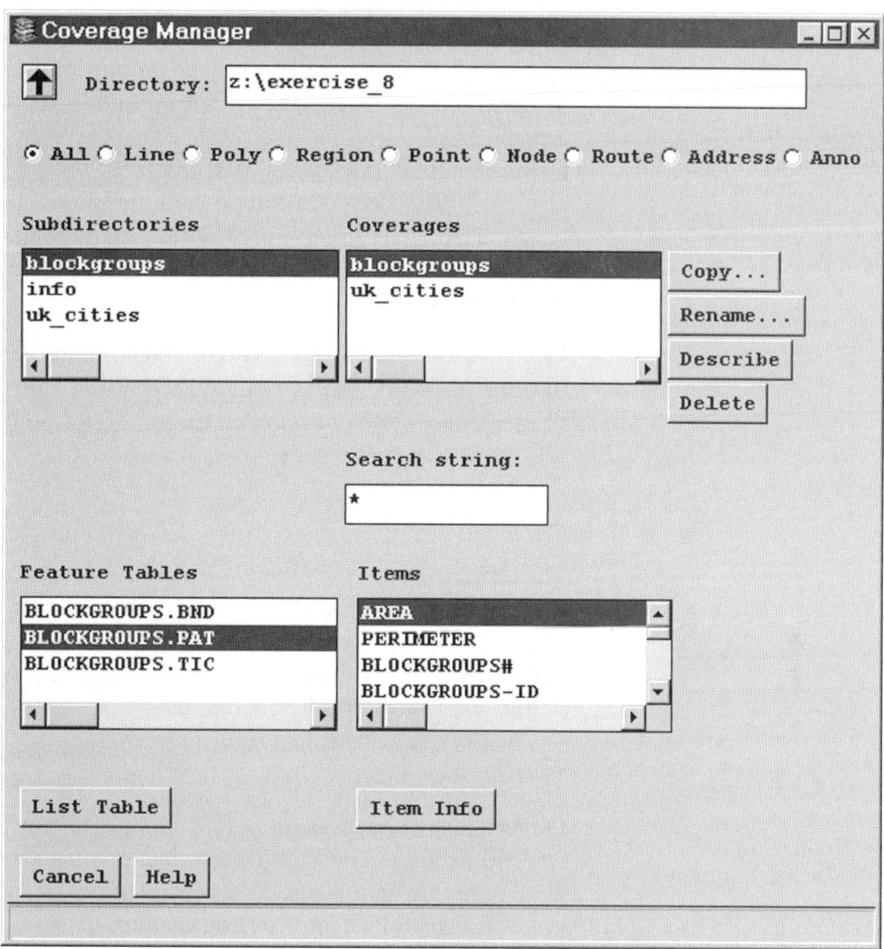

Coverage Manager menu.

The Coverage Manager menu displays the coverages present in your current directory. Buttons are available to Copy, Delete, Rename, and Describe coverages. The Describe button displays a text screen of the coverage's geographic projection, how many features it contains, its geographic extent, and so on.

☛ **WARNING:** *Be careful! Once deleted, coverages cannot be "undeleted." Remember, it is best to always work with copies of your data.*

Two additional functions available from this menu are the List Table and Item Info buttons. The List Table button lists the INFO data for the selected coverage. The Item Info button displays a list of a coverage's items and their definitions. You can navigate through directories to access coverages, and even list their features and feature attributes. The other data type options under ArcTools ➡ Manage (grids, TINs, INFO files) offer similar management menus.

Tool Browser

Use Tool Browser to look at the AMLs "behind" the ArcTools menu options (these are AMLs written by ESRI). To view a text file of one of the AMLs, select it and click on the Menu button. Remember that you can incorporate these tools into your own AML programs.

Applications

The Applications selection is similar to Tool Browser; the difference being that with Applications you can access your custom AMLs from useful categories (for example, "Analysis"). This can help you organize your AMLs by function. Select an AML and it will run from within the Command Tools menus.

Quit

Quit dismisses both Command Tools menus and returns you to the ArcTools menu. The foregoing selections under ArcTools are available in all traditional module menus. That is, the pull-down menu systems for ArcPlot and ArcEdit contain these same options.

Command tools...
Units calculator...

Tools selection under Command Tools.

The next primary Command Tools selection, Tools (shown in the illustration at left) contains two options. The first option launches the secondary Command Tools menu (shown previously), which is dedicated to geoprocessing commands. Its options are discussed in detail in the next section. The other option, a useful "units calculator" called the Unit Conversion Tool, brings up the Unit Conversion Tool menu, shown in the following illustration.

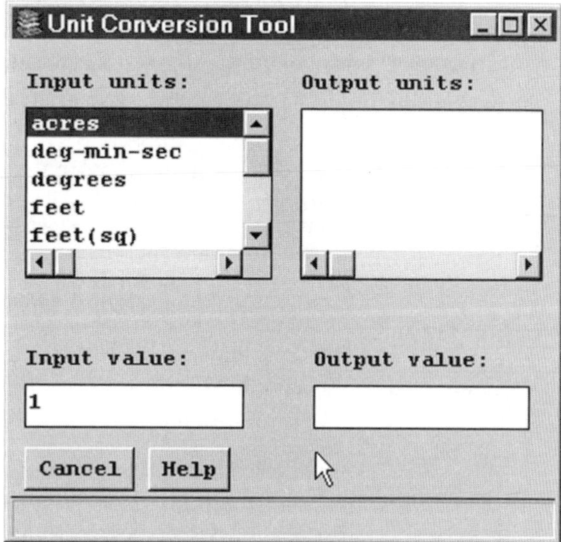

Unit Conversion Tool menu.

The Unit Conversion Tool menu lets you quickly and easily convert map units: inches to meters, hectares to acres, miles to kilometers, and so on. The final selection on the primary Command Tools menu is Help. Use it to access both descriptions of the Command Tools menu choices and ArcInfo's on-line help.

The Secondary Command Tools Menu: Geoprocessing Tools

Geoprocessing tools allow you to manipulate and analyze spatial (geographic) data. The secondary Command Tools menu, shown in the following illustration, contains the bulk of these analysis and spatial data processing tools. Its three pull-down menu are each organized around a particular category.

Command tools for geoprocessing.

The three categories are Edit, Analysis, and Conversion. Each category contains many commands. (For detailed information on any one command, see Chapter 4 or the on-line help.) The following sections provide general descriptions of the commands available in

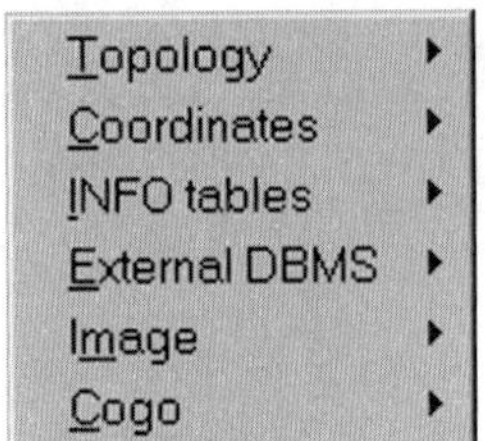

Edit Tools menu.

each category. Again, you may find it helpful to follow along on-screen. Begin by examining the Edit tools. The illustration at left shows the options available under the Edit Tools menu of the secondary Command Tools window.

The Edit Tools menu contains commands grouped around six task headings. These task headings are described in the section that follows.

Command Groups Under the Edit Tools Menu

The following task headings are found under the Edit Tools menu. The task headings are Topology, Coordinates, "INFO tables," External DBMS, Image, and Cogo.

Topology: Under the Topology heading you will find several commands for creating and modifying coverages' topology. Commonly used commands such as Build and Clean (which create topology) and Eliminate and Dissolve (which remove polygons) can be found here. You can also perform various other functions, such as adding polygon labels, creating regions, and renumbering nodes.

Coordinates: The Coordinates commands are used to perform operations on coverages' coordinates. For example, you can convert a coverage from one projection to another.

INFO tables: These commands access your INFO databases. You can add and delete items, modify their values, create new files, and even add values from a text file (ASCII) to an INFO file.

External DBMS: This set of commands allows you to access an external SQL-based database. Here you can establish the connection, list tables, and issue commands to external databases.

Images: Use these commands to rectify and register an image from a satellite or aerial photo for display with coverages.

Cogo: This grouping contains commands used to enter spatial data using surveyor's measurements. COGO is a separately purchased module from ESRI and is thus beyond the scope of this book.

Command Groups Under the Analysis Tools Menu

Next, follow along with the descriptions of the Analysis Tools. The real power of ArcInfo lies in its analytical capabilities, and the

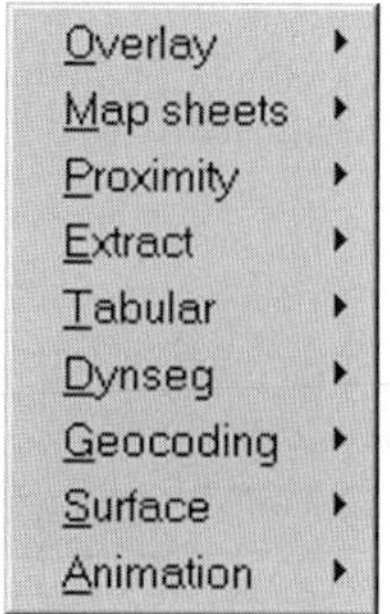

Analysis Tools in ArcMap.

options under the Analysis Tools menu make accessing these capabilities easier. The menu contains nine groups of tools, shown in the illustration at left, each with a particular focus. These are described in the material that follows.

Overlay: This group contains spatial overlay commands, such as Identity, Intersect, and Union. Each of these combines two coverages to create a third coverage containing information about overlapping features. Also included are commands such as Pointnode, which transfers feature attribute information from one coverage to another.

Map sheets: This group contains commands used to join coverages (Append) or to eliminate areas in one coverage based on the features of another (Clip and Erase). The commands available here are Append, Clip, Erase, Mapjoin, and Split.

Proximity: Use this group of commands to calculate the proximity of features. Included commands are Buffer, Near (computes point distances), and Thiessen (converts point coverage to a polygon coverage).

Extract: These commands extract features from one or more coverages to create a new coverage. Included commands are Reselect (named Select in the menu), "Area query," "Region query," and "Select region."

Tabular: These commands access some basic statistical reporting commands available in ArcInfo, such as Frequency. The "Report writer" option helps you create a report from an INFO file, and "Summary statistics" calculates means, standard deviations, and so on from an INFO file.

Dynseg: These commands are used to create and maintain events in a network coverage. Working with network coverages requires the purchase of the Network module.

Geocoding: This group contains commands for creating and maintaining address ranges from a line coverage. This also requires the purchase of the Network module.

Surface: These commands construct viewsheds, contour maps, and "lines of sight" from TIN and raster data sets. These capabilities require the purchase of the ArcInfo modules TIN and Grid.

Animation: This group contains commands that allow you to create an MPEG animation file consisting of graphic images. Once created, you can then play back the animated MPEG file. Only Sun raster files are supported.

As mentioned previously, these commands are explained in depth in the reference section of Chapter 4. The intent of this chapter is to introduce you to the pull-down menu systems available for use in Arc.

Command Groups Under the Conversion Menu

The next option available under the secondary Command Tools is the Conversion menu, shown in the illustration at left.

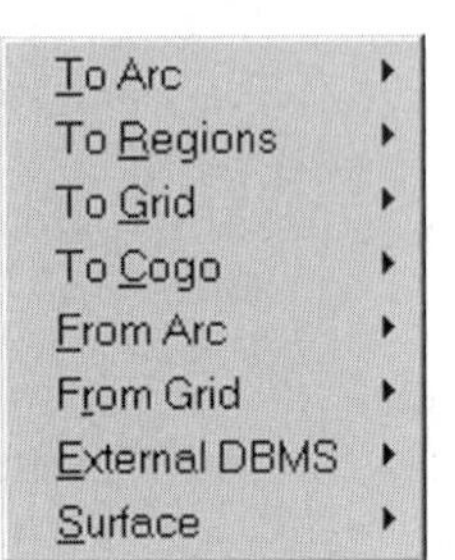

Conversion menu.

The Conversion menu contains commands for converting other file types (such as DIME, DXF, or shapefiles) to Arc coverages, and vice versa. You can also export coverages to other file formats using the Export and Ungenerate commands. The commands fall into eight groups, described in the following material.

To Arc: This group contains many commands for importing spatial data from other software formats into ArcInfo. The formats include DXF (AutoCAD Drawing files), DLG (Digital Line Graphs), and shapefiles (used in ESRI's ArcView software). A more substantial list of importable file choices can be found in ArcToolbox under Conversion Tools.

To Regions: This group contains commands for converting arc or polygon coverages to Region coverages.

To Grid: This group of commands creates grids from ArcInfo coverages, DEM, GRASS, or from an ASCII file. The Grid module, a separately purchased ArcInfo module, is required for using these tools.

To Cogo: These commands convert ArcInfo coverages to COGO coverages and can create a COGO coverage from a file of surveyor's measurements. COGO is a separately purchased module.

From Arc: Use these commands to convert ArcInfo coverages to file formats (such as DXF, DLG, and shapefiles) used by other programs. A lengthier list of conversion commands is available in ArcToolbox under Conversion Tools.

From Grid: These commands convert grids to ArcInfo coverages or to lattices. The Grid module, a separately purchased ArcInfo module, is required for using these tools.

External DBMS: Use these commands to convert INFO files to tables in an external database, and vice versa. Note that you must have an ESRI-supported SQL database.

Surface: Use these commands to convert a TIN to a lattice or a DEM to a lattice. The TIN module is required for using these commands.

The Help Option

The last option available on the secondary Command Tools menu is the Help option. From here you can access ArcTools help. To access the on-line documentation for ArcInfo, remember to select Help ➡ "ARC/INFO on-line doc" from the primary Command Tools menu.

To dismiss the Command Tools menu bars, select ArcTools ➡ Quit from the primary Command Tools menu. This returns you to the main ArcTools menu. Here, click on the Quit button to return to Arc's command line prompt.

This section has taken you through the main menu choices of the ArcTools Command Tools pull-down menus. These are the menus you use to access Arc commands. In the following section you will practice accessing and using Arc's menus and commands through a spatial analysis exercise.

Spatial Analysis Using the Menu Commands

Exercise 8-1, which follows, will help familiarize you with Arc's commands and menus. The best way to learn the menu commands is to use them. This exercise demonstrates a simple spatial analysis process.

Generally, spatial analysis involves looking for patterns among features over an area. Often such analyses require looking at the features in one coverage to see how they relate to features in another coverage. Sometimes the process also involves extracting data from one coverage to add to or create a new coverage. For this exercise you will perform such an extraction: selecting data from one coverage and creating a new coverage from it.

Exercise 8-1: Stepping Through a Spatial Analysis

Before beginning, be sure the data for this exercise has been loaded from the companion CD-ROM to your workspace (assumed to be *D:\ai80exercises\chap_8*). You should have also browsed the first section of this chapter, and you should be familiar with the basic concepts of GIS and ArcInfo Workstation (see chapters 1 and 5).

This exercise uses two coverages, *uk_eire_bdrs* (a polygon coverage) and *uk_eire_cty* (a point coverage). You may be familiar with these if you have already worked through the exercises in Chapter 7. You will use options from the Command Tools menus to extract the borders of Ireland from the *uk_eire_bdrs* coverage and to then extract cities in Ireland from the *uk_eire_cty* coverage.

Remember that the Arc module has no graphics display capability. You will have to use ArcMap, ArcEdit, or ArcPlot to graphically view the coverages. For this exercise, some command-line entry ArcPlot commands are included, but feel free to use the graphics module with which you are the most familiar.

Take a moment now to view the coverages you will be using. To do this in ArcPlot, first start ArcPlot by clicking on Start ➥ Programs ➥ ArcInfo ➥ ArcInfo Workstation ➥ ArcPlot. A blank, black background window pops up containing the copyright information and the "arcplot" command line prompt. This window is used to execute editing commands. A few moments later another blank, black background windows appears. This is your display canvas—where the graphics will appear. You can resize and move these windows to suit your working habits and needs. One way is to have the command window be long and narrow and at the bottom of the screen, with the canvas window a large rectangle above it. This is shown in the following illustration.

✓ **TIP:** *You can also start ArcPlot from the Arc prompt by typing the command* arcplot *(or simply* ap*). This also works for starting ArcEdit (*arcedit *or* ae*) and INFO (*info*).*

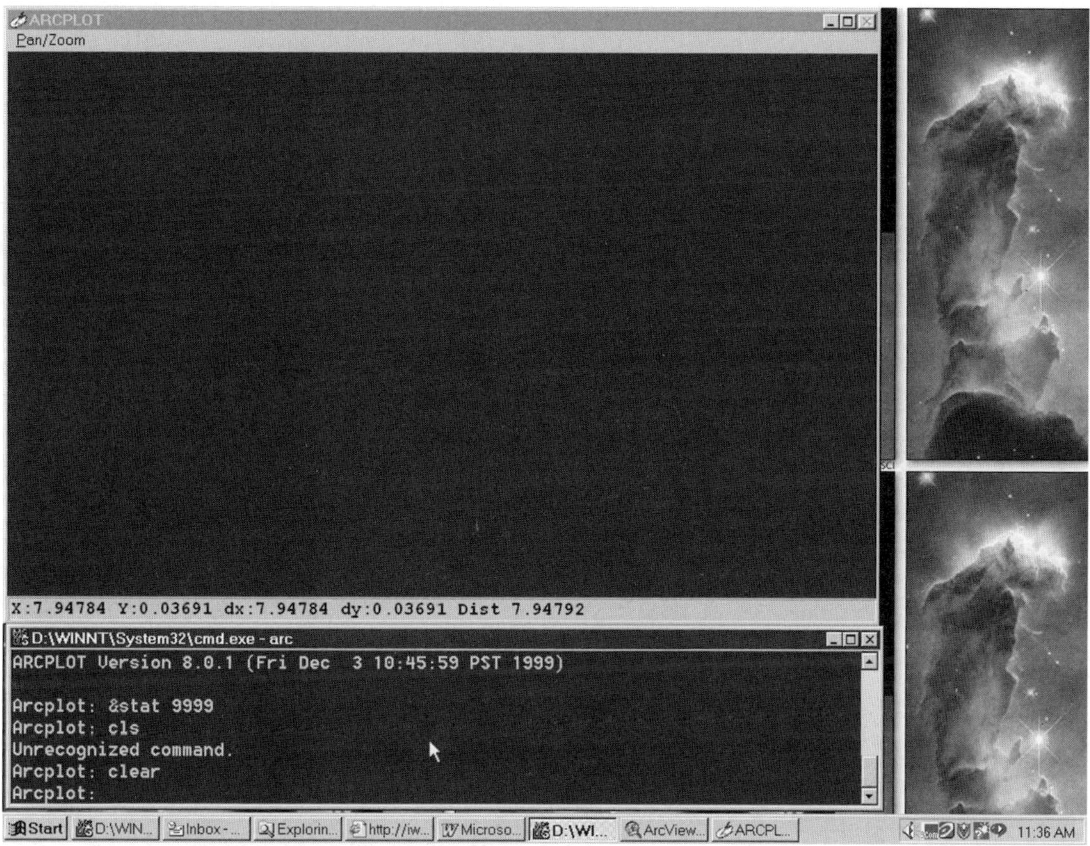

One way of arranging your ArcPlot command and canvas screens.

Now type in the following commands at the ArcPlot command and canvas screens.

```
Mapextent uk_eire_bdrs
Arcs uk_eire_bdrs
Points uk_eire_cty
Pointtext uk_eire_cty city_name
```

Your ArcPlot display canvas should look like that shown in the following illustration. Notice that the coverages include the boundaries, cities, and city names of the United Kingdom and the Republic of Ireland. Now, quit ArcPlot by typing *quit* at the ArcPlot prompt. You will return to ArcPlot later to view the results of your analysis.

The uk_eire_bdrs *coverage showing Ireland and the United Kingdom.*

Step 1: Extracting Ireland from the uk_eire_bdrs *Coverage*

Start Arc by selecting Start ➥ Programs ➥ ArcInfo ➥ ArcInfo Workstation ➥ Arc. Launch the Command Tools menus by first typing *arctools* at the Arc prompt and selecting Command Tools from the pop-up menu.

First, you will extract the polygons representing Ireland's borders from the *uk_eire_bdrs* coverage. Use the Reselect (named Select in the menu) command to do this. From the secondary Command Tools menu, click on Analysis ➥ Extract ➥ Select. The Reselect menu is shown in the following illustration.

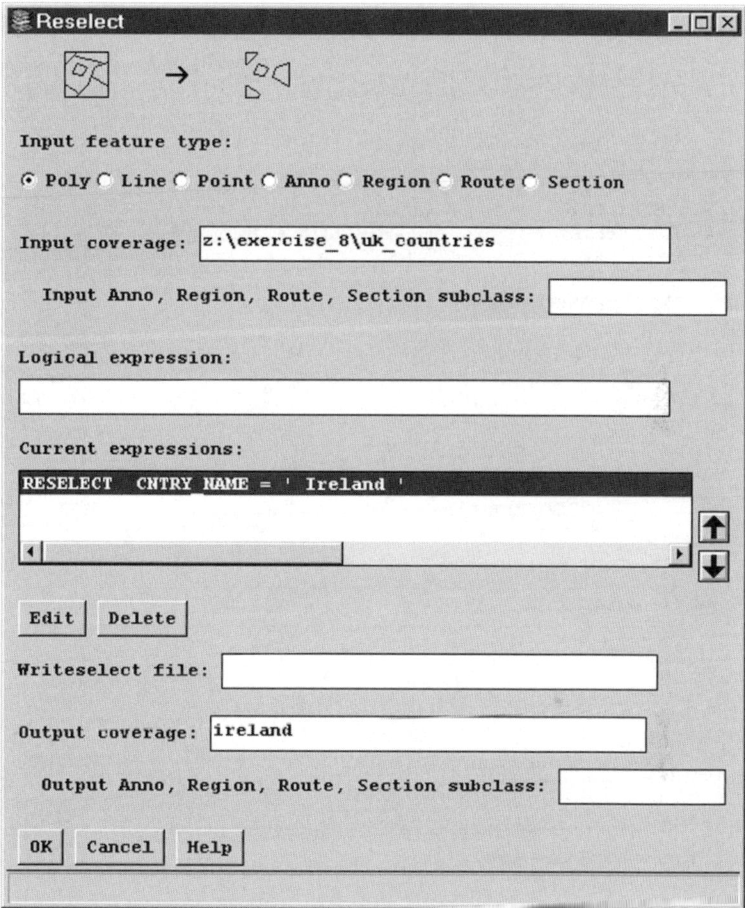

Reselect menu.

First, select the Poly button, because *uk_eire_bdrs* is a polygon coverage. Next, using the mouse, right click in the "Input coverage" text box. This brings up the "Select A Coverage – Type: POLY" window, which contains a list of polygon coverages present in the workspace. From this list, select the *uk_eire_bdrs* coverage and click on OK. Next, right click in the "Logical expression" text box to bring up the Logical Expression menu, shown in the following illustration.

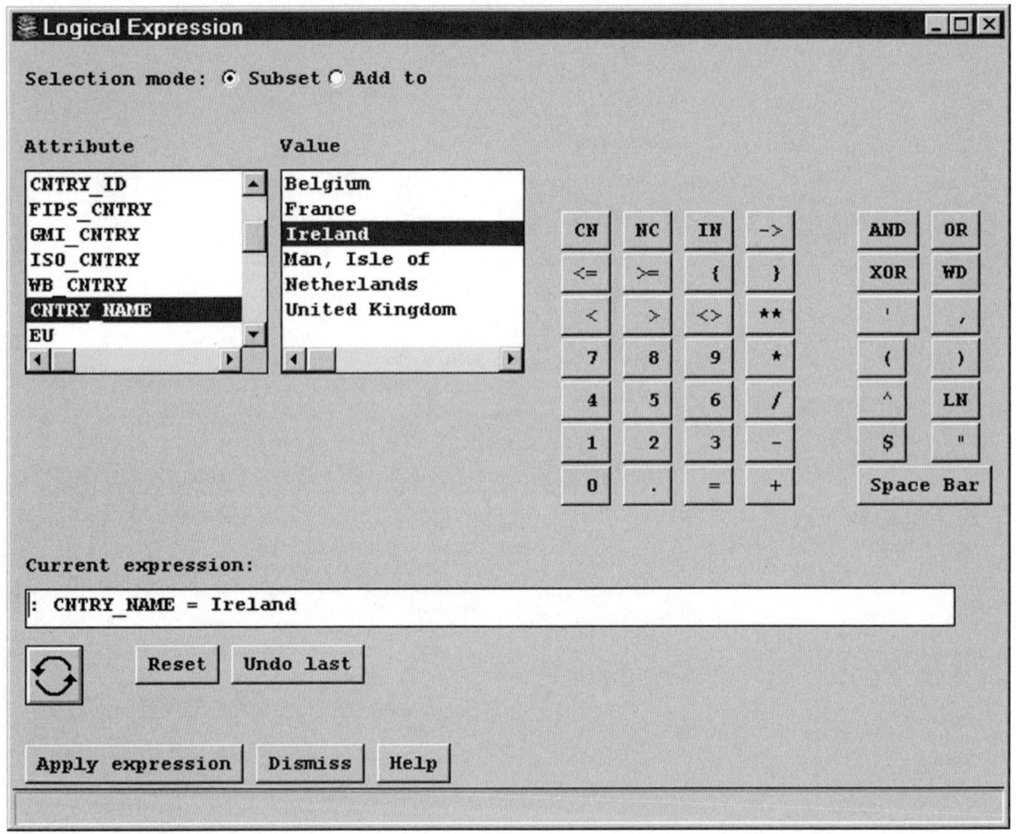

Logical Expression menu.

In the scrollable Attribute window, select the item CNTRY_NAME. The scrollable Value window now lists the country names present in the coverage. Before selecting *Ireland*, click on the "=" sign in the menu's keypad to continue forming your logical expression. Because *Ireland* is a text string, you will need to enclose it in single quotation marks. Do so by clicking on the "single quote" button (under the XOR button) on the keypad.

Now select *Ireland* in the Value window, and add a final single quotation mark. Remember, quotation marks are essential because you are asking ArcInfo to search for a text string. Finally, click on the "Apply expression" button. Dismiss this window, and you will see your newly created logical expression in the "Current expression" text box.

Now, type *Ireland* as the new coverage in the Output text box. To execute your reselection request, click on the OK button in the Reselect menu. ArcInfo will extract all polygons that have a CNTRY_NAME of "Ireland" and put them into a new coverage named Ireland. As the command executes, messages will appear in the Arc dialog box. When it is complete, the message "Reselect command tool completed" appears, shown in the following illustration.

Return to ArcPlot now to view your new Ireland polygon coverage, shown in the following illustration. As before, click on Start ➥ Programs ➥ ArcInfo ➥ ArcInfo Workstation ➥ ArcPlot. Arrange the windows to suit your preferences. At the ArcPlot prompt, type the following commands. Press <Enter> after each one.

```
Mapextent ireland
Arcs ireland
```

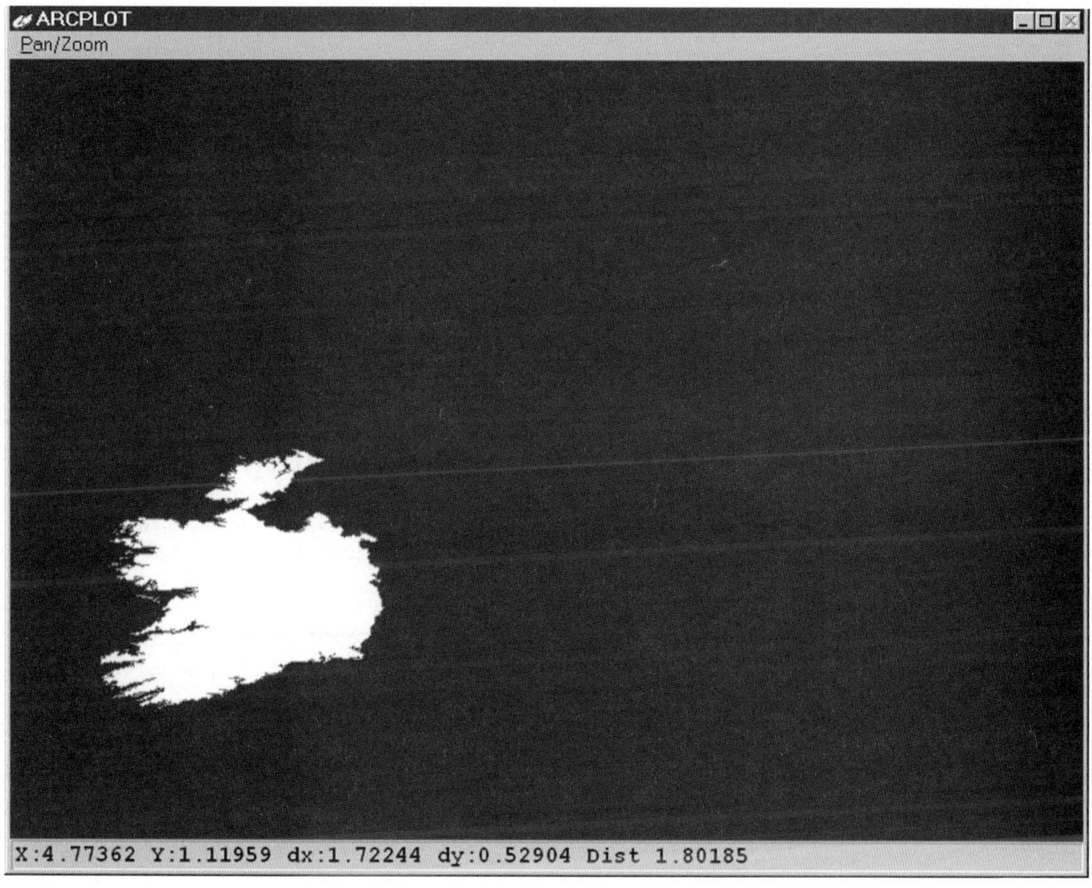

IRELAND coverage created using Reselect menu.

Step 2: Extracting Irish Cities from the uk_eire_cty Coverage

To extract Irish cities, you could use the same Reselect command as used in step 1. For this step, however, you will use the Intersect overlay command instead.

Access this command by selecting Analysis ➡ Overlay ➡ Intersect from the secondary Command Tools menu. This brings up the Intersect window, shown in the following illustration.

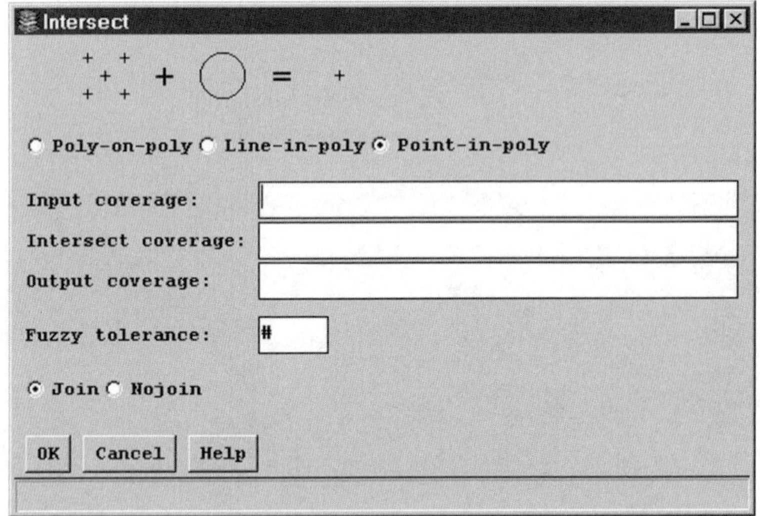

Intersect window.

Note that this window has a small graphic to demonstrate the concept behind the Intersect command. The Intersect command takes the features from one coverage that are in the area common to both coverages and creates a new coverage from these features. You will be using a polygon coverage, *Ireland*, to extract the cities from *uk_eire_cty* that are within the boundaries of *Ireland*.

First, specify that you are intersecting a point coverage and a polygon coverage. Do this by selecting the "Point-in-poly" option. Next, right click on the "Input coverage" text box to bring up the "Select A Coverage – Type: POINT" window. Select *uk_eire_cty*. Next, supply a polygon coverage that will be used to perform the intersect command. For this command, the intersect coverage must be a polygon coverage. Right click in the "Intersect coverage" text box and the "Select A Coverage –Type: POLY" window will appear. Select *ireland*.

Finally, specify the name for the new coverage by typing *irish_cities* in the "Output coverage" text box. The completed Intersect menu is shown in the following illustration.

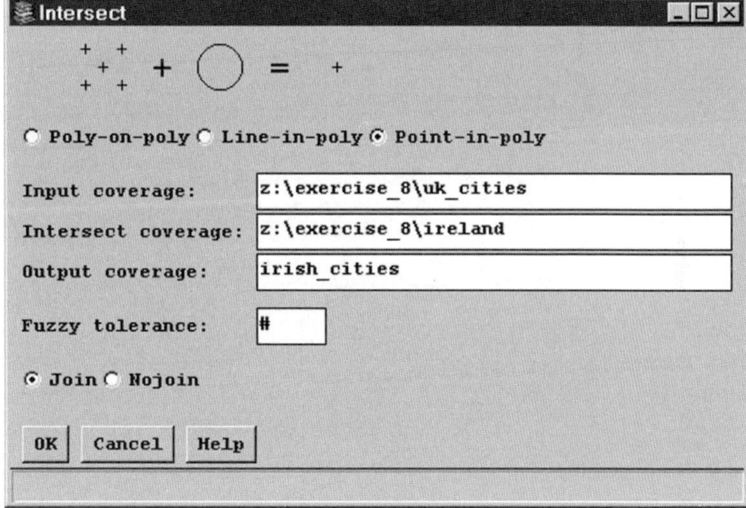

Completed Intersect menu.

Execute the Intersect command by clicking on the OK button. Once again, return to ArcPlot to view the results, shown in the following illustration. Type in the following commands, and press <Enter> after each one.

```
Mapextent ireland
Arcs ireland
Points irish_cities
Pointtext irish_cities city_name
```

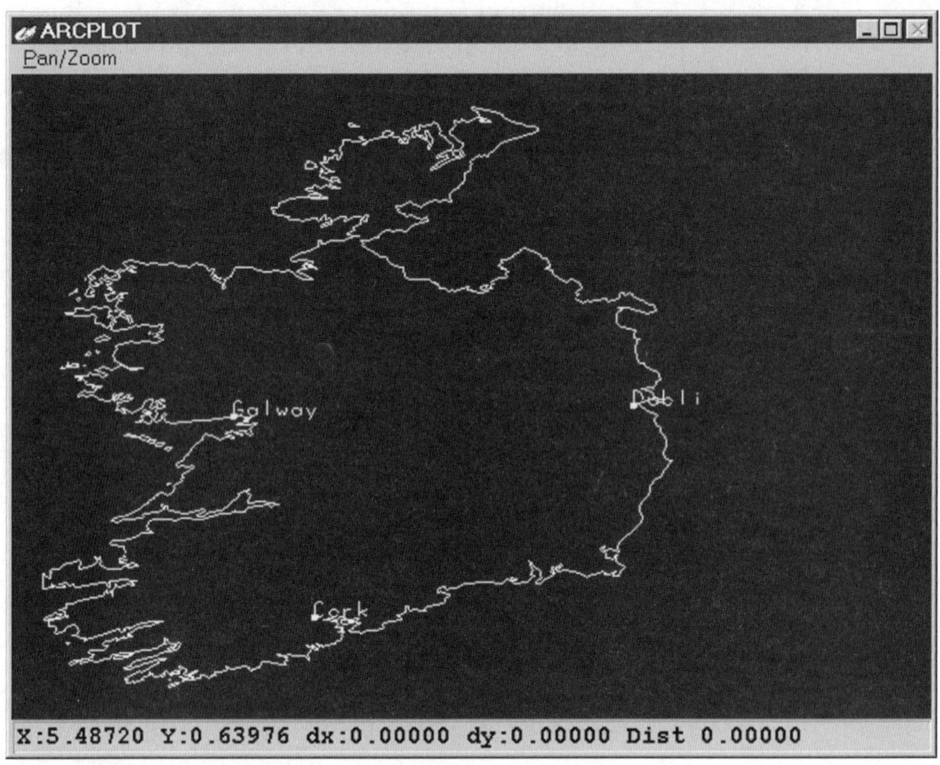

Result of irish_cities *intersection as IRELAND coverage.*

Using the polygon boundaries of the IRELAND coverage, the three Irish cities in the *uk_eire_cty* coverage have been extracted and put into their own coverage.

This exercise has shown you two ways of creating a new coverage by using a simple spatial analysis. Using the Command Tools pull-down menus, you extracted data from two coverages, *uk_eire_bdrs* and *uk_eire_cty*, and created two new coverages, *ireland* and *irish_cities*, respectively. You also viewed your newly created coverages.

Summary

Arc is one of ArcInfo's traditional modules. As such, its commands can be accessed using both command line entry and pull-down menus. This chapter introduced you to the primary and secondary

Command Tools menus of ArcTools. Remember that these menus can be incorporated into custom AML programs.

ArcInfo's spatial analysis and geoprocessing power originally resided in the Arc module, but with Version 8.0 much of Arc's functionality has been replaced by ArcToolbox and ArcCatalog. Users will find the newer interfaces easier to understand and use, and they are encouraged to rely on the new modules as their primary spatial analysis and data management tools.

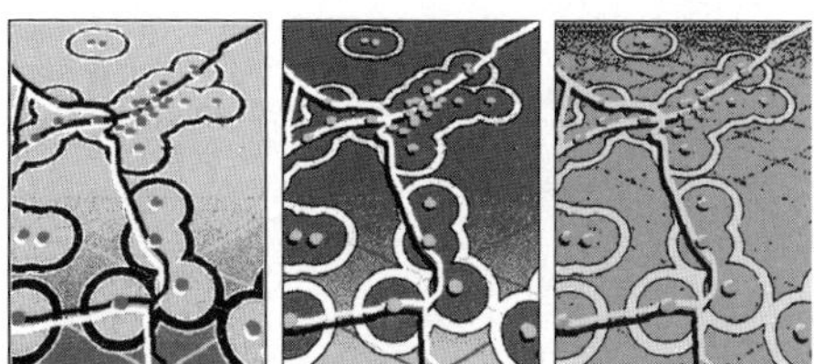

Chapter 9

Managing Data Using INFO

An Introduction to the INFO Database

INFO is ArcInfo's built-in database management software. A powerful relational database, INFO provides the information component to Arc's geographic component. Its interface may lack the sophistication common to today's databases, but INFO's simplicity and power more than compensate. Although ArcInfo works with a number of databases (such as Oracle, Sybase, Access, and Informix), many users rely on INFO as their primary GIS database.

INFO uses what is today considered nonstandard terminology in its database. The advent of SQL (Structured Query Language) as an independent database language, used by many database software programs, has made some of INFO's terminology obsolete. Table 9-1, which follows, provides a quick rundown of terms used by INFO and their corresponding SQL equivalents. The terms used in this chapter are INFO terms.

Table 9-1: INFO Terms and SQL Equivalents

INFO	*SQL*	*What It Is*
Table	Table	A database
Item	Column	The vertical column in a database
Record	Row	The horizontal row in a database
Item value	Datum	A value in an item (or column)
Relate	Join	Relating two or more tables

Other relational databases function much like INFO files. If your GIS accesses files from a third-party database program such as Oracle, keep these two points in mind: those database files are stored separately from INFO files, and most databases use the SQL standard programming language previously mentioned, whereas INFO does not.

This chapter covers the basics of working with INFO databases. The following are the topics covered.

- Attribute data
- Starting and quitting INFO
- Adding, deleting, and modifying records
- Arc commands that modify databases
- Creating and deleting databases
- Relating data files
- Creating reports and printing

The exercises in this chapter pretend that you have been hired to maintain data files and ArcInfo coverages about sales representatives and sales figures for National Widgets, a distributor of imported widgets. Your company maintains its data in INFO data files, and you must learn the basics about creating data files; adding, changing, and deleting data records; and creating custom reports.

Attribute Data

ArcInfo stores attribute data in two ways: in the feature attribute tables of coverages and as INFO database files. Coverages contain geographic data (points, lines, polygons) that represent features on the face of the earth, and they also can contain attribute data to better describe those features. For example, a coverage of streets will contain arcs representing real-life streets. These streets are each linked to data (street name, route number, and so on) in an Arc Attribute Table (signified by the *.AAT* suffix) describing them.

INFO files are database files that contain non-geographic information. They are not part of a coverage but may, however, be used in conjunction with coverages. They can hold additional descriptive information about the features in a coverage. This provides an efficient means of accessing and storing a lot of information. As long as a coverage's feature attribute table has one item in common with an INFO file, the information in that INFO file can be related to the coverage features.

This is the power of relational databases. Many coverages can "share" information in a single INFO file. Without the relational capability, that same information would have to be stored in each coverage. Similarly, INFO files can also be linked to other INFO files. For example, A nonprofit group could have a polygon coverage, named "habitat," of endangered species habitats. The *habitat.pat* table contains two items: the habitat's name and the U.S. state it is in. The nonprofit group also has a database file of members' addresses, including the state. By linking the two database files on the item "state," information pertaining to the habitats can be sent to members in that state.

The street addresses of members could be linked to a file containing Congressional district information, thus producing letters asking the group's members to petition their particular representatives, and providing the correct representative addresses. These powerful processes of relating and joining files are discussed in more detail later in this chapter. For now, use exercise 9-1, which follows, to get a feel for INFO's interface.

Exercise 9-1: Exploring INFO

This exercise gets you used to starting INFO and introduces a few basic INFO commands. You will start INFO from the Arc prompt, view which data files are in a directory, open and list a file, and quit INFO.

Step 1: Starting Arc and Info

Unlike the other modules in ArcInfo, INFO cannot start independently. You must first start Arc. Do this by clicking on Start ➡ Programs ➡ ArcInfo ➡ ArcInfo Workstation ➡ Arc. Change to the directory in which you have copied the sample data from the companion CD-ROM. This exercise assumes that the data has been copied to a folder named *D:\ai80exercises\chap_9*. To change to that directory, type the following command at the Arc prompt.

```
d:\ai80exercises\chap_9
```

✓ **TIP:** *Remember to use the "w" command to change workspaces when using ArcInfo.*

Now, list any coverages that may be in this workspace, by typing *lc* at the Arc prompt ("lc" is the abbreviation for the "listcoverages" command). Do not worry what the coverage looks like, because this chapter is principally concerned with data files.

You will see that there is one coverage, named SALES, present. Now, start INFO. At the Arc command prompt type *info*. The screen shown in the following illustration appears.

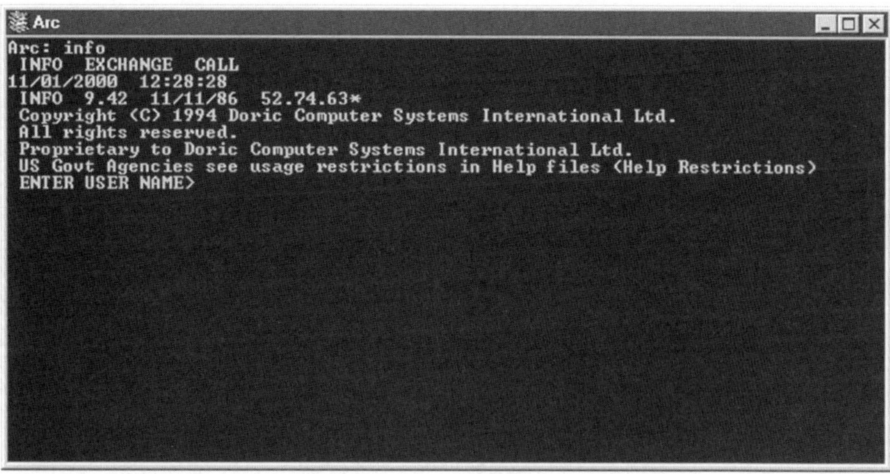

Starting an INFO session.

After starting INFO, you will always be prompted to enter a user name. This is a throwback to INFO's very early days. All you need to know is that for ArcInfo users the USERNAME is always ARC.

NOTE: *INFO is case sensitive. It is common convention that everything in INFO be typed using capital letters. Make life easier on yourself by adopting this convention.*

Now type *ARC* and press <Enter>. Your screen should look like that shown in the following illustration. Notice that the command line prompt for INFO is ENTER COMMAND >.

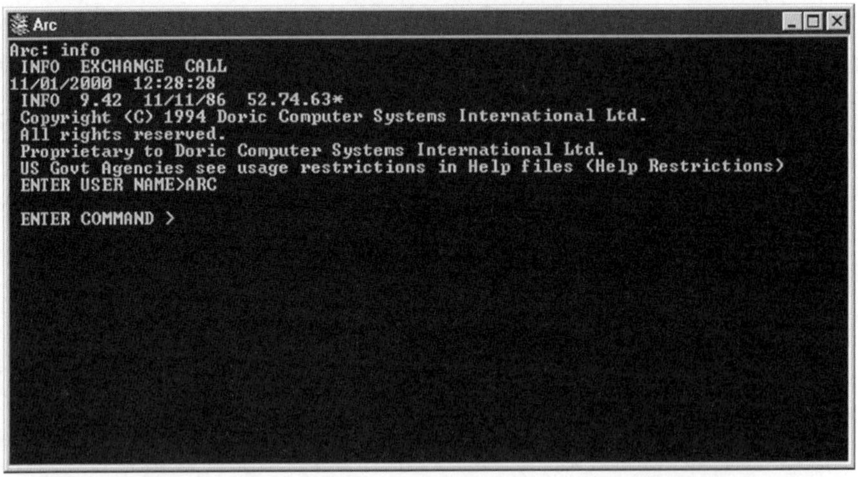

```
Arc
Arc: info
 INFO  EXCHANGE  CALL
11/01/2000  12:28:28
 INFO  9.42  11/11/86  52.74.63*
 Copyright (C) 1994 Doric Computer Systems International Ltd.
 All rights reserved.
 Proprietary to Doric Computer Systems International Ltd.
 US Govt Agencies see usage restrictions in Help files (Help Restrictions)
 ENTER USER NAME>ARC

 ENTER COMMAND >
```

Beginning INFO window.

✓ **TIP:** *Use capital letters for everything you do in INFO.*

Step 2: Listing Files in the INFO Directory

Now, type *DIR* (an abbreviation for DIRECTORY) to see a list of files in this directory. The results are shown in the following illustration.

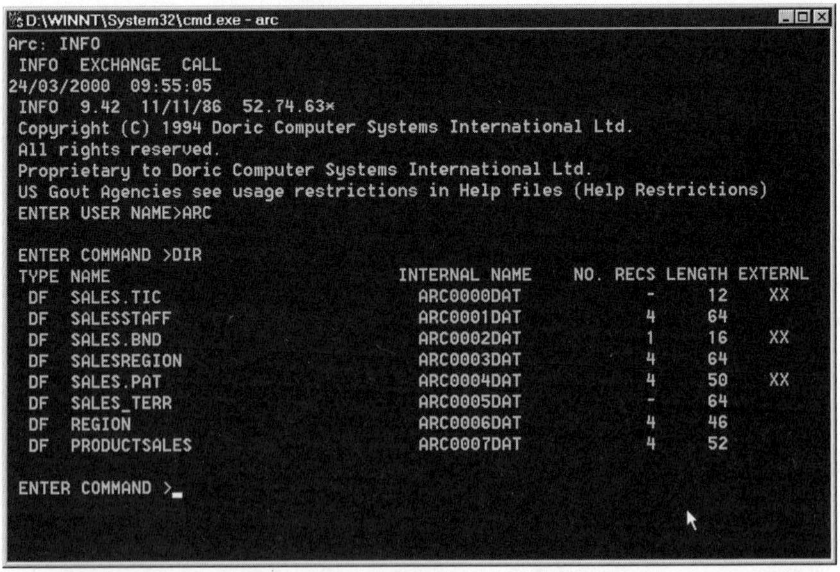

```
D:\WINNT\System32\cmd.exe - arc
Arc: INFO
 INFO  EXCHANGE  CALL
24/03/2000  09:55:05
 INFO  9.42  11/11/86  52.74.63*
 Copyright (C) 1994 Doric Computer Systems International Ltd.
 All rights reserved.
 Proprietary to Doric Computer Systems International Ltd.
 US Govt Agencies see usage restrictions in Help files (Help Restrictions)
 ENTER USER NAME>ARC

 ENTER COMMAND >DIR
 TYPE NAME                              INTERNAL NAME    NO. RECS LENGTH EXTERNL
  DF  SALES.TIC                          ARC0000DAT             -     12     XX
  DF  SALESSTAFF                         ARC0001DAT             4     64
  DF  SALES.BND                          ARC0002DAT             1     16     XX
  DF  SALESREGION                        ARC0003DAT             4     64
  DF  SALES.PAT                          ARC0004DAT             4     50     XX
  DF  SALES_TERR                         ARC0005DAT             -     64
  DF  REGION                             ARC0006DAT             4     46
  DF  PRODUCTSALES                       ARC0007DAT             4     52

 ENTER COMMAND >_
```

Using the DIR command to display files in INFO.

Notice that there are eight files present. There are three files associated with the SALES coverage: *SALES.TIC*, *SALES.BND*, and *SALES.PAT*.

These files are important to ArcInfo because they contain topology information. The other five are INFO files that contain tabular data.

Step 3: Selecting One File and Listing Its Items

Next, you will select the *SALES.PAT* file and display its items on the screen.

✓ **TIP:** *Unlike some other database software, INFO only allows one database to be open at a time.*

Select the file by using the SELECT command followed by the file name. This command can be abbreviated as SEL. Then list the file's items using the ITEMS command. Type the commands in as follows.

```
ENTER COMMAND > SEL SALES.PAT

ENTER COMMAND > ITEMS
```

✓ **TIP:** *In INFO, the SELECT command is used only for specifying which data file you want to view or modify. The RESELECT command (abbreviated as RESEL) is used to select records within the selected file.*

The results of the previous two commands are shown in the following illustration.

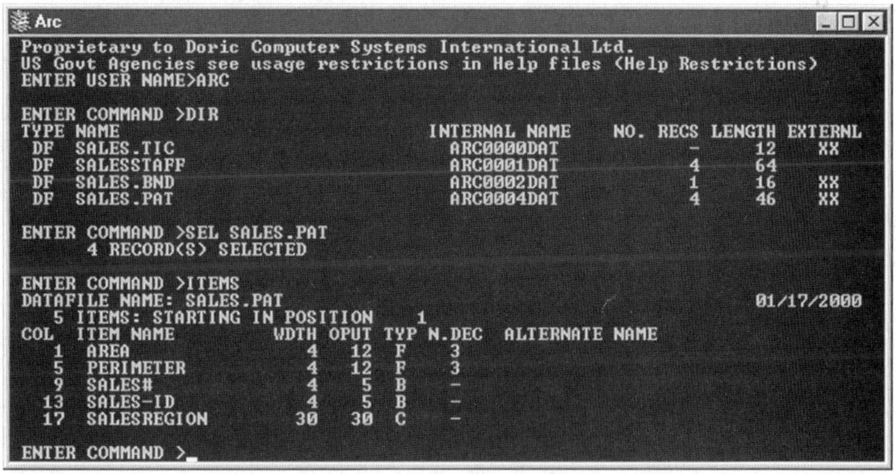

```
Arc
Proprietary to Doric Computer Systems International Ltd.
US Govt Agencies see usage restrictions in Help files (Help Restrictions)
ENTER USER NAME>ARC

ENTER COMMAND >DIR
TYPE NAME                          INTERNAL NAME    NO. RECS LENGTH EXTERNL
 DF  SALES.TIC                       ARC0000DAT           -      12     XX
 DF  SALESSTAFF                      ARC0001DAT           4      64
 DF  SALES.BND                       ARC0002DAT           1      16     XX
 DF  SALES.PAT                       ARC0004DAT           4      46     XX

ENTER COMMAND >SEL SALES.PAT
      4 RECORD(S) SELECTED

ENTER COMMAND >ITEMS
DATAFILE NAME: SALES.PAT                                            01/17/2000
   5 ITEMS: STARTING IN POSITION    1
COL  ITEM NAME          WDTH OPUT TYP N.DEC  ALTERNATE NAME
   1  AREA                 4   12  F     3
   5  PERIMETER            4   12  F     3
   9  SALES#               4    5  B     -
  13  SALES-ID             4    5  B     -
  17  SALESREGION         30   30  C     -

ENTER COMMAND >_
```

Selecting a datafile and listing its items.

Notice that *SALES.PAT* contains five items. As you may recall from Chapter 1, the first four items in the *.PAT* file (in this example, area, perimeter, sales#, and sales-id) are used exclusively by ArcInfo and cannot be changed by the user. You will destroy your coverage if you modify any of those four items.

☛ **WARNING:** *Never change the area, perimeter, covername#, or covername-id items in a coverage's* .PAT *file. Doing so will destroy your coverage.*

The fifth item, REGION, is a descriptive item for each polygon in this coverage. In this example, every polygon is a record. To view complete information on every record in the data file, type *LIST.* Each record, and every item in that record, will scroll across the screen. The prompt "More?" appears at the bottom of your screen. Type *Y* or press <Enter> to continue scrolling; type *N* to stop. The illustration that follows shows part of the content of the SALES.PAT database.

```
Arc
ENTER COMMAND >LIST
               1
AREA           =       -1.526
PERIMETER      =        6.171
SALES#         =    1
SALES-ID       =    3
SALESREGION    =C
               2
AREA           =        0.593
PERIMETER      =        3.378
SALES#         =    2
SALES-ID       =    2
SALESREGION    =B
               3
AREA           =        0.628
PERIMETER      =        3.038
SALES#         =    3
SALES-ID       =    1
SALESREGION    =A
               4
AREA           =        0.305
PERIMETER      =        2.192
SALES#         =    4
SALES-ID       =    4
MORE?_
```

Listing the content of the SALES.PAT database.

As you can see, INFO incorporates easy-to-use commands for selecting files, viewing the items in a file, and listing records on the screen. The next section discusses adding, modifying, and deleting records. First, however, you must end your INFO session.

Ending an INFO session is similar to starting one: you need two commands. When you are ready to quit, type *QUIT.* This gets you back to the ENTER USER NAME prompt. Again, this is a throwback to INFO's early days. Theoretically you could quit working under one user name and then begin working under another without leaving INFO. Now type *STOP*. The STOP command actually exits INFO and puts you back at the Arc prompt. These two commands are abbreviated as the following command: Q STOP. Typing *Q STOP* at the ENTER USER NAME prompt immediately returns you to the Arc prompt.

INFO's basic commands for adding, deleting, and modifying records are also easy to use. Exercise 9-2 explores how to perform these operations, using the *SALESSTAFF* file.

Exercise 9-2: Adding, Deleting, and Modifying Records

In this exercise, you are asked to update data in National Widget's data file of sales representatives. A sales person has left the company and two new ones have been hired. Use the INFO commands presented in this exercise to update the database.

Using the same steps as in exercise 9-1, start Arc and go to the folder where your Chapter 9 exercises are stored. Now start INFO. Open the *SALESSTAFF* file by typing the following command.

```
SEL SALESTAFF
```

Remember that SEL is short for SELECT. You will add two records to this file. They are listed in table 9-2, which follows.

Table 9-2: Records to Be Added

Sales Person	*Territory*	*Annual Sales*
Lasagna	B	23,456.90
Howard	R	14,789.00

Step 1: Entering Record Data

Using the ADD command, you will be prompted to enter information for each new record. Examine the screen shown in the following illustration to see how information for the two records is entered.

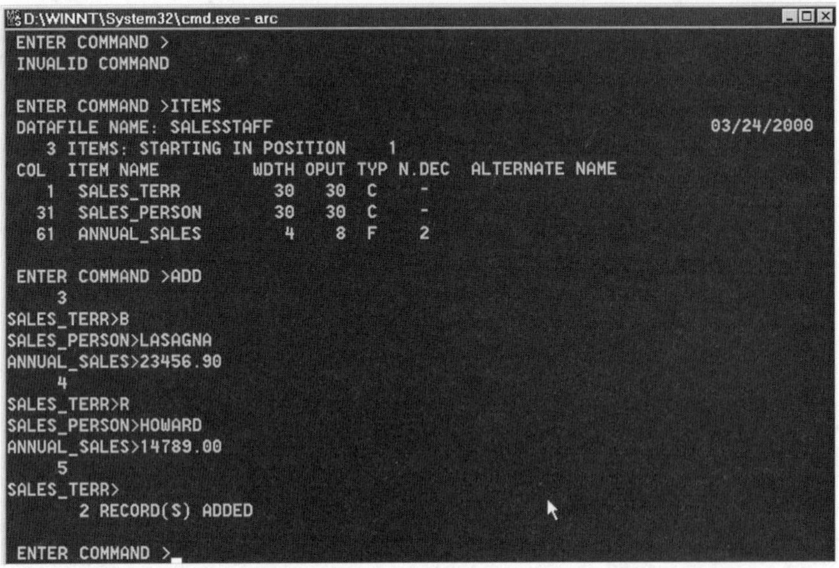

```
D:\WINNT\System32\cmd.exe - arc
ENTER COMMAND >
INVALID COMMAND

ENTER COMMAND >ITEMS
DATAFILE NAME: SALESSTAFF                                           03/24/2000
   3 ITEMS: STARTING IN POSITION     1
COL  ITEM NAME          WDTH OPUT TYP N.DEC  ALTERNATE NAME
   1  SALES_TERR          30   30  C    -
  31  SALES_PERSON        30   30  C    -
  61  ANNUAL_SALES         4    8  F    2

ENTER COMMAND >ADD
    3
SALES_TERR>B
SALES_PERSON>LASAGNA
ANNUAL_SALES>23456.90
    4
SALES_TERR>R
SALES_PERSON>HOWARD
ANNUAL_SALES>14789.00
    5
SALES_TERR>
      2 RECORD(S) ADDED

ENTER COMMAND >_
```

Adding two records to the SALESSTAFF file.

Now type *ADD* at the ENTER COMMAND prompt. Follow the on-screen prompts and use the previous illustration to enter the two records. Once entered, pressing the <Enter> key once when prompted to enter data again will tell INFO that the data entry session is complete. Now use the LIST command to examine the data just entered, as shown in the following illustration.

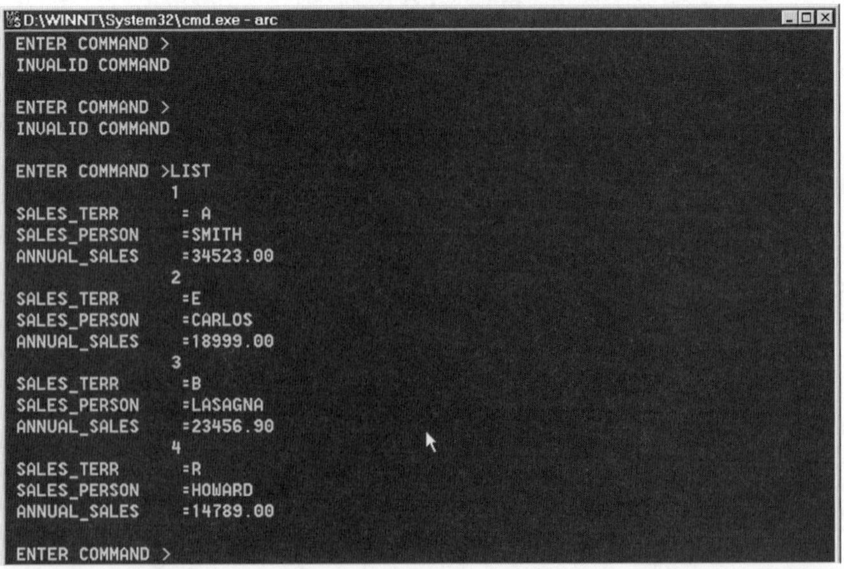

```
D:\WINNT\System32\cmd.exe - arc
ENTER COMMAND >
INVALID COMMAND

ENTER COMMAND >
INVALID COMMAND

ENTER COMMAND >LIST
               1
SALES_TERR      = A
SALES_PERSON    =SMITH
ANNUAL_SALES    =34523.00
               2
SALES_TERR      =E
SALES_PERSON    =CARLOS
ANNUAL_SALES    =18999.00
               3
SALES_TERR      =B
SALES_PERSON    =LASAGNA
ANNUAL_SALES    =23456.90
               4
SALES_TERR      =R
SALES_PERSON    =HOWARD
ANNUAL_SALES    =14789.00

ENTER COMMAND >
```

Using LIST to view records.

Step 2: Deleting a Record

Next, you will delete a record. Say that responsibility for Region E in the SALESSTAFF database has been shifted to another sales group, so that record must be deleted from your group's copy of the database.

Use the RESELECT command (abbreviated as RESEL) to select the record whose SALES_TERR item equals E. Type the following .

```
ENTER COMMAND > RESEL SALES_TERR = 'E'
```

Notice that the "E" is in single quotes. This tells INFO to search for a string of characters in the item SALES_TERR named "E." INFO will report that one record is now selected. Use the LIST command to verify that it is the record that contains E, as shown in the following illustration.

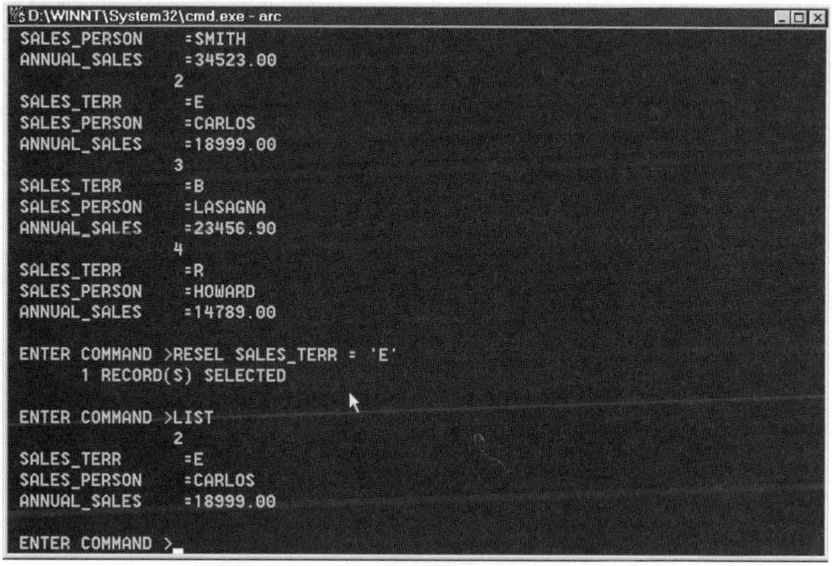

```
D:\WINNT\System32\cmd.exe - arc
SALES_PERSON     =SMITH
ANNUAL_SALES     =34523.00
                2
SALES_TERR       =E
SALES_PERSON     =CARLOS
ANNUAL_SALES     =18999.00
                3
SALES_TERR       =B
SALES_PERSON     =LASAGNA
ANNUAL_SALES     =23456.90
                4
SALES_TERR       =R
SALES_PERSON     =HOWARD
ANNUAL_SALES     =14789.00

ENTER COMMAND >RESEL SALES_TERR = 'E'
      1 RECORD(S) SELECTED

ENTER COMMAND >LIST
                2
SALES_TERR       =E
SALES_PERSON     =CARLOS
ANNUAL_SALES     =18999.00

ENTER COMMAND >_
```

One selected record in the SALESSTAFF database.

Now delete this record by using the PURGE command. Type the following.

```
ENTER COMMAND > PURGE
```

You will be prompted to verify that you want this record removed. This is a handy safeguard that makes you look twice at the record or records you are about to delete. Take advantage of it, and be sure you want to delete the selected records. For this exercise, type *Y*.

✓ **TIP:** *When deleting INFO file records, it is best to first make the deletions to a copy of the file. That way, important data will not be lost by mistake. When the edits are complete, delete the original and rename the copy.*

The selected record is now deleted and the *SALESSTAFF* file now has four records in it.

☛ **WARNING:** *Using the PURGE command in an AML (Arc Macro Language) program or any other program will cause the records to be deleted without prompting the user for verification. Be cautious when using PURGE in programs.*

Step 3: Modifying a Record

Now you will modify a record in an INFO file. Use the LIST command to view the current content of *SALESSTAFF*. Notice that the salesperson for region A is named Smith. She has recently left your organization and her responsibilities have been transferred to a Ms. Jones. You will change that record to reflect the new salesperson. To do this, use the UPDATE command followed by the item name.

```
ENTER COMMAND > UPDATE SALES_PERSON
```

INFO now prompts you for the record number you want to update. The record number is just the sequential number of all records in the database, starting at 1 and going, in this case, to 4. You can find the RECNO by reselecting for and listing a record. INFO displays the following prompt.

```
RECNO?>
```

Type *1* as a response and you should see the screen shown in the following illustration.

Updating a record in INFO.

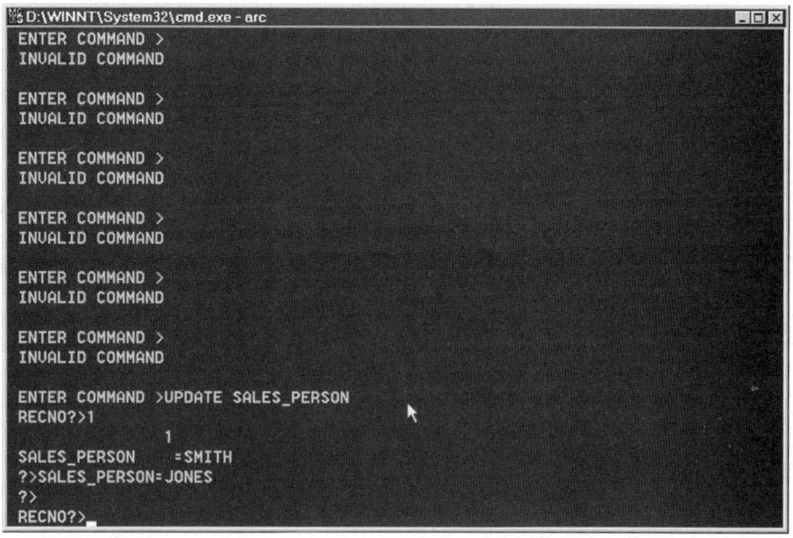

```
D:\WINNT\System32\cmd.exe - arc
ENTER COMMAND >
INVALID COMMAND

ENTER COMMAND >
INVALID COMMAND

ENTER COMMAND >
INVALID COMMAND

ENTER COMMAND >
INVALID COMMAND

ENTER COMMAND >
INVALID COMMAND

ENTER COMMAND >
INVALID COMMAND

ENTER COMMAND >UPDATE SALES_PERSON
RECNO?>1
                  1
SALES_PERSON     =SMITH
?>SALES_PERSON=JONES
?>
RECNO?>
```

Notice that INFO displays the current value stored in the SALES_PERSON field. To change this value, type the following.

```
SALES_PERSON = JONES
```

Press the <Enter> key again to stop the updating process. Now use the list command, and you will see that the file has been amended to reflect the new salesperson for Region A. The following illustration shows the updated file.

The updated SALESSTAFF file.

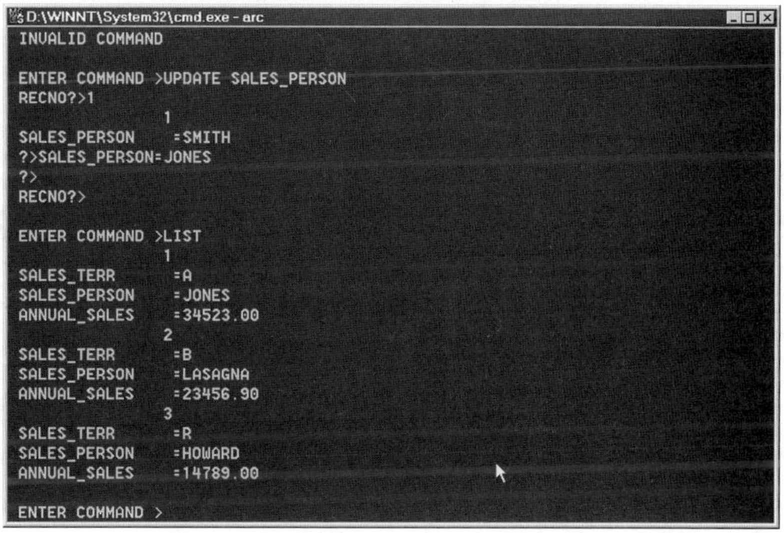

```
D:\WINNT\System32\cmd.exe - arc
INVALID COMMAND

ENTER COMMAND >UPDATE SALES_PERSON
RECNO?>1
                  1
SALES_PERSON     =SMITH
?>SALES_PERSON=JONES
?>
RECNO?>

ENTER COMMAND >LIST
                  1
SALES_TERR       =A
SALES_PERSON     =JONES
ANNUAL_SALES     =34523.00
                  2
SALES_TERR       =B
SALES_PERSON     =LASAGNA
ANNUAL_SALES     =23456.90
                  3
SALES_TERR       =R
SALES_PERSON     =HOWARD
ANNUAL_SALES     =14789.00

ENTER COMMAND >
```

With this exercise you have learned to add, delete, and modify a record. These are all INFO commands. The next section introduces you to adding and deleting items. You may be surprised to learn that the commands for doing this are actually executed at the Arc prompt.

Arc Commands That Modify INFO Files

You have seen that selecting data files and editing their records in INFO is fairly simple. Adding or deleting items to INFO files is also simple, but these operations are executed at the Arc prompt instead of at INFO's "ENTER COMMAND >" prompt. You can modify an existing file, or add or delete items to it and then save it as another file. In a way, these commands allow your database to grow as your organization's projects grow. This section describes how to modify database files using those Arc commands.

Adding Items

Arc's Additem command can be used on all INFO database files, including feature attribute tables (*.AAT*, *.PAT*, and so on). The syntax for the command is shown in the following illustration.

✓ **TIP:** *Typing a command name and pressing <Enter> gives you the command's syntax.*

```
Arc
Arc: additem
Usage: ADDITEM <in_info_file> <out_info_file> <item_name> <item_width>
        <output_width> <item_type> <decimal_places> <start_item>
Arc: _
```

Syntax for the Additem command.

The arguments for the command are straightforward. You must supply the name of the INFO file to which you will add an item; a name for an output file (it can have the same name as the original file) that contains all original items and data plus the new item's name; the new item name, width (maximum number of characters or numbers), and type (character, binary, and so on); and the item you want to be placed after in the file. Consider the example discussed in the material that follows.

The command Items at the Arc prompt works just like the ITEMS command in INFO; it displays on the screen the items in an INFO file. Use it to display the items in the file named *SALESREGION* by typing *items SALESREGION* at the Arc prompt, as shown in the following illustration.

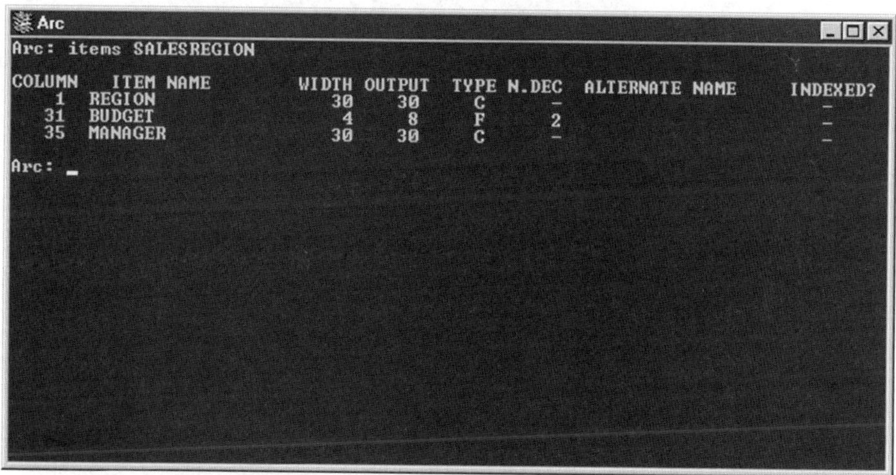

Listing items in SALESREGION *file.*

Now add a new item named POPULATION and place it after the item REGION. Do this by using the Additem command. For this example, type the following.

```
Arc: additem salesregion salesregion population 4 5 B # region
```

The following illustration shows the executed Additem command and the resulting change to the file *SALESREGION*.

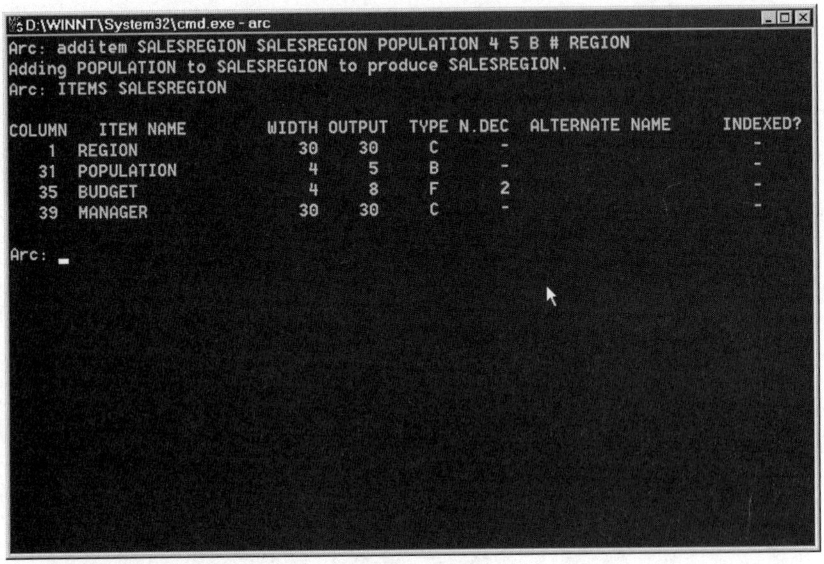

```
D:\WINNT\System32\cmd.exe - arc
Arc: additem SALESREGION SALESREGION POPULATION 4 5 B # REGION
Adding POPULATION to SALESREGION to produce SALESREGION.
Arc: ITEMS SALESREGION

COLUMN   ITEM NAME          WIDTH OUTPUT  TYPE N.DEC  ALTERNATE NAME     INDEXED?
     1  REGION                 30     30     C     -                         -
    31  POPULATION              4      5     B     -                         -
    35  BUDGET                  4      8     F     2                         -
    39  MANAGER                30     30     C     -                         -

Arc: _
```

Results of Additem command on the SALESREGION *file.*

Notice one thing in particular about this example. Because POPULATION is a binary item and thus has no numbers to the right of the decimal place (binary items only store integer numbers), you must use a pound sign (#) in the command's parameters where number of decimal places would be specified. The # acts as a placeholder and tells ArcInfo to skip it and move on to the next parameter, which in this case is "starting item."

Deleting Items

Deleting an unwanted item in a file can be done using the Dropitem command. This command requires that you supply the name of the file to use, the name of the output file, and which item(s) to drop. The syntax is shown in the following illustration.

Using the Dropitem command.

Type the following at the Arc prompt.

```
Arc: dropitem salesregion salesregion population
```

Next, to view the results of the Dropitem command, use the Items command, as follows.

```
Arc: items salesregion
```

The following illustration shows the result.

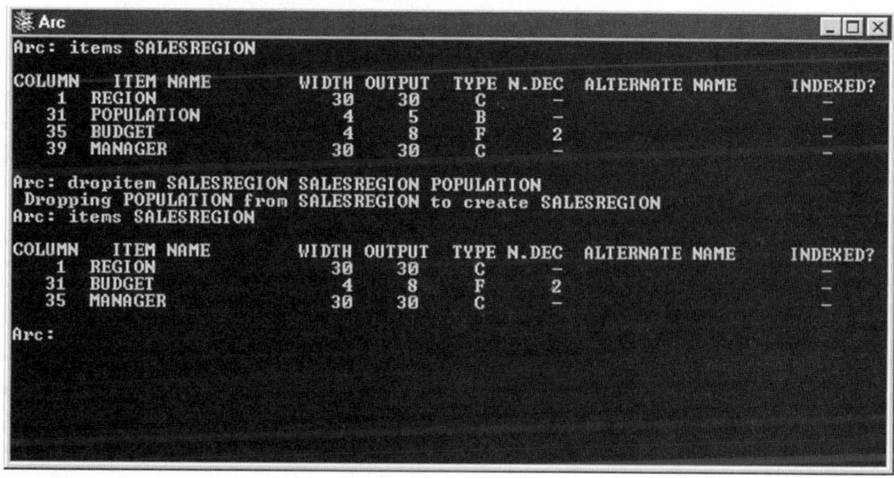

Result of Dropitem command on SALESREGION.

☛ **WARNING:** *Be careful when deleting items from a feature attribute table. Remember from Chapter 1 that the first few items in an attribute table are reserved for ArcInfo's use. If you delete any of them using Dropitem, you will destroy your coverage!*

This section introduced you to two commands that operate on INFO files but are executed from the Arc prompt. Additem and Dropitem are used to modify items in INFO files. Another command, named Joinitem, is also executed at the Arc prompt. It allows you to combine information in two INFO files into one file. This command is especially useful for adding items from one coverage to another, or for adding items from a text file (perhaps output by another database program) to a coverage. The next section discusses the steps necessary to join information from a text file to an INFO file.

Adding Records from a Text File

At first glance, adding information from a text file to an INFO file may seem a bit complex, but mastery of this process will save you much time and frustration. It will also add more flexibility to your GIS as data from other sources (typed lists, output from other database programs, and so on) can be readily incorporated into your coverages. Finally, human error and proofreading will be greatly reduced by having ArcInfo do this for you, instead of typing in the information by hand.

Exercise 9-3 introduces the Arc commands Joinitem and List and takes you through the process. As usual, be sure you have loaded the companion CD-ROM files for the Chapter 9 exercises into your workspace. This exercise assumes those files are stored in *D:\ai80exercises\chap_9*, and that you have worked through the previous exercises.

Exercise 9-3: Adding Data from a Text File

The research department has handed you a computer file with sales figures for four sales regions. The computer file was produced by their database software and is a comma-delimited text file. This exercise shows you how to read a text file into INFO, where it can then be joined to an ArcInfo coverage. This highlights how tabular data from other databases can be used with ArcInfo coverages.

Getting data from a text file to an INFO file requires four basic steps. First, be sure the text file items are in a format acceptable to INFO. A comma between each item is one example of an acceptable format. (Other formats are possible; refer to INFO's DELIMTER command for more information.) Also be sure the text file has at least one item in common with the INFO file to which you want to add the text file's information.

Second, create an INFO file into which you will load the text file's information. Third, load the information from the text file into the new INFO file. Fourth, join the two INFO files. This exercise demonstrates these four steps.

Step 1: Verifying Compatibility

Start Arc and navigate to your work directory. Use your system's file browser to look at the text file named *customer.txt*. You will see that it contains the following comma-separated information.

```
A,1000

B,34912

C,1800

D,451
```

Here, the first column is the sales region and the second is the number of customers in that region. You want to add the customer data to your *SALES.PAT* file. Both this text file and the *SALES.PAT* file have an item in common: region. Again, the text file must have at least one item in common with the INFO file to which you want to add the text file's information.

Step 2: Creating an INFO File

Create an INFO file into which you will copy the records in *customer.txt*. You will use INFO's DEFINE and ADD FROM commands. For now, follow the steps listed here; the next section contains more information and an exercise about creating INFO files.

From Arc, start INFO. Use the DEFINE command to create a new INFO file named CUSTOMER. Do this by typing *DEFINE CUSTOMER* at the "ENTER COMMAND >" prompt. You will be prompted to enter information about the items in the new CUSTOMER file. Your file will have two items: REGION and NUM_CLIENTS (for number of clients). Respond to the prompts so that the item named REGION has a width of 30, an output width of 30, and a type of C (for character).

The NUM_CLIENTS items should have a width of 4, an output width of 5, and be a B (for binary) type. This syntax is shown in the following illustration. Do not worry about the meaning of these, which is covered in this chapter's next section. Note, however, that you defined your INFO file's items to correspond to the *customer.txt* file's items. That is, the first column in both files contains the region, and the second column in both files contains the number of customers in that region.

✓ **TIP:** *When defining an INFO file for the ADD FROM operation, be sure the items in both files are in the same order and have the same parameters (widths, type, and so on).*

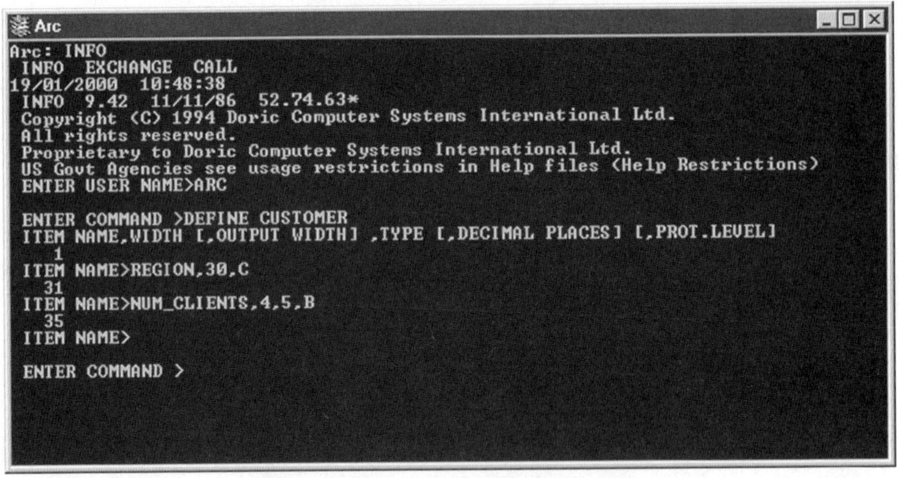

```
Arc
Arc: INFO
 INFO  EXCHANGE  CALL
19/01/2000  10:48:38
 INFO  9.42  11/11/86  52.74.63*
 Copyright (C) 1994 Doric Computer Systems International Ltd.
 All rights reserved.
 Proprietary to Doric Computer Systems International Ltd.
 US Govt Agencies see usage restrictions in Help files (Help Restrictions)
 ENTER USER NAME>ARC

 ENTER COMMAND >DEFINE CUSTOMER
 ITEM NAME,WIDTH [,OUTPUT WIDTH] ,TYPE [,DECIMAL PLACES] [,PROT.LEVEL]
    1
 ITEM NAME>REGION,30,C
   31
 ITEM NAME>NUM_CLIENTS,4,5,B
   35
 ITEM NAME>

 ENTER COMMAND >
```

Creating an INFO file named CUSTOMER using the DEFINE command.

Step 3: Reading Information into an INFO File

Next, use the ADD FROM command to read the information in the *customer.txt* file into your newly defined CUSTOMER file. ADD FROM reads each line of the text file as a record, and each comma-separated item as an item for that record. The syntax for the command is as follows.

```
ENTER COMMAND > ADD FROM ..\CUSTOMER.TXT ..\CUS.REJ
```

Notice the "..\" parameter in the command. It tells INFO to look one directory level up for the file. This is necessary because not only is

INFO its own module in Arc, it operates in a subdirectory of Arc. In other words, while working in INFO you are actually working within the INFO directory (within Arc). Test this for yourself by now performing a DIR command while in INFO. The *customer.txt* file is not listed. To access the *customer.txt* text file, you must use the "..\" parameter.

The "..\CUS.REJ" parameter sets up a file, also one directory up, that stores any records that did not properly load from the text file into the INFO file. The usefulness of this function cannot be underestimated. The text file in this exercise contains only four records, and if one or two did not load it would be easy to find them, examine them, and figure out the error. For text files containing hundreds or thousands of records, spotting "load" errors is not as easy.

The reject file provides a way to search through the rejected records to find a common denominator. One common error is that of an item in the target file (the defined INFO file) not being wide enough. A character item may be defined as only six letters long, whereas the source file (the text file) contains records with that item having seven or more characters. Develop the habit of always specifying a reject file name.

Now type *LIST* to view the newly added information. The following illustration shows the results of the ADD FROM command. All four records loaded perfectly, thus leaving the *cus.rej* file empty. You have now recreated the *customer.txt* file as an INFO file and are now ready to join it to SALES.PAT, the polygon attribute table of the SALES coverage.

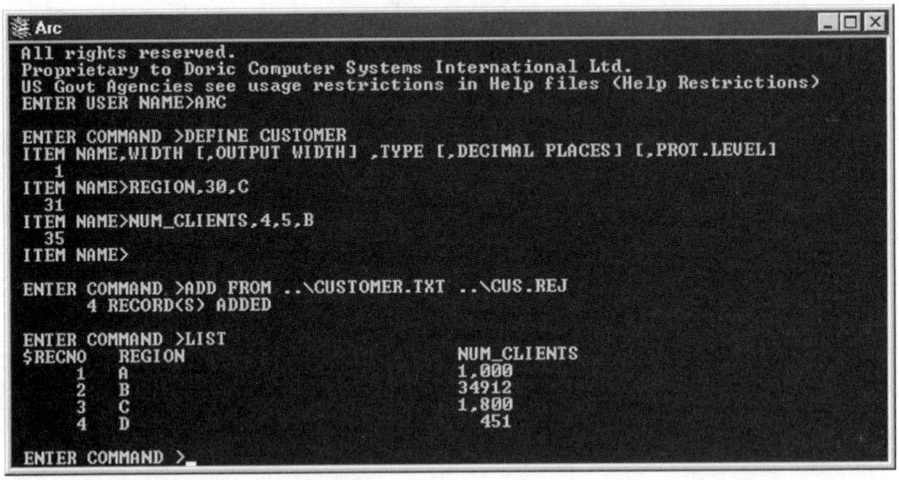

```
Arc
All rights reserved.
Proprietary to Doric Computer Systems International Ltd.
US Govt Agencies see usage restrictions in Help files (Help Restrictions)
ENTER USER NAME>ARC

ENTER COMMAND >DEFINE CUSTOMER
ITEM NAME,WIDTH [,OUTPUT WIDTH] ,TYPE [,DECIMAL PLACES] [,PROT.LEVEL]
   1
ITEM NAME>REGION,30,C
  31
ITEM NAME>NUM_CLIENTS,4,5,B
  35
ITEM NAME>

ENTER COMMAND >ADD FROM ..\CUSTOMER.TXT ..\CUS.REJ
      4 RECORD(S) ADDED

ENTER COMMAND >LIST
$RECNO   REGION                          NUM_CLIENTS
     1   A                               1,000
     2   B                               34912
     3   C                               1,800
     4   D                                 451

ENTER COMMAND >_
```

Listing results of the ADD FROM command.

Step 4: Joining Two Files

The final step is to use Arc's Joinitem command to merge the *CUSTOMER* file with the *SALES.PAT* file. This will effectively add the item NUM_CLIENTS to your SALES coverage. Remember that both files must have one item in common in order to use the Joinitem command. In this exercise the common item is REGION. Your resulting file will not have two items named REGION in it; rather, the REGION item from the "in" file will be kept and the REGION item from the join file will be dropped. Refer to the on-line documentation on the Joinitem command for detailed information.

Quit INFO to return to the Arc prompt. Type *joinitem sales.pat customer sales.pat region*. This specifies *sales.pat* as the "in" file, *customer* as the join file, *sales.pat* as the out (resulting) file, and *region* as the relate item (the item in common). Notice that you could create a new INFO file by specifying the "out" file as a different name from the "in" file. The following illustration shows the Joinitem command on screen.

```
Arc
Arc: joinitem
Usage: JOINITEM <in_info_file> <join_info_file> <out_info_file> <relate_item>
               <start_item> <LINEAR | ORDERED | LINK>
Arc: joinitem sales.pat customer sales.pat region
Joining sales.pat and customer to create sales.pat
Arc: _
```

Using Joinitem to combine items from two INFO files.

Check your "out" file to make sure the Joinitem operation worked. Do this using Arc's List command by typing *list sales.pat.* The screen should look as it does in the following illustration.

```
Arc
Joining sales.pat and customer to create sales.pat
Arc: list sales.pat
              1
AREA                       =          -1.526
PERIMETER                  =           6.171
SALES#                     =      1
SALES-ID                   =      3
REGION                     = C
NUM_CLIENTS                =   1800
              2
AREA                       =           0.593
PERIMETER                  =           3.378
SALES#                     =      2
SALES-ID                   =      2
REGION                     = B
NUM_CLIENTS                = 34912
              3
AREA                       =           0.628
PERIMETER                  =           3.038
SALES#                     =      3
SALES-ID                   =      1
REGION                     = A
NUM_CLIENTS                =   1000
              4
Continue?
```

Using the List command to view records in SALES.PAT.

You can see in the previous illustration that the NUM_CLIENTS item has been copied from the *CUSTOMER* file to SALES.PAT. You will find this *Joinitem* an efficient method for adding items to coverages and INFO files. It allows you to use text files from other sources without having to first hand-enter the text files' information into

INFO. Use this method whenever you have text file data you need to incorporate into your GIS.

Creating and Deleting INFO Files

There are as many reasons for creating INFO files as there are GIS. You could create files on sales information for a national retailer, population information for a health care provider, or wildlife inventories for nature preserves. The previous section and exercise introduced you to creating an INFO file using the DEFINE command. This section explores INFO items in detail, and shows you how to create an INFO file and how to delete INFO files.

First, examine and understand how INFO stores item data. Item data can be stored as numbers (binary, floating point, and so on), characters, or dates. It is important to define items as comprehensively as possible. For example, you do not want to define the item named VA_CITIES (Virginia cities) as having a width (a number of characters) of ten, only to find later that "Charlottesville" does not fit. The following are descriptions of INFO item characteristics.

✓ **TIP:** *Put some thought into the future needs, not just the present needs, of your GIS database when creating new INFO files. This will save you time in modifying files later on.*

ITEM NAME: The name you want to give the data field (the item). It cannot contain blank spaces, and it must be no longer than 16 characters.

WIDTH: This is the maximum width of the data field. If you specify 20 and you make the field a character field, the maximum number of letters this field can hold is 20.

OUTPUT WIDTH: This field, enclosed in brackets, is optional. This controls how many spaces are set aside when the field is displayed on the screen or printed. You should specify a value slightly larger than the WIDTH so that fields do not run together when printed.

TYPE: This specifies the type of data the field holds. The following are acceptable values.

- *C:* Signifies that the field will contain alphanumeric data. The maximum width is 320 characters.

- *I:* Use this for integer data (numbers with no decimal places). The maximum width is 16.
- *F:* Use this for floating point numbers, which you may be more familiar with as real numbers. The width for this item must be either 4 or 8, 4 designating single-precision storage and 8 double precision. Single-precision numbers store about 7 places of accuracy, and double precision stores up to 14. Note that you must specify an output width when using floating point numbers!
- *B:* Designates the field as binary data. Binary fields hold integer data in binary format, which takes up less storage place (memory) than "I" (integer) format.
- *N:* Used to store numeric data with one character per digit. It is not used very often because it requires more computer storage than F.
- *D:* Use this to store date information in year-month-day format. It is Y2K compliant and thus uses the YYYYMMDD format.

Greater detail on these characteristics can be found in the on-line help. See the INFO command DEFINE for additional information. Now that you are more familiar with the characteristics of INFO items, exercise 9-4 walks you through creating an INFO file.

Exercise 9-4: Creating an INFO File

After becoming familiar with the data used by National Widgets, you determine that a new file that tracks annual sales by sales region would be helpful. This exercise shows how to create that INFO data file.

This exercise continues with the sales and territory sample data found in the previous exercises. As always, be sure the exercises for Chapter 9 are loaded into your workspace. This exercise assumes the workspace is named *D:\ai80exercises\chap_9*. It also assumes you have completed the previous exercises in this chapter.

Step 1: Collecting Items into a File

The first file you create will contain data for three items: sales territory, salesperson, and annual sales. You will name these items SALES_TERR, SALES_PERSON, and ANNUAL_SALES.

First, start INFO. Once started, use the DEFINE command to create the file and name it SALES_TERR (for sales territory). Do this by typing *DEFINE SALES_TERR* at the "ENTER COMMAND >" prompt. INFO first provides a list of the item parameters (characteristics), as follows.

```
ITEM NAME, WIDTH [, OUTPUT WIDTH], TYPE, [DECIMAL PLACES] [PROT.LEVEL]
```

Although there are six parameters listed, you can actually only use the first five. PROT. LEVEL is a leftover from the 1980s, when INFO was marketed as an independent product. It is no longer used in ArcInfo today. Underneath this list is the "ITEM NAME>" prompt, asking you to begin entering item characteristics. You will be prompted to enter information for each characteristic (name, width, and so on). The following illustration shows what your screen should look like.

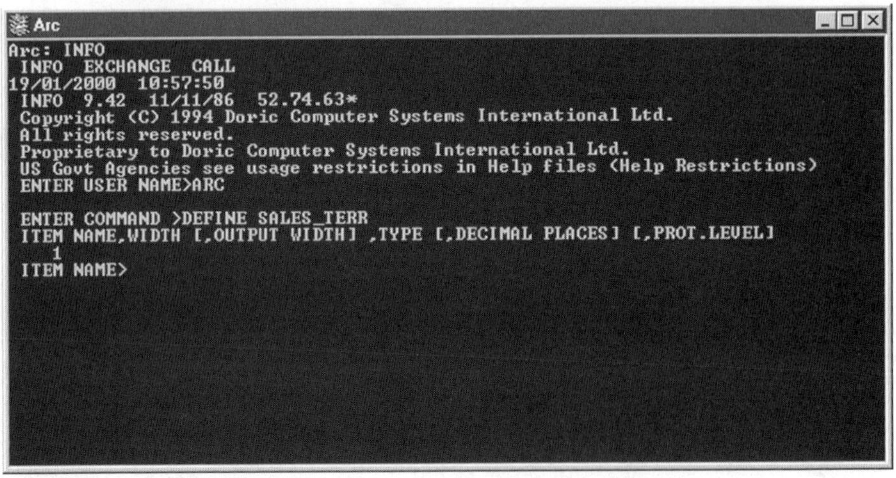

```
Arc
Arc: INFO
 INFO  EXCHANGE  CALL
19/01/2000  10:57:50
 INFO  9.42  11/11/86  52.74.63*
 Copyright (C) 1994 Doric Computer Systems International Ltd.
 All rights reserved.
 Proprietary to Doric Computer Systems International Ltd.
 US Govt Agencies see usage restrictions in Help files (Help Restrictions)
 ENTER USER NAME>ARC

 ENTER COMMAND >DEFINE SALES_TERR
 ITEM NAME,WIDTH [,OUTPUT WIDTH] ,TYPE [,DECIMAL PLACES] [,PROT.LEVEL]
    1
 ITEM NAME>
```

Defining the INFO file SALES_TERR.

Now enter the three items previously listed. Table 9-3, which follows, lists the characteristics to use with each item. The illustration that follows shows how the information can be typed in.

Table 9-3: SALES_TERR File Entries

Item Name	Width	Output Width	Type	Decimal Places
TERRITORY	30	32	C	0
SALES_PERSON	30	32	C	0
ANNUAL_SALES	4	8	F	2

```
Arc
Arc: INFO
 INFO  EXCHANGE  CALL
19/01/2000  10:57:50
 INFO  9.42  11/11/86  52.74.63*
 Copyright (C) 1994 Doric Computer Systems International Ltd.
 All rights reserved.
 Proprietary to Doric Computer Systems International Ltd.
 US Govt Agencies see usage restrictions in Help files (Help Restrictions)
 ENTER USER NAME>ARC

 ENTER COMMAND >DEFINE SALES_TERR
 ITEM NAME,WIDTH [,OUTPUT WIDTH] ,TYPE [,DECIMAL PLACES] [,PROT.LEVEL]
    1
 ITEM NAME>TERRITORY,30,32,C
   31
 ITEM NAME>SALES_PERSON,30,32,C
   61
 ITEM NAME>ANNUAL_SALES,4,8,F,2
   65
 ITEM NAME>

 ENTER COMMAND >
```

Defining items in an INFO file.

When you are finished, type *ITEMS* to see a listing of the items in the *SALES_TERR* file. Your screen should look like that shown in the following illustration.

```
Arc
 All rights reserved.
 Proprietary to Doric Computer Systems International Ltd.
 US Govt Agencies see usage restrictions in Help files (Help Restrictions)
 ENTER USER NAME>ARC

 ENTER COMMAND >DEFINE SALES_TERR
 ITEM NAME,WIDTH [,OUTPUT WIDTH] ,TYPE [,DECIMAL PLACES] [,PROT.LEVEL]
    1
 ITEM NAME>TERRITORY,30,32,C
   31
 ITEM NAME>SALES_PERSON,30,32,C
   61
 ITEM NAME>ANNUAL_SALES,4,8,F,2
   65
 ITEM NAME>

 ENTER COMMAND >ITEMS
 DATAFILE NAME: SALES_TERR                                            01/19/2000
    3 ITEMS: STARTING IN POSITION     1
 COL  ITEM NAME         WDTH OPUT TYP N.DEC  ALTERNATE NAME
   1  TERRITORY           30   32  C    -
  31  SALES_PERSON        30   32  C    -
  61  ANNUAL_SALES         4    8  F    2

 ENTER COMMAND >_
```

Using ITEMS to display fields in an INFO file.

Step 2: Adding Records to a File

At this point you are be ready to add records to your *SALES_TERR* file. You will recall from exercise 9-2 that the ADD command in INFO lets you add records interactively: you are prompted for item information for each record. Exercise 9-3 demonstrated how to add and drop items, and how to load data into an INFO file from a text file. For this exercise, however, you will not need to add any records.

As part of your GIS data management you will also sometimes need to delete INFO files. INFO has no recycling bin, or trashcan, or "undo" command, so be very sure you are deleting the correct file. The INFO commands DELETE and ERASE both delete INFO files. Do not confuse these two commands with PURGE. PURGE removes records only; DELETE and ERASE remove entire files.

To use DELETE you must first SELECT the file. Then type *DELETE <filename>*. Info responds with the following.

```
THIS COMMAND WILL ERASE THE SPECIFIED DF

DO YOU WISH TO CONTINUE (Y OR N) >
```

Note that in the INFO response, DF stands for "data file." You respond with Y (yes) or N (no) as appropriate; any answer other than a Y is interpreted as an N.

This section discussed creating INFO files, the item characteristics of those files, and how to delete INFO files. Exercise 9-4 took you step by step through creating an INFO file.

In the following section you will learn how to relate your INFO files to one another. Relating files allows you to expand the capabilities of your GIS by allowing coverages to access data without having to actually store the data.

Relating Data Files

INFO incorporates powerful, easy-to-use commands for relating (linking) files. Relating files can enhance your analyses by providing more information. Relates are also an efficient means of storing information. Relying on relates can make updates easier, as one file can be updated and several coverages can access that file. This is a much easier method than updating several coverages.

The general requirement for relating two files is that they have an item in common. Ideally, these items have the same name and same definitions (both are binary and have the same widths, for example). This is not a hard and fast requirement, however, and files may be related on items with different names. Those types of relate operations are beyond the scope of this chapter. If you need to perform such operations, refer to the RELATE command (under the alphabetical list of INFO commands) and the "Relating data files" section (under the functional list of INFO commands) in the on-line documentation.

In keeping with our sales and territory examples in this chapter, exercise 9-5 shows you how to relate two files, one with sales persons and one with product information. Relating the two will let you use both for analysis, which will reveal, in this case, information about which sales person sold the most widgets. Again, be sure you have copied the exercises for this chapter from the companion CD-ROM to your workspace.

Exercise 9-5: Relating INFO Files

INFO is a relational database. This means that it has the ability to link data files based on common items. For this exercise, say that your boss has asked you to determine how many widgets are sold per customer. Use INFO's capabilities to determine the answer by linking two files.

In this exercise, you will relate two INFO files, CUSTOMER and PRODUCTSALES. (You will recall that in exercise 9-3 you created the INFO file CUSTOMER and that you loaded data into it from a text file.) You will use the results of the relate operation to calculate a ratio of how many widgets per customer were sold in each region. The higher the ratio, the more widgets sold per customer. To store the results of this calculation, you must first add a new item to the PRODUCTSALES file. This new item will store the calculated ratio.

Step 1: Adding an Item to the PRODUCTSALES Data File

Start Arc and navigate to your workspace. Use the Additem command, as discussed in exercise 9-3, to add an item named SALES_RATIO to the *PRODUCTSALES* file. Because this item will hold a real number, specify its type as floating point (F). Do this by typing the following.

```
Additem PRODUCTSALES PRODUCTSALES SALES_RATIO 4 8 F 2
```

Next, type *items productsales* to view the PRODUCTSALES items. As the following illustration shows, you should see the new item SALES_RATIO.

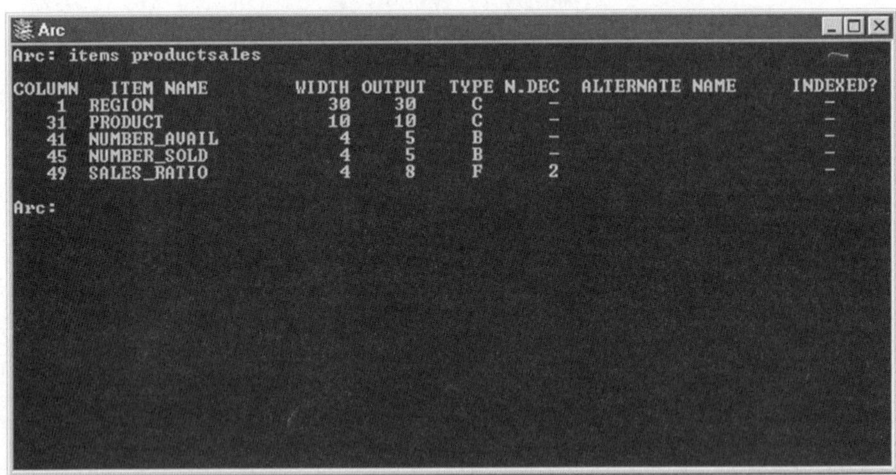

Arc
Arc: items productsales
COLUMN ITEM NAME WIDTH OUTPUT TYPE N.DEC ALTERNATE NAME INDEXED?
1 REGION 30 30 C - -
31 PRODUCT 10 10 C - -
41 NUMBER_AVAIL 4 5 B - -
45 NUMBER_SOLD 4 5 B - -
49 SALES_RATIO 4 8 F 2 -
Arc:

Using the List command to view the added item SALES_RATIO.

Step 2: Relating Two Files

Start INFO and select the *PRODUCTSALES* data file by typing *SEL PRODUCTSALES* at the "ENTER COMMAND >" prompt. Next, use the RELATE command to connect the two files. Remember, the files must have an item in common; that is, an item with the same name and same characteristics. In this case, the identical item is named REGION. The RELATE command can become quite complex in execution, as up to nine files can be related at once.

You can study the syntax of the RELATE command by referencing it in the on-line help. For now, keep it simple by typing the following.

```
ENTER COMMAND > RELATE CUSTOMER BY REGION
```

Your two files are now related. Use a special form of the LIST command to see this for yourself.

Step 3: Viewing Data in Both Files

To see the power of the RELATE command in action, use a form of the LIST command in INFO. Notice that to list an item in the related file, you must preface it with "$1," which signifies that you are listing an item in the first related data file. You can relate many files at once, up to nine, and you would access items in each by prefacing the item names with $2 (for the second related file), $3 (for the third), and so on.

Now type the following.

```
ENTER COMMAND > LIST REGION,$1 NUM_CLIENTS,NUMBER_SOLD
```

Spaces between the items are not important, but commas are. You should see items from both files, as shown in the following illustration.

```
Arc
ENTER COMMAND >
INVALID COMMAND

ENTER COMMAND >
INVALID COMMAND

ENTER COMMAND >
INVALID COMMAND

ENTER COMMAND >
INVALID COMMAND

ENTER COMMAND >SEL PRODUCTSALES
     4 RECORD(S) SELECTED

ENTER COMMAND >RELATE CUSTOMER BY REGION

ENTER COMMAND >LIST REGION,$1NUM_CLIENTS,NUMBER_SOLD
$RECNO   REGION                          NUM_CLIENTS NUMBER_SOLD
     1   A                               1,000        12000
     2   B                               34912           89
     3   C                               1,800          491
     4   D                                 451        5,400

ENTER COMMAND >
```

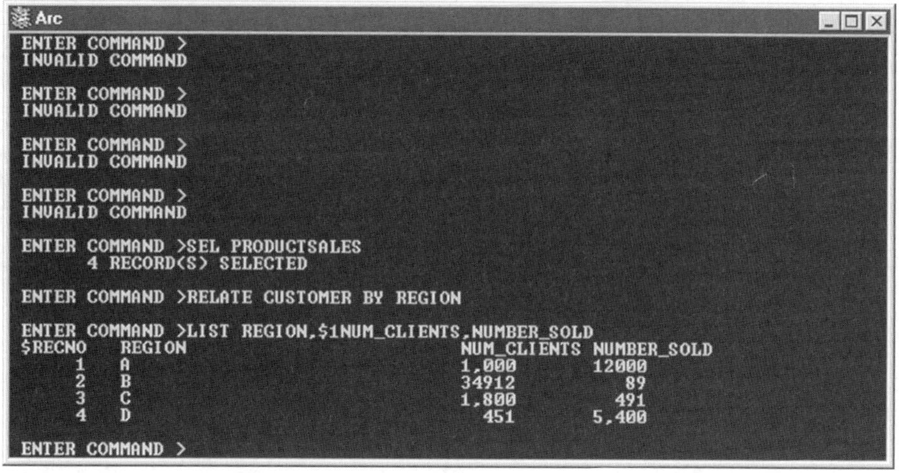

Viewing items from two related files.

Step 4: Calculating Sales of Widgets per Customer

Finally, calculate a value for the newly added item SALES_RATIO. Its value will be the result of dividing the NUMBER_SOLD item in the *PRODUCTSALES* file by the NUM_CLIENTS item (number of customers) in the CUSTOMER file. Do this by typing the following.

```
ENTER COMMAND > CALC SALES_RATIO = NUMBER_SOLD / $1NUM_CLIENTS
```

Your screen should look like that shown in the following illustration.

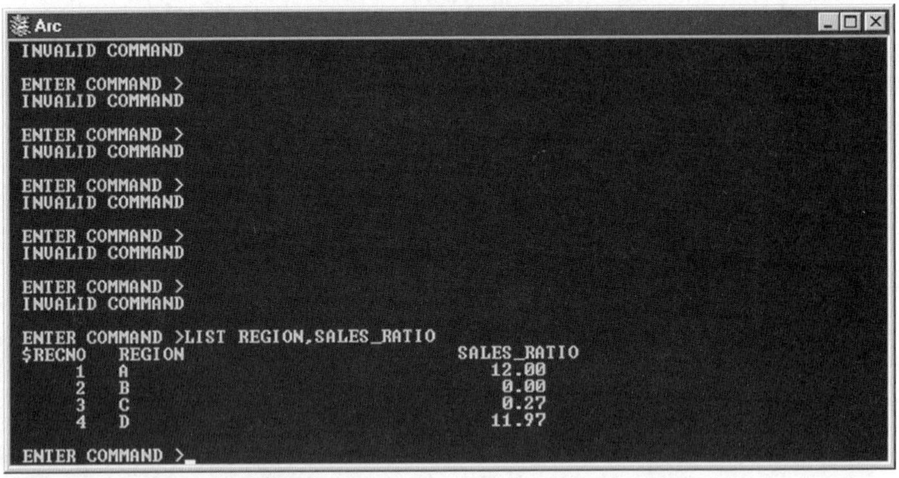

```
INVALID COMMAND

ENTER COMMAND >
INVALID COMMAND

ENTER COMMAND >
INVALID COMMAND

ENTER COMMAND >
INVALID COMMAND

ENTER COMMAND >
INVALID COMMAND

ENTER COMMAND >
INVALID COMMAND

ENTER COMMAND >LIST REGION,SALES_RATIO
$RECNO   REGION                        SALES_RATIO
     1   A                                   12.00
     2   B                                    0.00
     3   C                                    0.27
     4   D                                   11.97

ENTER COMMAND >_
```

Listing REGION and SALES_RATIO.

CALC is an abbreviation for CALCULATE, the INFO command for performing mathematical operations. As you can see, regions A and D sell a good number of widgets per customer, whereas regions B and C lag far behind.

By using the RELATE and CALCULATE commands you were able to perform an analysis using items from different INFO files. Relating files provides a powerful means of accessing data across multiple files for performing complex and powerful analyses. The next section shows you how to share the results of your analyses by creating reports from INFO.

Creating a Report

INFO has a report-generating function you can use for creating and printing reports. A report contains formatted information that you select from a file's records and related records. Reports can contain totals, averages, calculation results, text fields, and so on. They can also include column headers, page numbers, and titles. You can route reports to your screen or to your spool file for printing.

This section introduces you to creating and displaying a simple report. There are many commands and options that are outside the scope of this section. For detailed information on all report and printing commands and their options, see "Creating Reports" and

"Printing from INFO" in the functional list of INFO commands in the on-line help.

To create a report in INFO, simply type *REPORT <filename>* at the "ENTER COMMAND >" prompt. An interactive dialog prompts you to enter information about what should appear on the report page. Exercise 9-6 takes you through the steps of creating a report for the results of the analysis you performed in exercise 9-5.

Exercise 9-6: Creating a Simple INFO Report

Your boss now wants to see a report of sales activity in each region. Follow the steps in this exercise to create a report.

Before beginning, be sure you have loaded the exercises for Chapter 6 from the companion CD-ROM to your workspace. Also be sure you have completed exercise 9-5, as this exercise builds on the results of that exercise. This exercise assumes your workspace is named *D:\ai80exercises\chap_9*. Start Arc, navigate to your workspace, and then start INFO.

Step 1: Defining the Report

First, decide what items you want to show in your report. For this exercise, you will show REGION, NUMBER_SOLD, $1CUSTOMER, and SALES_RATIO—items found in the PRODUCTSALES file and its related file, CUSTOMER. Now use the REPORT command to set up a report named SALES.RPT by typing the following.

```
ENTER COMMAND > REPORT SALES.RPT
```

INFO now prompts you to enter the first item. Type *REGION*.

INFO then prompts you to enter "report options" and "column headings." Skip over these options by pressing the <Enter> key. When the "ENTER COLUMN HEADINGS>" prompt appears, type *SALES REGION*. SALES REGION will be the heading for the column that holds the REGION item. You will again be prompted to enter a column heading. This is INFO's way of allowing you two rows to enter a column heading. Because you are only using one column, just press the <Enter> key.

You are now prompted to enter the second item. Type *NUMBER_SOLD* and skip through the rest of the prompts by pressing the <Enter> key. Do not add column headings for it. Repeating this same step, add *$1CUSTOMER* and *SALES_RATIO* (also without column headings).

Once you have entered all four items, you will be prompted for a report name. This name will be displayed on the top of your report. Name this report 'SALES REPORT,' being sure to enclose SALES REPORT in single quotes. This tells INFO to display the words as text. If you do not use single quotes, INFO will try to find an item in the database called SALES REPORT, and an error message will appear.

INFO now asks if you want to execute this report. Press the <Y> key. When prompted to send this report to the printer, press the <N> key. This will route your report display to the screen.

Now INFO asks you to specify how many lines per page to display. Because you are displaying this on the screen, type *20.* If you were sending this to a printer, typing a number around 60 sends a full page of text to a standard letter-size printer.

Finally, you will be prompted for report options. For this exercise, just press the <Enter> key again. The report will now display on the screen. It should look like that shown in the following illustration.

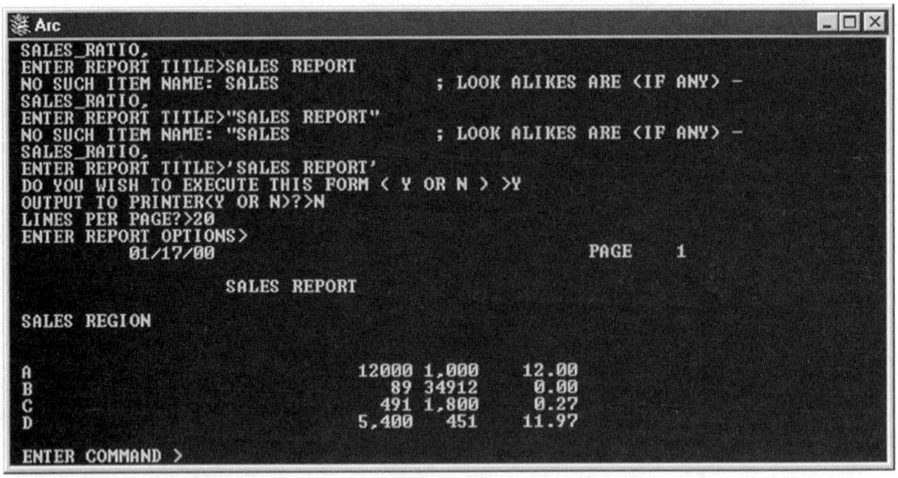

```
Arc
SALES_RATIO,
ENTER REPORT TITLE>SALES REPORT
NO SUCH ITEM NAME: SALES                ; LOOK ALIKES ARE (IF ANY) -
SALES_RATIO,
ENTER REPORT TITLE>"SALES REPORT"
NO SUCH ITEM NAME: "SALES               ; LOOK ALIKES ARE (IF ANY) -
SALES_RATIO,
ENTER REPORT TITLE>'SALES REPORT'
DO YOU WISH TO EXECUTE THIS FORM < Y OR N > >Y
OUTPUT TO PRINTER(Y OR N)?>N
LINES PER PAGE?>20
ENTER REPORT OPTIONS>
       01/17/00                                   PAGE    1

                SALES REPORT

SALES REGION


A                              12000 1,000    12.00
B                                 89 34912     0.00
C                                491 1,800     0.27
D                              5,400   451    11.97

ENTER COMMAND >
```

Displaying the SALES.RPT report.

Immediately you notice a problem. The only column to have a heading is REGION, where you typed *'SALES REGION*'. The other three columns should have displayed their item names by default, but they do not appear at all. Why? Because all three items are defined as having a smaller width than their item names.

For example, the item name SALES_RATIO has 11 characters in it, but the item itself is defined as having an output width of only 8 characters. Because INFO cannot display the entire item name in eight characters, it decides to print no name at all. Keep this in mind when deciding on widths for item names, as well as for items. This is an example of an annoying problem that crops up just as you are trying to meet an important deadline.

✓ **TIP:** *You may want to make item names smaller than the item widths so that they appear when using INFO's report function.*

One way to fix this problem is to type a column header that is less than the width of the item. Remember that you have two lines to enter a column heading, so you can abbreviate as well. Examine the screen shown in the following illustration to see an alternative report format.

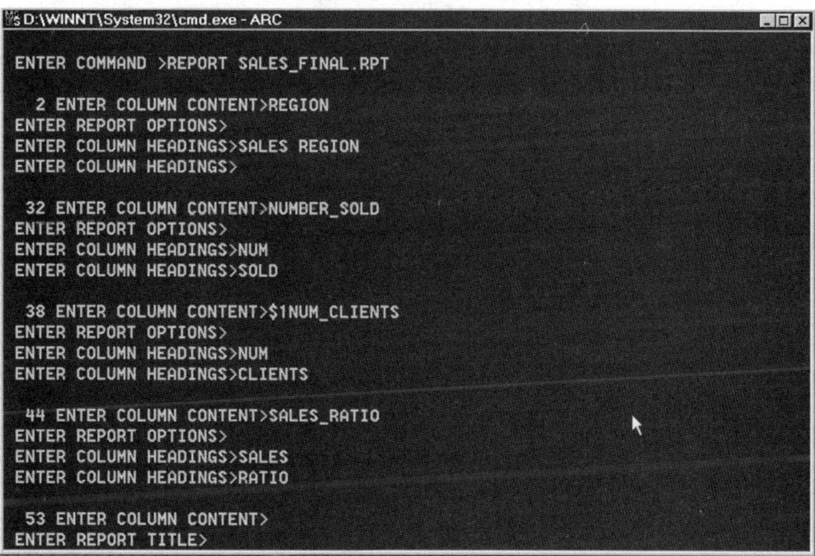

FIXED_SALES.RPT report format.

Using the example shown in the previous illustration and the steps previously outlined, create your own FIXED_SALES.RPT. Or, if you prefer, continue with this exercise, using the SALES_FINAL.RPT report included.

Step 2: Using an Existing Report

To view an example of a corrected report, type *SALES_FINAL.RPT.* This executes the report and displays it to the screen. Your screen should look like that shown in the following illustration.

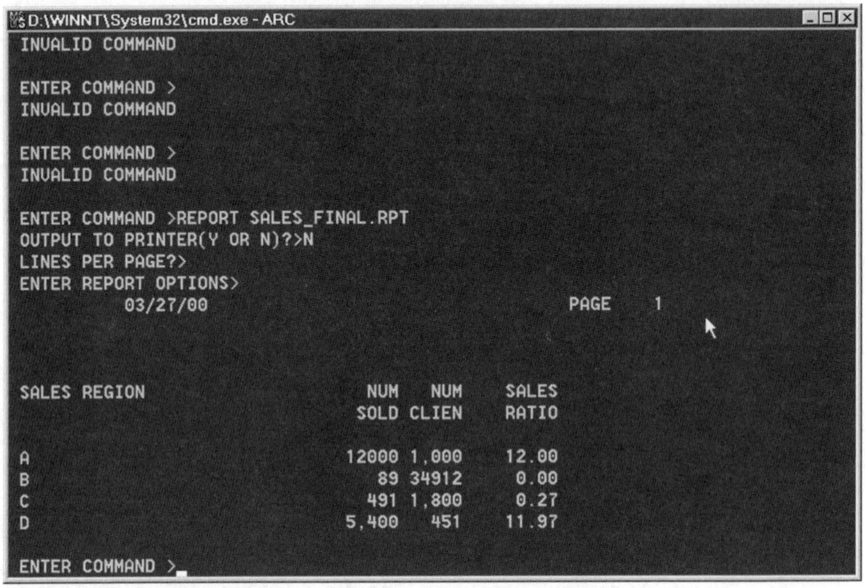

```
D:\WINNT\System32\cmd.exe - ARC
INVALID COMMAND

ENTER COMMAND >
INVALID COMMAND

ENTER COMMAND >
INVALID COMMAND

ENTER COMMAND >REPORT SALES_FINAL.RPT
OUTPUT TO PRINTER(Y OR N)?>N
LINES PER PAGE?>
ENTER REPORT OPTIONS>
        03/27/00                                  PAGE    1



SALES REGION                          NUM   NUM     SALES
                                     SOLD CLIEN     RATIO

A                                   12000 1,000     12.00
B                                      89 34912      0.00
C                                     491 1,800      0.27
D                                   5,400   451     11.97

ENTER COMMAND >_
```

View of the corrected sales report.

Step 3: Printing a Report

To print a report, first create the report, as previously shown. When prompted to send the report to the printer, press the <Y> key.

The report will be generated and kept in INFO's spool file. This file holds everything you have printed so far in INFO and stores it there until you issue a command to send it all to the printer. Use the SPOOL command at the "ENTER COMMAND >" prompt to send the report to the printer. This will send the report (plus any other reports you have run but have not yet sent to the printer) to the printer.

Reports are an essential tool for displaying data. They make it presentable and easier to understand. INFO, although an older database, still has a very good report generator that will serve most users' day-to-day needs. You are encouraged to take some time to explore the other report and printing options, as mentioned at the beginning of this section. You will uncover even more useful commands and options for dressing up your reports.

Summary

This chapter makes accessible some of INFO's power and highlights. INFO is ArcInfo's database component. A full-featured relational database, it has served ArcInfo users well since the software's inception. It uses older, non-SQL standard terms but offers power and ease of use. This chapter introduced you to some basic but very useful INFO operations. You learned how to add, delete, and update records; how to create and modify INFO files; how to add information from a text file to an INFO file; how to relate INFO files to each other; and even how to create and display reports.

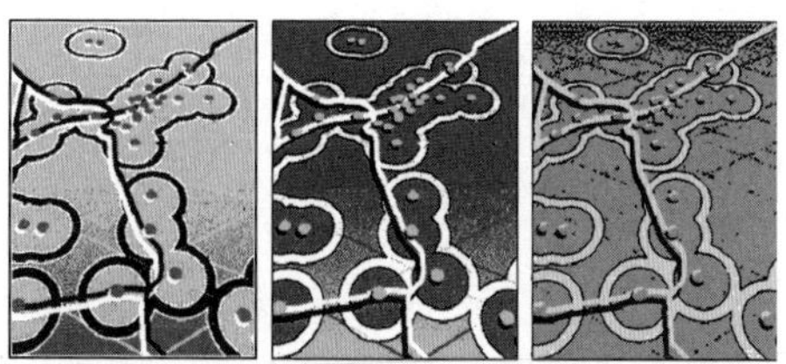

Chapter 10

The Basics of Arc Macro Language

Arc Macro Language (AML) is ArcInfo's scripting language. Based originally on the command language used by PrimOS, the operating system used by Prime minicomputers in the 1980s, AML has evolved to include many programming functions found in computer languages today. You can use AML to create simple programs that run one ArcInfo command after another, relieving you, for example, of having to type the same list of commands each time you need to create a map file in ArcPlot.

You can also use AML's more advanced functions to open and read from text files, and to perform numerical calculations and advanced text string manipulation. As your ArcInfo experience grows, AML will become an essential companion.

This chapter introduces you to the basics of AML. You do not need to be a programmer to use AML. However, some knowledge of basic programming concepts will be helpful. Note one convention among ArcInfo users: although AML is an acronym for the programming language, it is also used to refer to the actual programs. For example, your boss might say "Run that AML to create our WWW maps," meaning "Execute that Arc Macro Language program to create our WWW maps."

AML commands can only be used with the ArcInfo Workstation modules. If you want to create programs that interface with ArcCatalog, ArcToolbox, or ArcMap, you will need an object-oriented programming language such as Visual Basic. Creating such interfaces is beyond the scope of this book, and beyond the capabilities of AML. AML can be used to create complex programs that utilize the power of ArcInfo's traditional modules.

AMLs can include any of the pull-down menus of the traditional ArcInfo modules. In fact, you can even build your own pull-down menus using FormEdit. Pull-down menu interfaces used with your custom AMLs can make using ArcInfo easier for persons in your organization.

This chapter assumes you have some familiarity with the traditional ArcInfo modules. If you are new to ArcInfo, work through the other chapters before beginning this one. You will need to know the commands before you can put them into programs. This chapter covers the following.

- Scripting and using *&watch* files
- AML basics: definitions, reading files, and looping
- Using FormEdit to create menus

Although setting up AMLs may seem time consuming, the effort it takes to write one saves you much more effort later. A simple example: say that every time you start an ArcInfo session, you always have to change workspaces to get to a particular folder, and then you always start ArcTools (the pull-down menu system). Thus, at the Arc prompt you are always typing the same three commands, which follow.

`Arc: workspace d:\ai80exercises\chap_10` [Move to a new folder]

`Arc: &stat 9999` [Initialize the graphics display]

`Arc: arctools` [Start ArcTools menu system]

Instead of typing these each time you start Arc, use a text editor to put these three commands into a text file named *start.aml.* This AML (program) contains the following three lines.

```
w d:\ai80exercises\chap_10

&stat 9999

arctools
```

Note that these commands are identical to what you would type. Now, by simply typing *&r start* (*&r* is the abbreviation for "&run"), you execute all three commands.

You can use a word processing utility such as Windows NT Notepad to write and edit your AMLs. You can also use word processing software (such as Microsoft Word or WordPerfect), but be sure to do two things: always save your AML using the *.aml* file extension, and save your file as a text file, *not* as a Word or WordPerfect file. ArcInfo only recognizes AML programs if they are text files and if they end with the *.aml* extension.

✓ **TIP:** *If using word processing software to write AMLs, be sure to save the files as text files with the* .aml *extension.*

The *&run* command is used to execute all AMLs. It tells ArcInfo to execute the commands in the file specified. If you forget to type *&run* (or its abbreviation *&r*) before the file name, ArcInfo will interpret the file name as a command, and return the "Unknown command" error message. Although AMLs must be named with the *.aml* extension, you can reference them with the *&run* command using just the prefix. For example, instead of typing *&run start.aml*, you can type *&r start.*

The *start.aml* example is indeed a simple one, but it gives you an idea of the convenience and efficiency AML programs can provide users. The next section includes a more complex AML for displaying a coverage on the screen, and introduces you to *scripting*, a simple method of creating AML programs.

Scripting and Using &watch Files

In the preceding ArcInfo Workstation chapters you learned how to use many commands in ArcEdit, ArcPlot, Arc, and to work with the INFO database. With each of these modules, however, you enter commands one at a time, whether by typing in the command at the module's command prompt or executing it via pull-down menus. Scripting allows you to store many commands in a file, and then sequentially execute them by typing just one command. Thus, you can automate many of the commands you previously entered one by one. *Start.aml*, previously discussed, is an example of scripting.

Scripting is a simple form of AML programming. There are no decision trees, "do" loops, or reading in or writing out of files associated with AML. Scripting is just many commands executed one after the other. Exercise 10-1, which follows, will give you a feel for scripting an AML.

Exercise 10-1: Drawing a Coverage in ArcPlot

This exercise introduces you to the power and convenience of placing many commands into one AML. You will create an AML named *cnty_map.aml*, which will draw in ArcPlot a map of the major roads in an Ohio county.

Before beginning this exercise, be sure you have copied the data from the companion CD-ROM to your workspace on your hard drive. The data for this exercise are the base layer data from the Blue-eyed Corn Moth project used in the Chapter 6 exercises. This exercise assumes that workspace has the name *D:\ai80exercises\chap_10*. You should also be familiar with the basic functions of ArcPlot, ArcInfo's traditional map composition module.

Step 1: Creating an AML

Using the word processing utility of your choice, create a new file named *cnty_map.aml*. Once you have created the new text file, type in the following lines exactly as shown.

```
&stat 9999
arcplot
reselect proj_cnties poly name = 'Madison'
mapextent poly proj_cnties
polygonshades proj_cnties 2
reselect proj_roads line overlap proj_cnties poly
arcs proj_roads
```

The *&stat* command tells ArcInfo that you are using a workstation, and *arcplot* (which could be abbreviated as *ap*) starts the ArcPlot module. The next five commands are ArcPlot commands. The *reselect* command selects Madison County from the larger multi-county coverage. The *mapextent* command (abbreviate it as *mape* if you wish) tells ArcPlot from which coverage to draw the extents of the map composition.

Notice that the syntax states *mapextent poly.* This option to the *mapextent* command sets the map limits to the named feature from the *proj_cnties* coverage, which, in this case, is Madison County. *Polygonshades* causes the selected polygons of the coverage specified *(proj_cnties)* to be shaded red (the number *2* here indicates the color red).

Another "reselect" command is issued, this time on the coverage of roads named *proj_roads.* This is done to draw just the roads that fall within Madison County or that intersect with it. The reselect command uses the overlap option to select those roads inside the *proj_cnties* coverage.

The *arcs* command is also an ArcPlot command. It tells ArcPlot to draw the arcs of the specified coverage, *proj_roads,* to the graphics screen. For more information on these and other ArcPlot commands, see Chapter 7 or the on-line help.

Now save this file, being sure to save it in your Chapter 10 workspace (where you copied the exercise data for Chapter 10 exercises: *D:\ai80exercises\chap_10*). You have just created an executable AML.

Step 2: Executing (Running) an AML

Start Arc and navigate to your workspace using the *workspace* (abbreviated as *w*) command: *w d:\ai80exercise\chap_10.* Execute the AML by typing *&r cnty_maps.*

The AML executes each command in turn. Watch it work, and when it is finished, the map shown in the following illustration should be displayed on your screen.

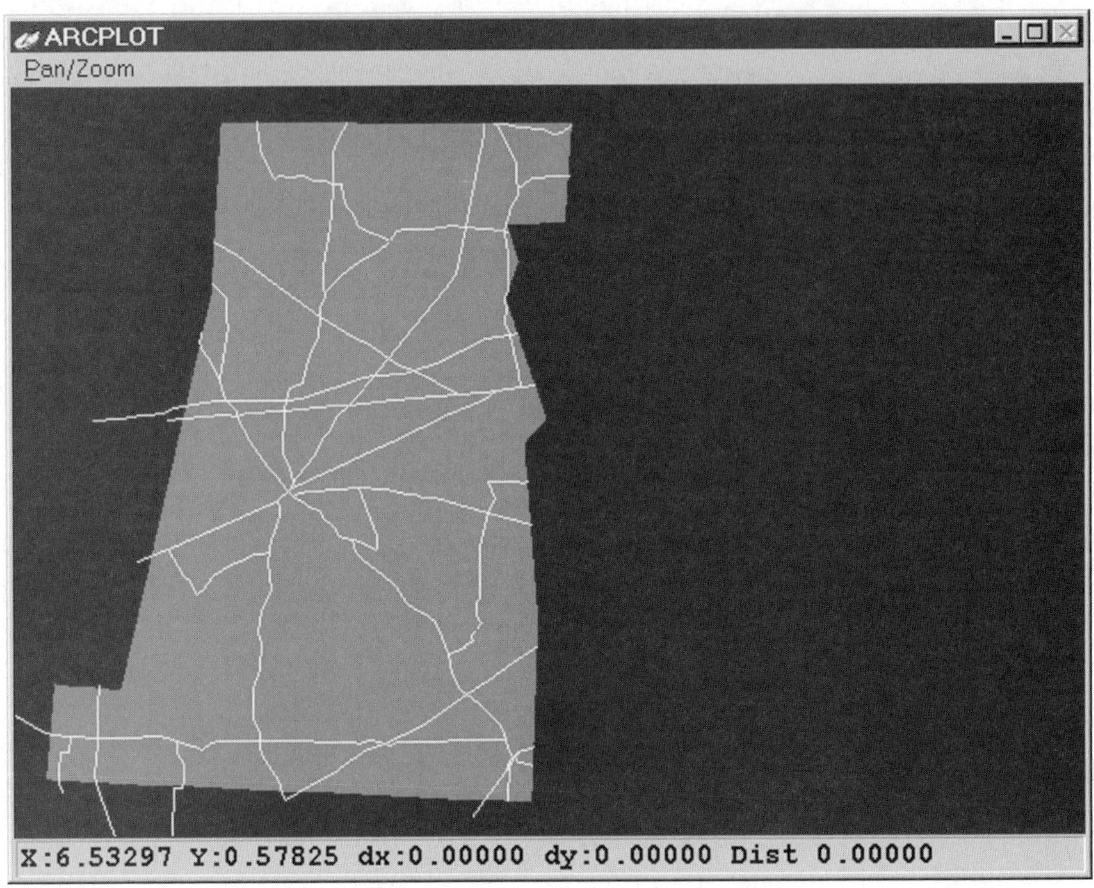

Results of cnty_map.aml.

Notice that when the AML ended, it left the ArcPlot graphics screen active. This is because AMLs do *only* what you tell them to, and *nothing more.* Go ahead and quit ArcPlot now, by typing *q* at the ArcPlot prompt.

By using *cnty_map.aml*, you now have to type just one command instead of five. AMLs such as this save time and reduce errors.

Step 3: Modifying an AML to Create a Plot File

You have seen this simple map on the screen. Now suppose you need to make a paper copy of the map created by *cnty_map.aml.* You can easily modify the AML so that it creates an ArcInfo graphics file suitable for printing. Use the Text Editor to make the following changes and additions to the *cnty_map.aml* file.

```
&stat 9999

arcplot

display 1040

madison.gra

reselect proj_cnties poly name = 'Madison'

mapextent poly proj_cnties

polygonshades proj_cnties 2

reselect proj_roads line overlap proj_cnties poly

arcs proj_roads

quit

draw madison.gra 9999
```

Notice that four new lines have been added. Examine each one closely. The *display 1040* command tells ArcPlot that you are creating a graphics file for printing, and *madison.gra* is the name of the graphics file (commonly called a "plot file"). The name of the file created by the *display 1040* command is always placed on the line after the command. See Chapter 7 for a refresher on creating plot files.

The third new line, *quit*, ends the ArcPlot session. The fourth added line, *draw madison.gra 9999*, draws the graphics file you just created to the screen. This allows you to preview your map composition before actually printing it.

Now save this amended AML file under the same name *(cnty_map.aml)*, and return to the Arc prompt and run the AML again. Your screen should now look like that shown in the following illustration. Notice that the map graphic appears on the Draw screen rather than on the ArcPlot display canvas screen.

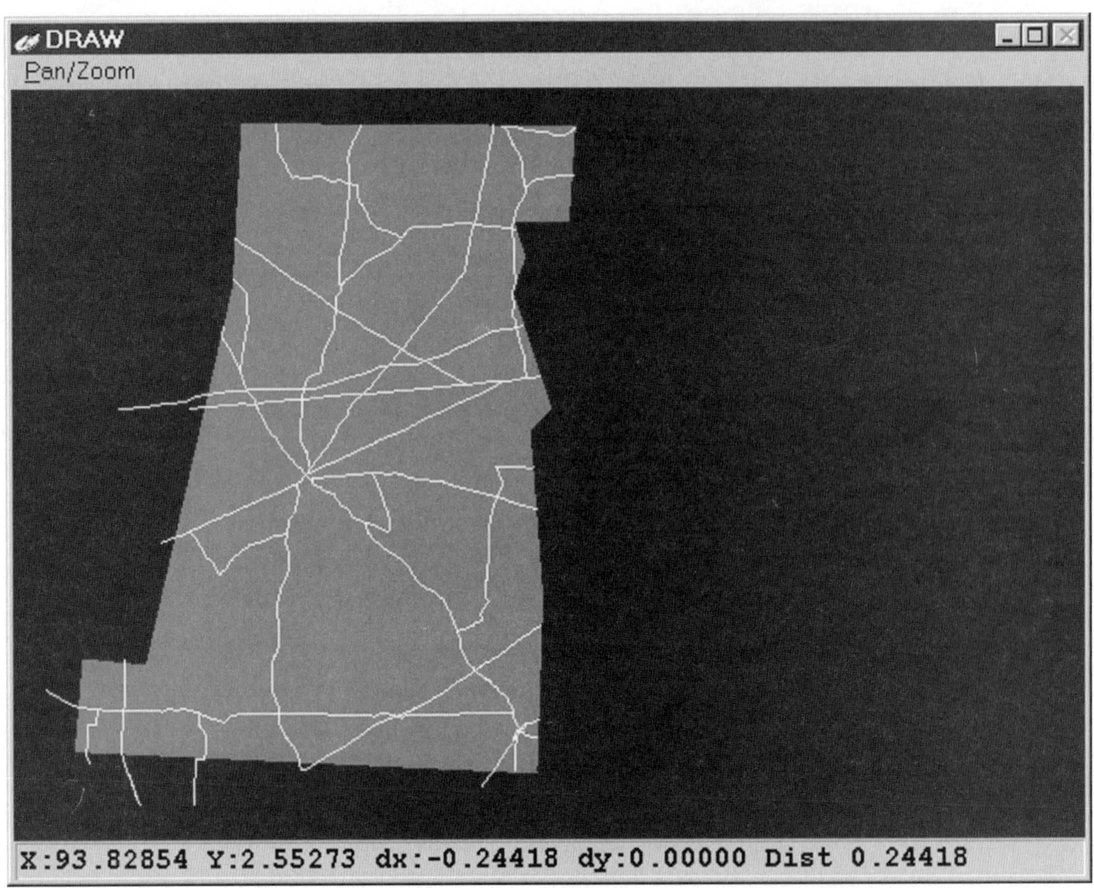

Running the edited cnty_map.aml.

Step 4: Adding Documentation to cnty_map.aml

It is always a good idea to document the AMLs you write. Many times you will examine an AML you wrote previously and will wonder why you did something the way you did. Or perhaps someone else will have to use your AML. Documentation is one of those steps that may seem time consuming up front, but always pays off with time saved later.

✓ **TIP:** *Always document your AMLs. Document them as if you are explaining their purpose to someone else. You might also consider developing a standard documentation format for your organization.*

In AML, you can add notes to yourself or comments about the program's purpose by prefacing the comments with the "/*" characters. These characters tell ArcInfo to skip anything that follows them and to proceed to the next line. The following is an example of how to document the *cnty_map.aml.*

```
/*****************************************************************
/* Program:          CNTY_MAP.AML
/* Purpose:          Draw map of county roads in project area
/* Date:             April 2000
/* Programmer:       Brendan Kyla
/*****************************************************************
/* Initialize graphics environment
&stat 9999
/* Start ArcPlot
Arcplot

/* Create plotfile called madison.gra
display 1040
madison.gra

/*************************

/* Reselect Madison County from proj_cnties county coverage
reselect proj_cnties poly name = 'Madison'

/* Set mapextent and draw Madison County
mapextent poly proj_cnties
polygonshades proj_cnties 2

/* Select and draw roads that fall within Madison County
reselect proj_roads line overlap proj_cnties poly
arcs proj_roads

/*************************
quit

/* Draw madison.gra to the screen
draw madison.gra 9999

/* End
```

Although this may seem like a lot of documentation for a short program, it is generally best to err on the side of including too much. Always include some information about the purpose of the AML, who wrote it, and when. This makes it much easier to use the AML at a later date, long after you have forgotten the details of how it was written.

Some programmers set off the header with rows of asterisks, or put their documentation notes in all capital letters. Blank lines can also be used to set off sections of your AML, which are simply skipped by ArcInfo. As you become proficient in AML you will develop your own documentation style. The important thing is to document, and to do so clearly.

Using &watch Files

ArcInfo provides an interesting directive called an *&watch file* that acts as a recorder for every command you enter. Once you are done, all commands you entered can be converted to an AML. You will, however, have to edit the AML to some degree, especially if you entered a command you did not mean to.

Exercise 10-2: Creating an AML from an *&watch* File

Just as in exercise 10-1, in this exercise you will create a map of the roads in Madison County. Enter all commands by typing them in, but this time, use an *&watch* file to record them.

Step 1: Starting an &watch File

To start an *&watch* file, first start Arc. At the Arc prompt, type the following.

```
Arc: &watch counties.wat
```

All commands you subsequently enter will be captured to the file named *counties.wat* until you give the command to stop recording.

Step 2: Entering Commands to Draw Madison County Roads

Now, type the following commands.

```
arcplot

reselect proj_cnties poly name = 'Madison'

mapextent poly proj_cnties
```

```
polygonshades proj_cnties 2
reselect proj_roads line overlap proj_cnties poly
arcs proj_roads
```

Step 3: Turning Off the &watch File

Now, turn off the *&watch* file by typing *&watch &off*. ArcInfo stops recording commands. Now exit ArcPlot by typing *quit*.

Step 4: Examining and Editing an &watch File

The final step is to convert the *&watch* file to an AML. At the Arc prompt, use the *&conv_watch_to_aml* directive (directives are discussed in the next section), as shown in the following.

```
Arc: &conv_watch_to_aml counties.wat counties.aml
```

Examine this new AML by opening *counties.aml* with a text editor. It should look like the code shown in the following illustration.

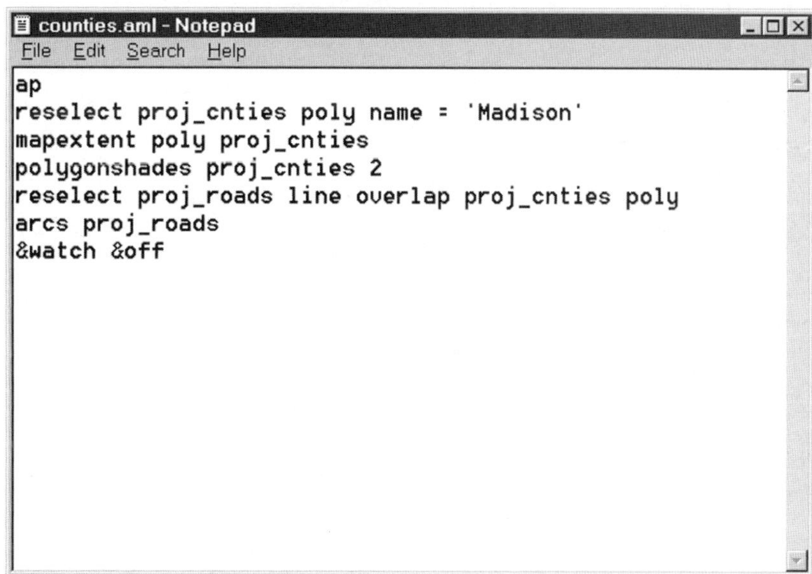

Counties.aml *code generated by using "&watch."*

The code should look identical to what you typed. Notice that the last line contains the command *&watch &off*; just delete this line. You have now created a short AML from a few typed commands.

You can see that using *&watch* files can be useful for capturing commands. It works as well in ArcEdit, Arc, and ArcPlot. Use it to capture your interactive sessions. The commands can become their own AML programs, or they can be incorporated into larger, even menu-driven, programs.

In this section, and by completing the exercises, you have learned the method behind scripting an AML. Scripting is really just stacking commands, one after the other, in a file. The commands are then executed sequentially by typing just one command (*&r* and the file name). You created, edited, and (hopefully) documented your short AML. The next section introduces you to some programming basics of AML.

AML Programming Basics

AML has powerful capabilities that let you add common programming functions to your scripts. There are commands to prompt a user for a response, that let you do things repetitively ("Do" loops), that read data in from a text file, and so on. Thoroughly discussing AML is, again, beyond the scope of this book. For readers interested in learning AML in depth, the authors recommend a visit to the ESRI Press web site. This section provides an introduction to some programming basics, and walks you through an exercise that puts AML to work.

AML Directives

A *directive* is any AML command that starts with an ampersand *(&)* sign. When ArcInfo sees a command that starts with "&," it recognizes it as being an AML command rather than an Arc or ArcPlot command. You have already used one AML directive, in exercise 10-1, when you issued the *&run CNTY_MAPS.AML* command. This told ArcInfo to execute an AML program.

The following is another simple directive. At the Arc prompt, type *&type Hi there!* ArcInfo knows that *&type* means to display the specified text to the screen. The following illustration shows the results of this command.

Results of using the &type directive.

Dozens of directives are available in AML. The following are examples of directives.

- **&LISTPROGRAM** [Lists the program's variables and values of those variables]
- **&POPUP** [Displays a scrollable text file to the screen]
- **&COMMANDS** [Lists AML directives and functions to the screen]
- **&PAUSE** [Pauses the program; user presses a key to restart]

For complete information on directives, see "Functional List of AML Directives and Functions" under "Command References for ArcInfo Prompts" (AML) in the on-line help. Also read through "Customizing ArcInfo" (AML). These sections provide excellent information about AML programming.

Variables

Like most programming languages, AML uses *variables* to store information. Variables represent information assigned to them through commands in an AML program. Variables may contain text, and numeric, date, or Boolean (True-False) data.

Assign information to a variable by using the *&setvar* directive. This directive can be abbreviated as *&s*. Examine the following lines to see how information is assigned to a variable.

```
&s landusetype := urban

&s landusetype = urban

&s landusetype urban
```

All three previous statements are equivalent. Each assigns the value *urban* to a variable named *landusetype*. Notice that after the variable *landusetype* there are ":=", "=", and no equal sign at all. Although the ":=" and "=" are optional in AML, they are important to use because they make reading and editing the programs much easier.

☛ **WARNING:** *Because the ":=" and "=" operators are optional in AML, variables cannot contain blank spaces. To do so would confuse AML as to what is part of the variable name and what is the value assigned to it.*

Once a variable is assigned a value, it keeps that value until it is reassigned a new one. Once a variable is assigned a value, it is referenced later in the program by surrounding it with "%" signs. For example, to reference the current value of *forest* of the variable *landusetype*, you would refer to it as *%landusetype%*. The following illustration shows an example of this methodology.

```
Arc
Arc: &s landusetype = Forest
Arc:
Arc: &type landusetype
landusetype
Arc:
Arc: &type %landusetype%
Forest
Arc:
Arc:
```

Referencing a variable in AML.

In the previous illustration, you see that a value *Forest* has been assigned to the variable *landusetype*. Using the *&type* directive introduced earlier, you see the difference between typing *&type landusetype* and *&type %landusetype%*. The former displays just the text "landusetype" on the screen, whereas the latter, using the % signs, prints the *value* stored in the *variable* named *landusetype*.

AML Functions

In addition to variables, AML also employs something called *functions*. A function is an AML command enclosed in brackets ([ ]); it returns a text string or the results of a calculation to the program. The following illustration shows an example of the *[response]* function used after the *&type* directive. The AML will print a message, incorporating a user's response, to the screen.

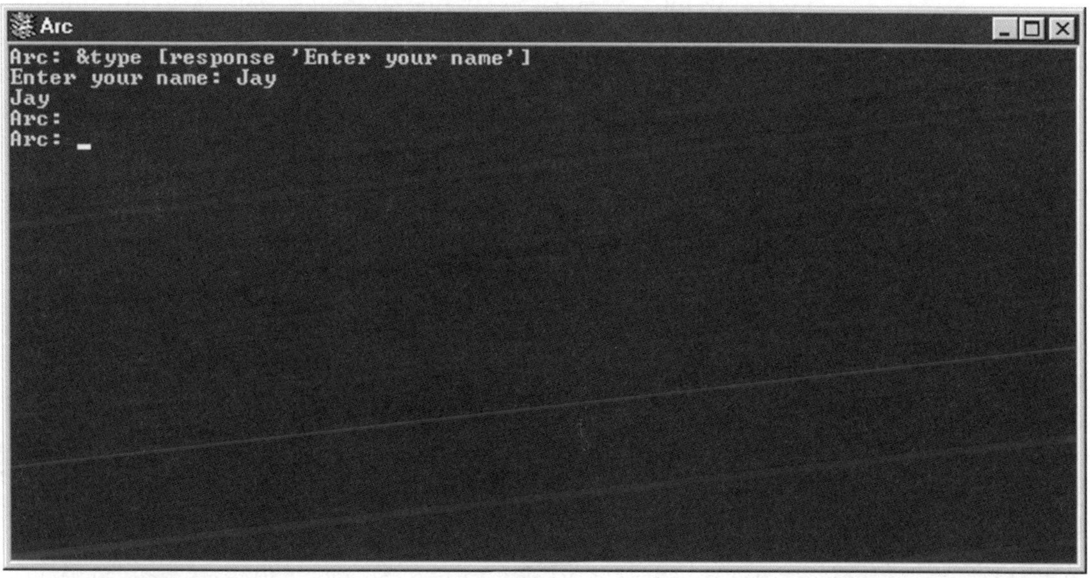

```
Arc: &type [response 'Enter your name']
Enter your name: Jay
Jay
Arc:
Arc:
```

Using a function in AML.

In this example, the function is *[response]*. The string of text, enclosed in single quotes, following it is displayed to the screen as a message prompting the user for a response. You see it in the previous illustration as "Enter your name." The *[response]* function requires the user to type a response. In this case, the user typed *Jay*. The *&type* directive preceding the *[response]* command will display the user's response to the screen. Thus, the message "Welcome Jay" would appear on the screen.

There are many very powerful functions in AML that perform mathematical calculations, geometry, and advanced text string manipulation. The following are examples.

- `[EXISTS]` determines whether an object (such as a coverage, a file, etc.) exists.
- `[FILELIST]` creates a list of files of a specified object type (such as coverages).
- `[EXTRACT]` extracts elements, based on a criterion, from a list of elements.
- `[INVDISTANCE]` calculates the distance between two points.

For complete information on the many functions, see "Functional List of AML Directives and Functions" under "Command References for ArcInfo Prompts" (AML) in the on-line help. Also read through "Customizing ArcInfo" (AML). Both on-line help sections will provide excellent information on using functions.

Reading In a File and Writing a Loop in AML

One main reason to create an AML is to automate the repeated performance of a task. You can capture commands in a *loop*. The program will then return to the loop each time it needs to execute those commands.

AML has a number of ways to let you loop through a section of programming code. One of the most used applications for a loop is to read in lines of information from a file. For each line of the file, the AML would perform the same action. The file could contain the names of coverages, data to be assigned to items in an INFO table, and so on. This section examines one method of creating such a loop.

In the *D:\ai80exercises\chap_10* directory there is a short, five-line text file named *cnty_names.txt*. The following is its content.

```
Clark

Clinton

Fayette
```

```
Greene

Madison
```

You will create an AML that reads in this file and performs the same function on each coverage. In this case, it draws a map displaying just that county and its roads. First, however, you need to understand the mechanics of opening a file and reading information from it.

AML provides the *[open]* and *[read]* functions to control opening and reading in files. Examine the syntax of each in the following.

```
&s file1 = [OPEN <file name> <status variable> <-READ | -WRITE | -APPEND>]
```

Because *OPEN* is a function, it and its *arguments* are enclosed in brackets. The function contains three arguments: *file name, status variable*, and either *READ*, *WRITE*, or *APPEND*. The *file name* is just the name of the text file being opened. The *status variable* shows whether or not the file opened correctly, which is indicated by a numerical code. (More details on the "status" are provided in exercise 10-3.)

READ, *WRITE*, and *APPEND* specify what you will do with the file. You may read information from it, write information to it, or add (append) information to it. Note that you can only do one of these actions at a time. In other words, you cannot both read from and append to the same file at the same time.

You must set a variable to "hold" the file and the action you wish to perform on it. Once the file is open, AML refers to this file by this variable name surrounded by "%" signs (in this case, *%file1%*). *Never* open an existing file with the *WRITE* option, or the data in it will be erased so that new data can be added.

☛ **WARNING:** *If you open an existing file with the* WRITE *option, the file will be erased before new data is written to it!*

Now that the file is open, you can read what is in it, line by line. For the purposes of looping, the value of *each line* is assigned to a variable. The variable value changes as the AML completes the loop and selects the next line in the file. To see how this works, look at the syntax for the *[read]* function. In the following example, remember that *&s* (or *&setvar*) sets a variable, the variable here being *curr_cnty* (for current county).

```
&s curr_cnty = [READ %file1% status]
```

The *[read]* function reads one line from the file, referred to by the variable *%file1%*. Recall that the *[open]* function opened the file, specified the type of operation to perform on it (read, write, or append), and assigned it to a variable name. The *[read]* function then needs just the variable name *%file1%* to know *which* file is open and *what* can be done with it.

You have looked at opening a file and reading one line from it. You will now learn how to set up a loop that reads each line, performs an action (such as creating a map composition in ArcPlot), and then repeats the action with information from the next line of a file.

Exercise 10-3: Looping with an AML

This exercise uses a loop to read a text file consisting of five county names. A map of each county and its roads will be composed. This exercise exemplifies the benefits of looping because the same AML code is used to create each map. Writing an AML to perform a task such as this also saves you from repeating the commands fives times.

Step 1: Typing Code for an AML

Using your word processing utility, create an AML file named *cnty_loop.aml*. Type in the following lines.

```
&s file1 = [open cnty_names.txt status -read]
&s curr_cnty = [read %file1% readstatus]
&do &until %readstatus% = 102
        arcplot
        display 1040
        %cover%.gra
        reselect proj_cnties poly name = [quote %curr_cnty%]
        mapextent poly proj_cnties
        polygonshades proj_cnties 2
        reselect proj_roads line overlap proj_cnties poly
        arcs proj_roads
        quit
        &s cover = [read %file1% readstatus]
&end
```

The first line opens the file named *cnty_names.txt* for reading, and assigns this information to a variable named *file1*. The second line reads the first line in the file and assigns it to a variable named *curr_cnty*. The *status* of the read function is stored in the variable named *readstatus*.

Line 3 contains all information necessary to start the looping. The *&do &until* tells the AML to continue repeating all commands listed between it and *&end* until the *readstatus* variable equals 102.

Readstatus variable? Look at line 2 again. This program line reads in a line from the file and maintains a status of its success in a variable named *readstatus*. The important thing to note here is that when the last record in *cnty_names.txt* is read (something called reaching "end of file" in computer programming circles), the variable is automatically assigned a value of 102 by ArcInfo. When this occurs, the lines between *&do &until* and *&end* are skipped and the program proceeds to the next line in the AML, if any.

The remaining lines are the same as the lines in *cnty_map.aml* from exercise 10-1, with one exception. The exception is that another *[read]* function is added as the last line before the *&end* directive. This also reads a line from the text file. Once the five maps are completed, this line will attempt to read the text file once more and will encounter an error because all of the file's lines have already been read. This is when the *readstatus* variable is assigned the value of 102 automatically by ArcInfo, which signifies that the end of the file has been reached. Again, once *readtstatus* equals 102, the loop terminates.

Step 2: Executing an AML

Now, run the AML by typing the following.

```
Arc: &r cnty_loop.aml
```

The AML will read each county name from the text file and create a new map of county roads for each. You will see ArcPlot start and stop for each map.

Step 3: Viewing the New Maps

The new maps are named *clark.gra*, *clinton.gra*, *fayette.gra*, *greene.gra*, and *madison.gra*. View them on the screen by using Arc's Draw command, input as in the following example.

```
Arc: draw clark.gra 9999
```

The illustrations that follow show the five maps composed using *cnty_loop.aml*.

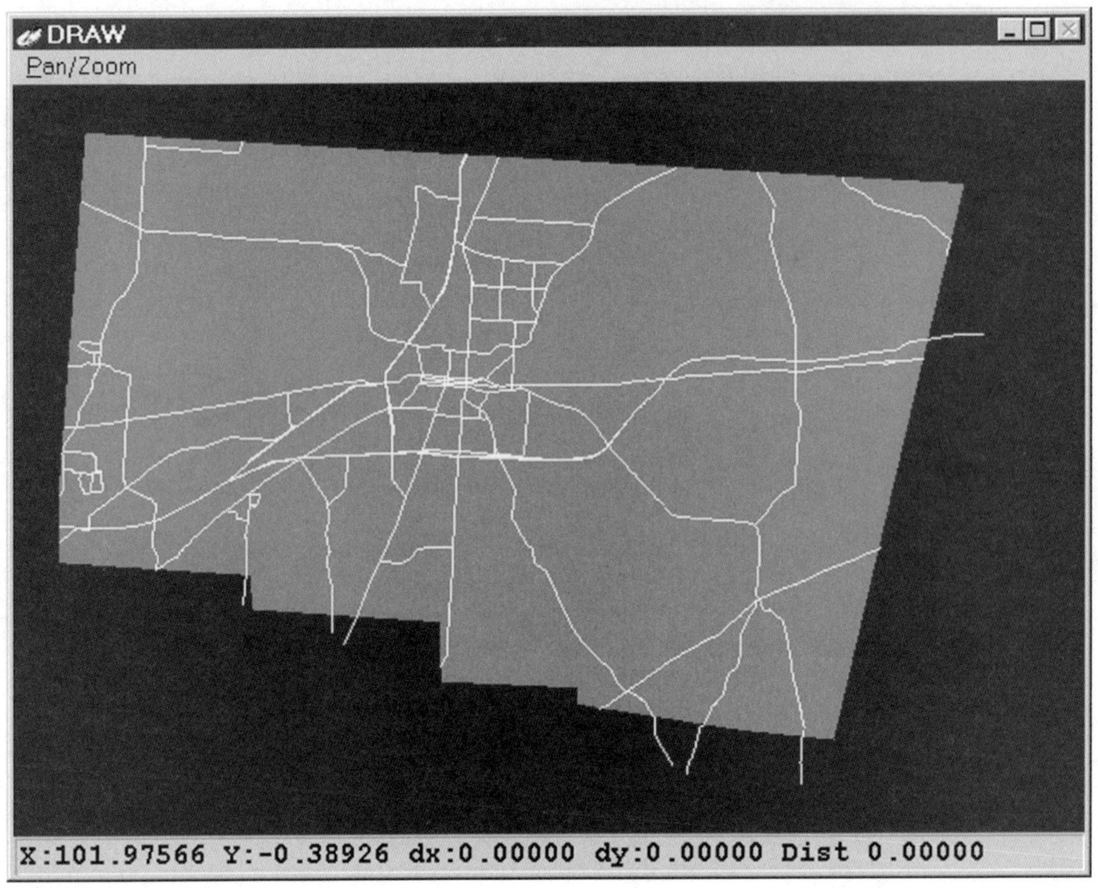

Map of roads in Clark County created with cnty_loop.aml.

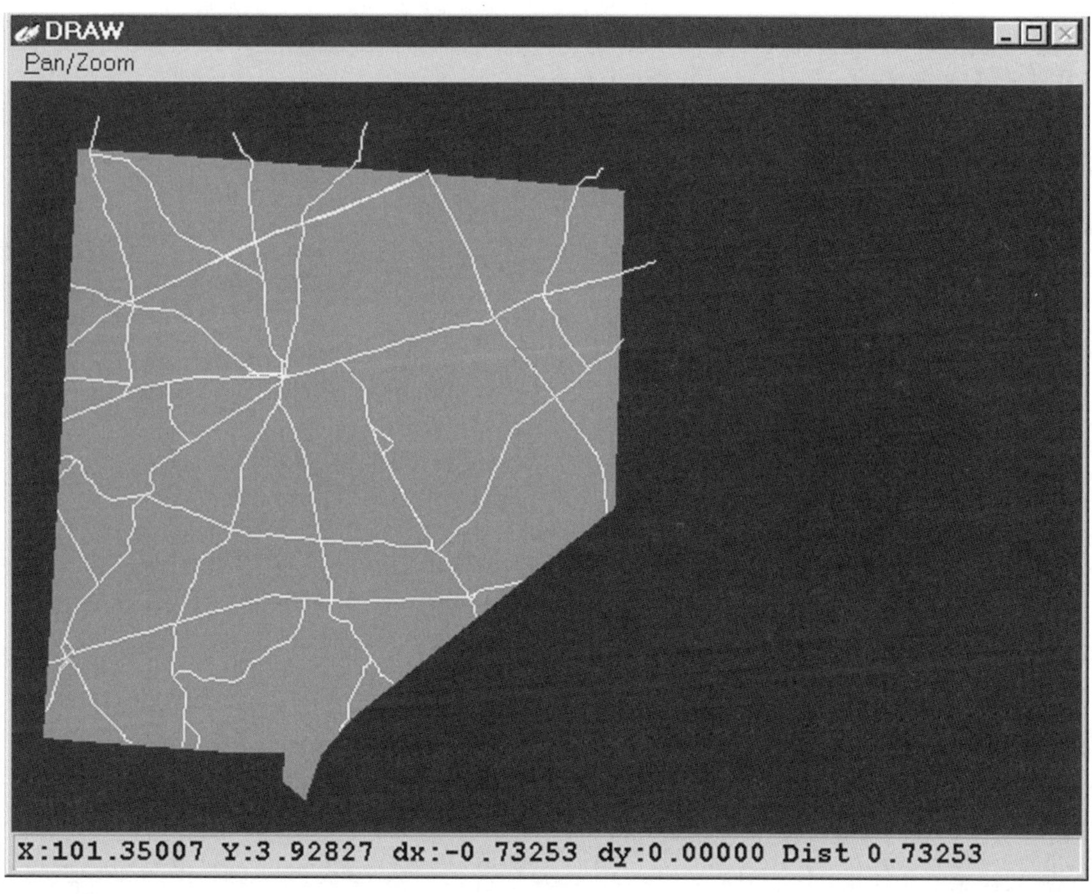

Map of roads in Clinton County created with cnty_loop.aml.

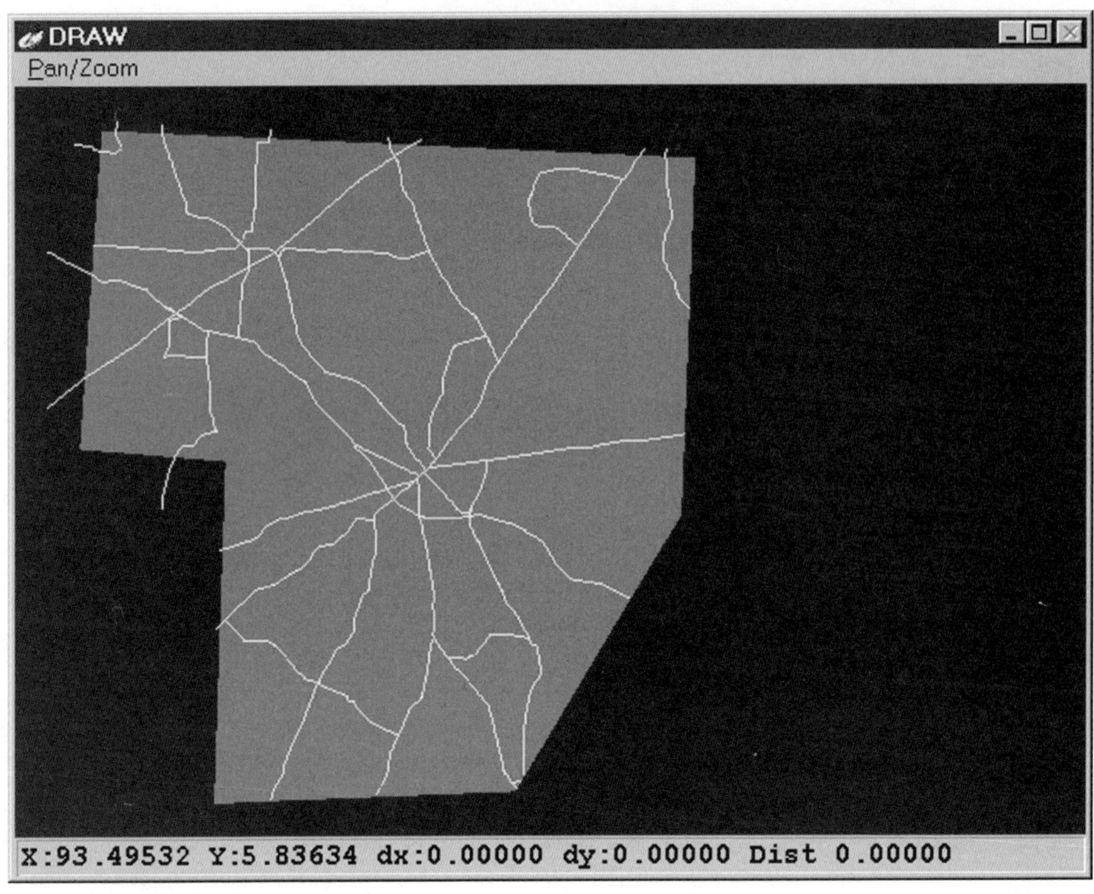

Map of roads in Fayette County created with cnty_loop.aml.

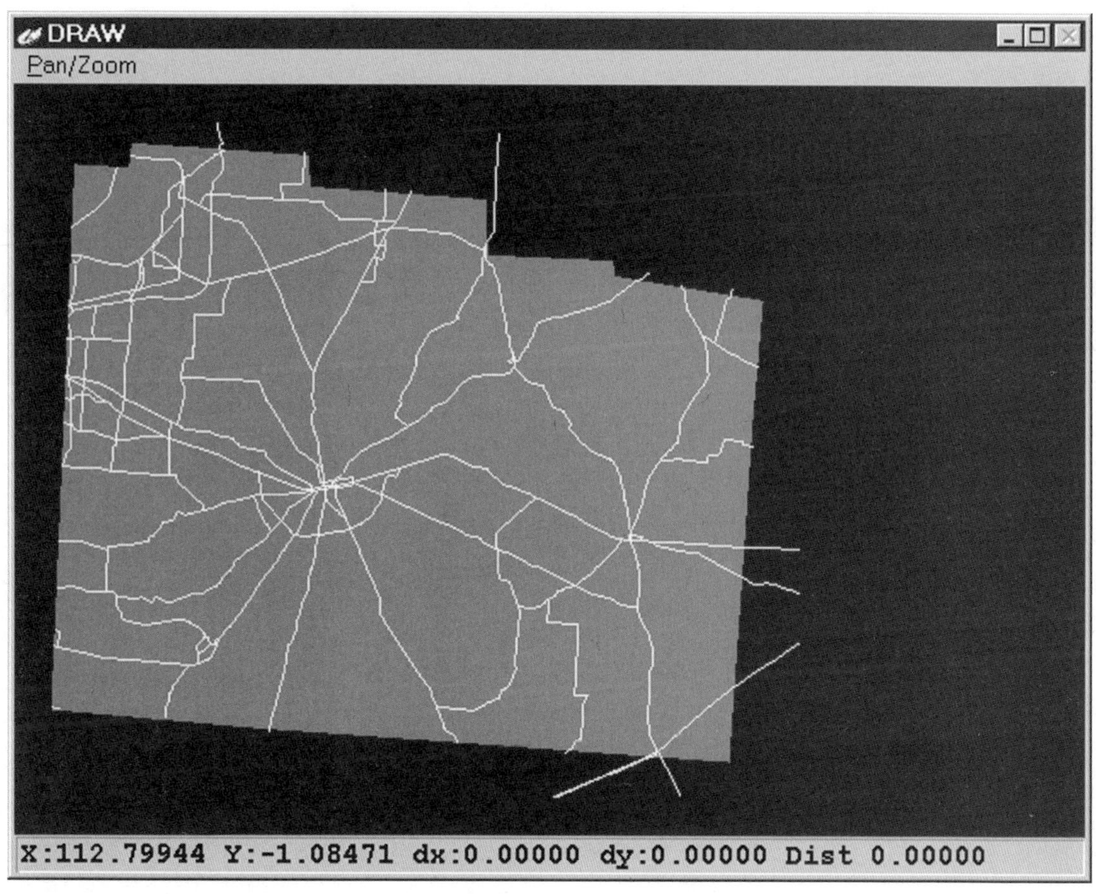

Map of roads in Greene County created with cnty_loop.aml.

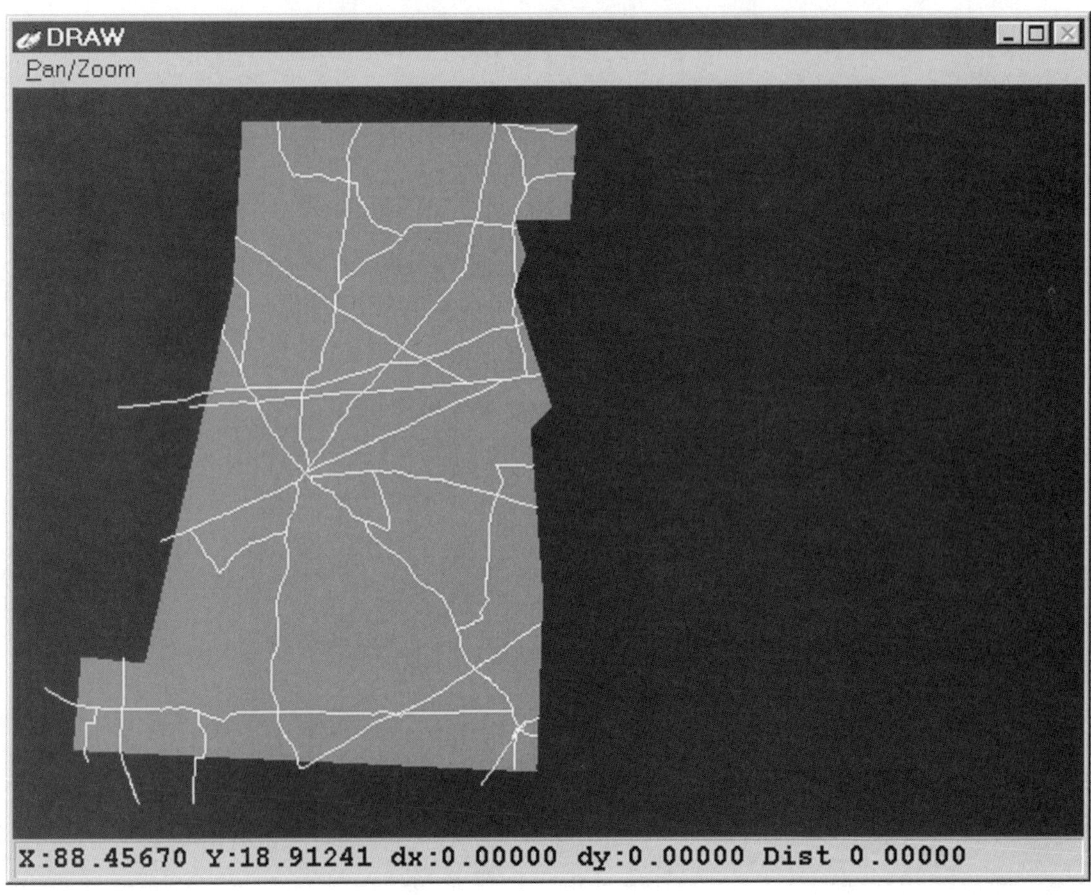

Map of roads in Madison County created with cnty_loop.aml.

Now that you are becoming proficient in authoring AMLs, you will need to organize your storage of them. The next section introduces *&ATOOL* and *&AMLPATH*, two directives for "pointing" to your AMLs.

&ATOOL *and* &AMLPATH

After you write an AML, generally the AML must be in the workspace in which you are working in order to run it. This can be good for AMLs that are specific to projects in that workspace. In the previous exercise, the AML you wrote resides in the *D:\ai80exercises\chap_10* workspace. If you change to another workspace and want to use *cnty_loop.aml* there, you must either copy the AML over or specify the entire path name to it.

ArcInfo lets you place often-used AMLs into a folder from which you can access them from any workspace. This provides the benefit of needing only one copy of an AML (in addition to its backup, of course!). Using this, you would not need to copy the AML from folder to folder. Anytime you edit that AML, you would be editing the *one* copy, which would be available from any workspace on your computer.

There are actually two means of accessing a "central repository" of AMLs. The first is to use the *&atool* directive, and the second is to use the *&amlpath* directive.

The *&atool* command is used to specify the directory in which your AMLs exist. Say, for example, you have created an AML named *jumpstart.aml* that you will use often. Instead of copying it from workspace to workspace, move it to a folder named *D:\amlscripts*. Then, when you start ArcInfo, type *&atool d:\amlscripts* at the Arc prompt.

Now you can run the *jumpstart.aml* AML from any directory simply by typing *jumpstart*. The *jumpstart.aml* executes immediately. Notice that you do *not* have to include the *&run* directive to run an AML kept in a specified *&atool* directory. Using *&atool* is a great way to keep your workspaces from becoming cluttered with several copies of the same programs.

✓ **TIP:** *You can turn AML names into commands by storing them in* &atool*-specified directories.*

You can specify up to 20 directories using the *&atool* command. Once you have many AMLs, organize them logically into subdirectories, like those that work in ArcPlot, ArcEdit, and so on. This way, the *names* of the AMLs used in those modules actually work like *commands*. Specify one name after another using the *&atool* command. For example, you would type *&atool d:\amlscripts\arcplot d:\amlscripts\arcedit*.

The *&amlpath* directive is similar to *&atool* except that it allows you to specify up to 30 directories instead of 20. In addition, you must use the *&run* directive when running an AML. To run an AML named *jumpstart.aml*, you would type the following.

```
Arc: &r jumpstart.aml
```

ArcInfo searches through the directories specified with the *&aml-path* directive for the AML and executes it once it finds it.

✓ **TIP:** *You can store these path names, along with system startup commands, in an* .arc *file. Commands stored in this file automatically execute each time you start the Arc module. ArcInfo contains a system* atool *directory under the ArcInfo install directory. There are also* .arcedit *and* .arcplot *files. These are usually hidden files, so you may not immediately see them in your directory. See your System Administrator's Guide for information about the location of your system's* atool *directory.*

This section has introduced you to AML directives, variables, and functions. Directives are commands exclusive to AML; that is, they are programming commands, not commands in the traditional sense, such as ArcPlot or ArcEdit commands.

Variables are used to store information in a program, and functions are commands that return a text string or the results of a calculation. Functions let you perform many advanced mathmatical calculations, as well as advanced text string operations. See ESRI's on-line ArcInfo help for a list of functions. You have learned about storing AMLs for access from any directory.

You have also learned how to read a file into an AML and how to create a simple loop. Keep in mind that you can also *write* files from an AML. For example, you could have an AML write out the name of every county in a state's county coverage, or write out the coordinates of every utility pole from a point coverage. The next section shows you how to create your own pull-down menus by putting AML commands behind a ready-to-use interface.

Using FormEdit to Create Menus

ArcInfo also lets you design and use your own pop-up menus. For example, you could use these menus, called *forms*, to select and edit feature attributes or to view information about selected features. Forms are built to access AML and ArcInfo commands; therefore, you should be very comfortable with both before working with forms.

Start FormEdit by clicking on Start ➡ Programs ➡ ArcInfo ➡ ArcInfo Workstation ➡ FormEdit. The FormEdit window, shown in the following illustration, appears on the screen.

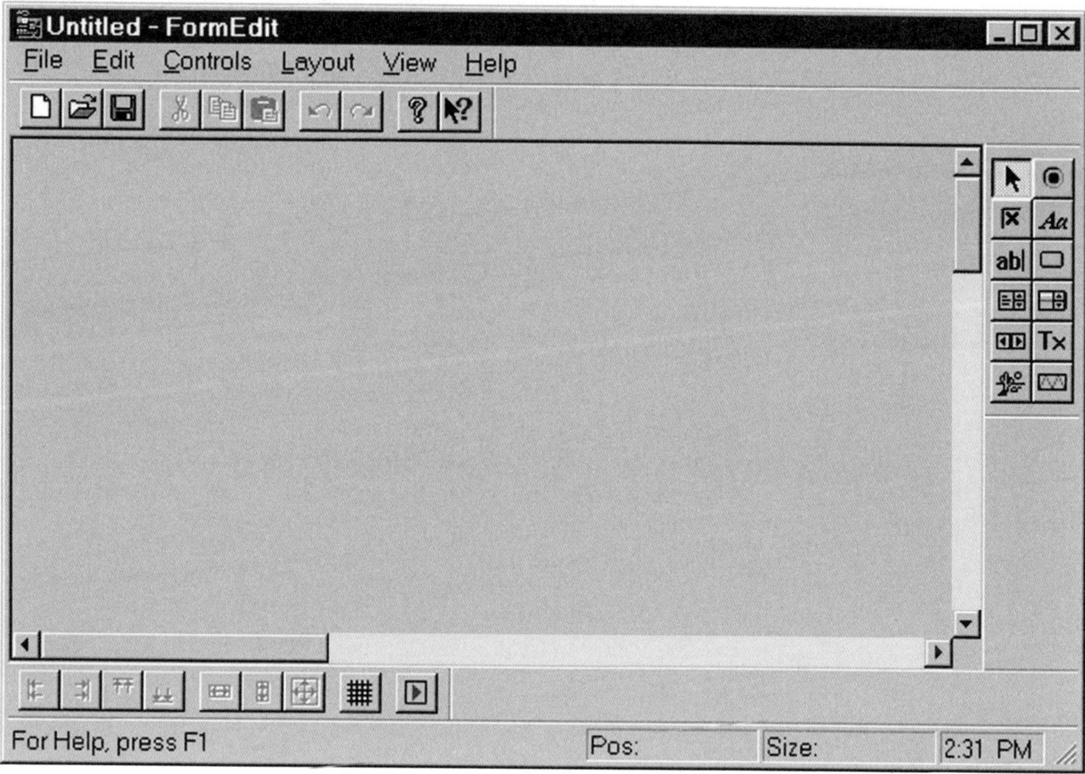

FormEdit window.

Along the top of the window are pull-down menus that let you add controls such as slider bars, buttons, and scrollable windows. The toolbar along the right of the screen duplicates the commands in the pull-down menus. The blank gray area is actually a template area. By activating Layout ➡ View Grid you can have a dotted line grid appear in its background, which you use to design your menus. To create menus, you would select a control from the toolbar and then use the mouse to place it on the template.

Creating forms can be an efficient method of interfacing with ArcInfo. It can be especially useful in helping non-technical users access ArcInfo data.

✓ **TIP:** *Create forms to help regular and occasional ArcInfo users alike in accessing data.*

The best way to learn about the controls and how they work is to actually create a form. Exercise 10-4, which follows, shows how to make a form that can be used to select a street name from a coverage and display it in red.

Exercise 10-4: Creating a Menu

For this exercise you will work for the Department of Public Works. You must design a menu that allows the person who fields phone calls about street locations to easily find and display streets using a GIS. To do this, she repeatedly displays and queries the city's street map using ArcInfo's ArcPlot module.

Use FormEdit to create a menu system that draws the map on the screen, lets her access any street from a pull-down menu, and then highlights the chosen street on her screen. You have a coverage named *Street* that will be used to display the city streets in ArcPlot.

Step 1: Creating an AML

Use a text editor to create a new AML named BEGIN.AML. Save it to the folder where you have copied the data for the Chapter 10 exercises. The following is the text of the AML. Type the following.

```
&stat 9999

arcplot

mapextent street

arcs street
```

Step 2: Using FormEdit to Create a Menu

The most important part about creating a form is first deciding what it should do. For this exercise, its purpose is to help you find and display information about any street in the city. The following capabilities will be needed in the form.

- A scrollable window of street names
- Allows the user to select a name from the street name list
- Draws the selected street in red
- Redraws to return the selected street to original color
- Dismisses the window

Because for every street selected you will perform the same actions (selecting it from the coverage, drawing it in red, and then redrawing it in its original color), you will create an AML to execute these actions for you. This AML will be incorporated into your menu. Use a text editor to create the AML that follows, named *showstreet.aml*. Type the following commands and documentation.

```
/**************************************************
/* Program:              SHOWSTREET.AML
/* Purpose:              To select a city street, and draw it in red,
/*                       then return it to original color
/**************************************************

/**** Select city street based on street selected in menu ****
reselect street line streetname = [quote %datalist0%]

/**** Draw selected street in red ****
arclines street 2

/**** List street info in the ArcEdit dialog window ****
list street line

/**** ASELECT to make all streets available for next selection ****
aselect street line

/**** End ****
& End
```

After creating the *showstreet.aml* file, start FormEdit. In the blank menu that appears, add a scrollable window by selecting Controls ➥ Data List from the pull-down menu. A data list is a scrollable text box, allowing you to view a list of files in a directory. Once you select a data list in the menu, a small icon representing it appears. Use the mouse to place it anywhere on the gray template area. A small, scrollable window appears, as shown in the following illustration.

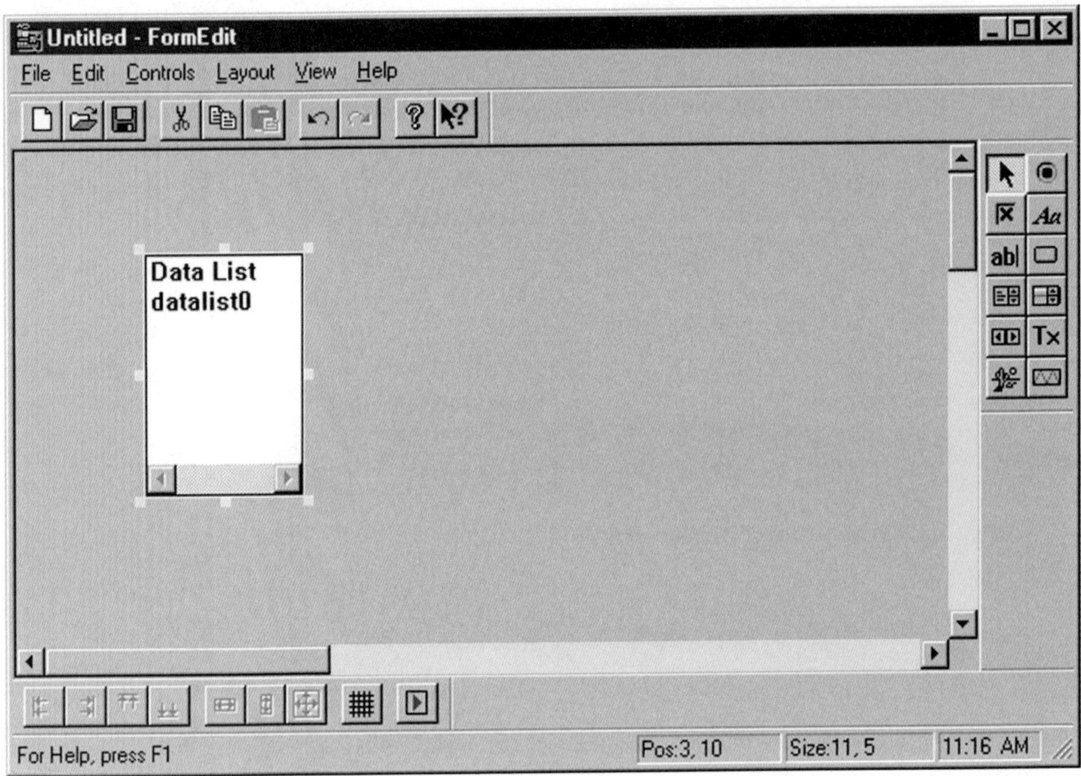

Adding a scrollable window to a new form.

Notice that the cursor now turns into a small hand. Use this hand to select and move the scrollable window around on the template. Place it in the upper left-hand area. Now make it larger by "grabbing" (clicking and holding down the left mouse button) its lower right corner. "Stretch" it diagonally to make it larger. Your form should look similar to that shown in the following illustration.

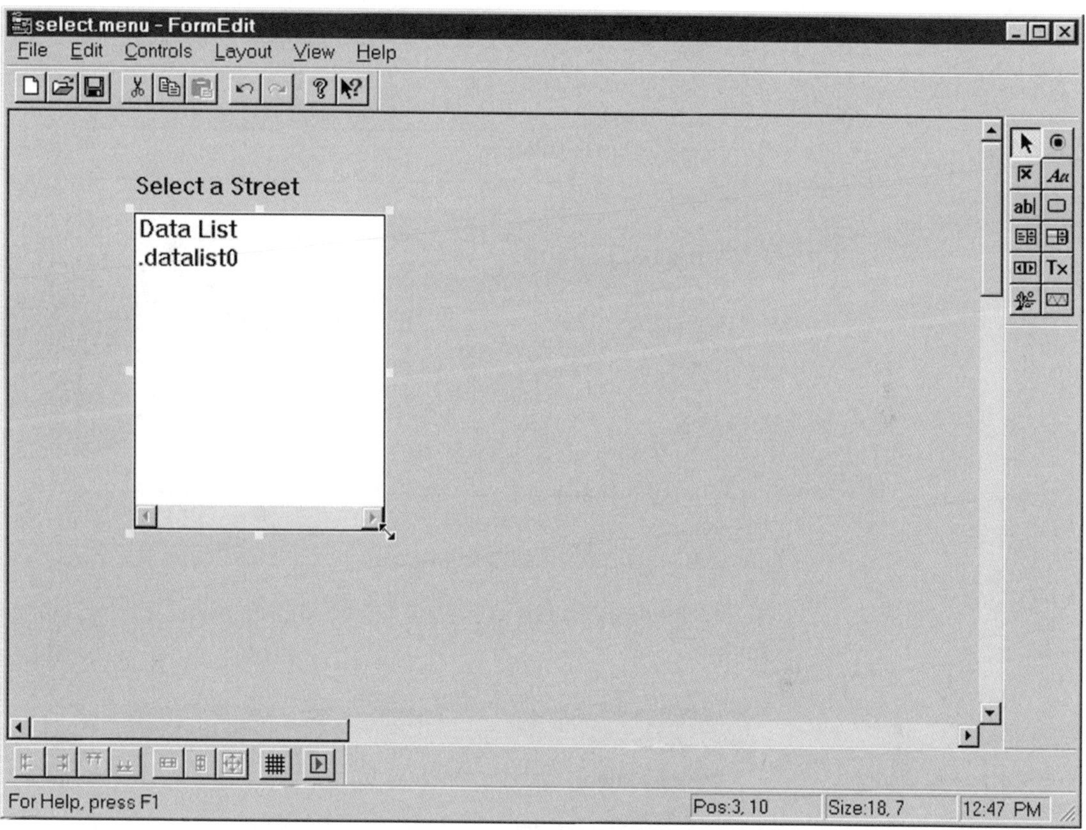

Moving and enlarging the scrollable window.

Now that you have placed the scrollable window, you need to "connect" it to the street coverage so that street names will appear in it. Double click on the scrollable window and you will see the Data List Properties dialog, shown in the following illustration, which allows you to modify the properties of the window.

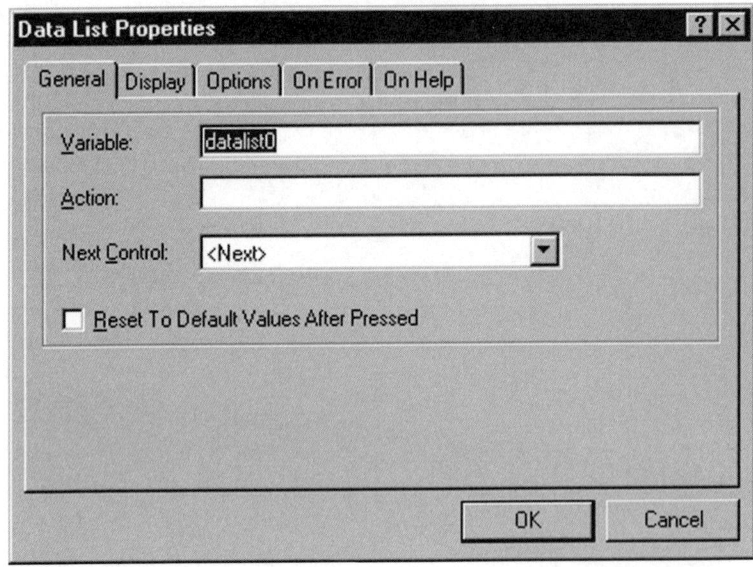

Data List Properties dialog.

Select the Options tab to get the screen shown in the following illustration.

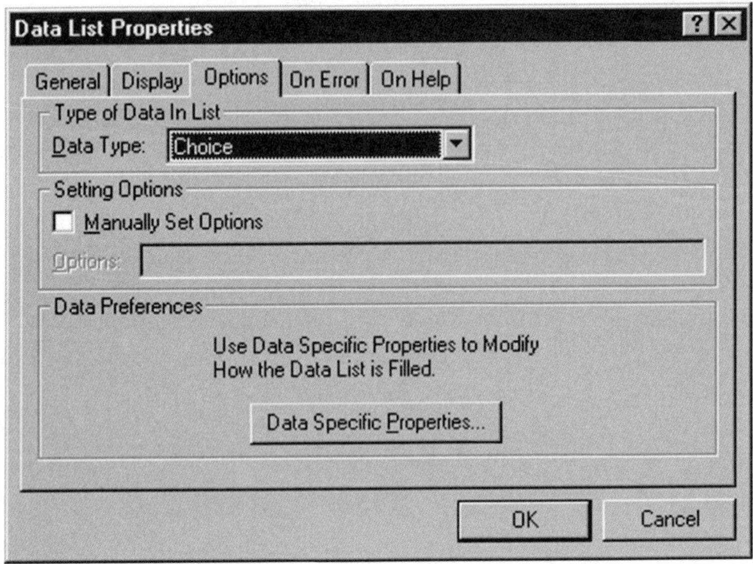

Options tab of the Data List Properties dialog.

In the Type of Data In List field, scroll down and select Unique. This will make a unique entry in the scrollable window for each street name. (Otherwise, because most of the streets consist of more than one arc, you may have each arc's name listed more than once.) Next, click on the Data Specific Properties button. The Unique Properties dialog appears, shown in the following illustration.

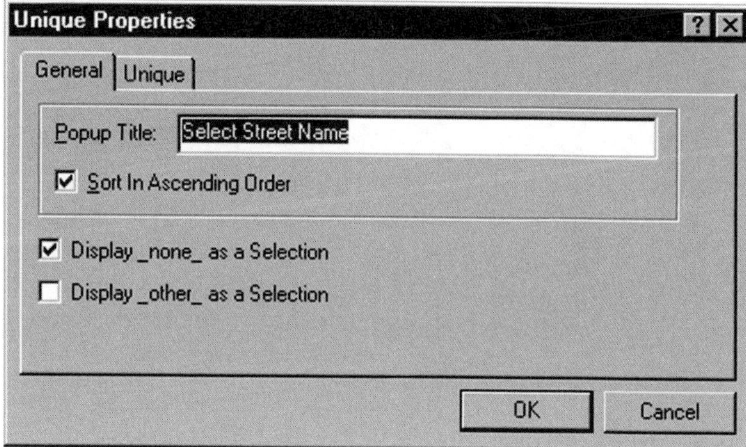

Unique Properties dialog.

Name your newly created form by typing *streetchoice* in the Popup Title box. Select the Sort In Ascending Order option to list the streets in alphabetical order. Then, turn on the "Display none as a Selection" option. This means that the choice of "none" will be added to the list of streets.

Now select the Unique tab on this menu. You will get the menu shown in the following illustration.

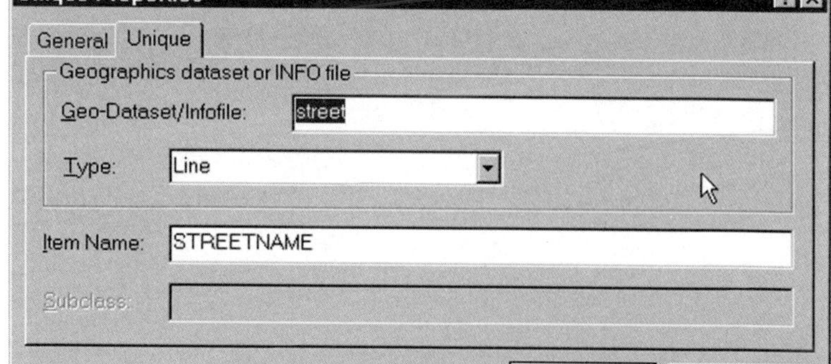

Unique tab of the Unique Properties dialog.

Type the coverage name *(street)* in the Geodatabase/Infofile text box. Then select the feature type "line" from the Type box (because *street* is an Arc coverage) and type the item name (STREETNAME) in the Item Name box. Click on the OK button in the Unique Properties menu to return to the Data List Properties window.

Now, incorporate *showstreets.aml* into your menu. In the box labeled Action, you can enter an ArcInfo command that will execute each time a street is selected. For this exercise, type in *show-street.aml.* It will run each time a street name is chosen.

As with most well-written menu systems, text appears somewhere (usually at the bottom) to educate or remind the user about what to do at that menu. Add this to your menu by selecting the Controls ➥ Text menu. Place the text box just below the scrollable window. Click on it twice to get its Text Properties menu. Type *Select a street* into the Text box and click on OK.

So far you have a scrollable list and an AML that will run each time a selection is made from the list. Now add two buttons. The first one will *redraw* the street coverage. The menu user will click on this after he or she is finished viewing the current street selection. This will redraw all streets in their original color (in this case, white). The second button will allow the user to *dismiss* (quit) the form.

To add a button that will redraw the street coverage, select the Controls ➥ Button menu. Using the button icon and your mouse, place the button to the bottom left of the scrollable window, as shown in the following illustration.

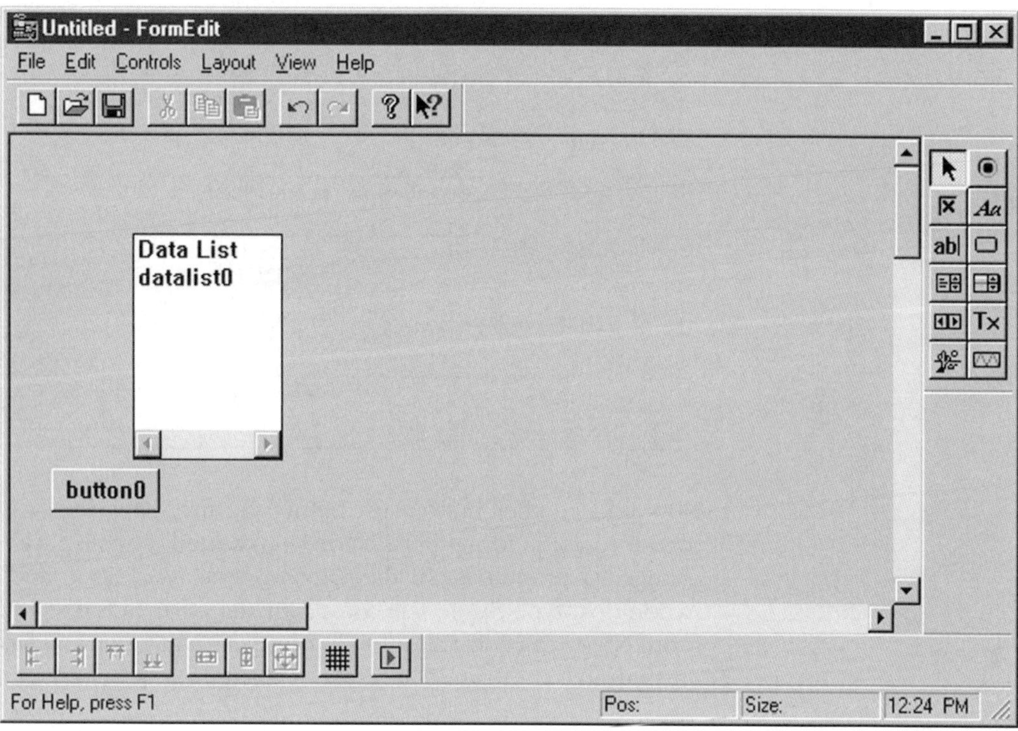

Adding a button to the form.

Click twice on the button to view its properties. The Button Properties menu is shown in the following illustration. Under Caption, type the word *Redraw*. This will appear on the button. In the Action text box here, type the following ArcPlot commands: *aselect street line; arclines street 1*. These are two commands separated by a semicolon. The first selects all arcs in the street coverage, and the second command draws them using symbol 1 (white).

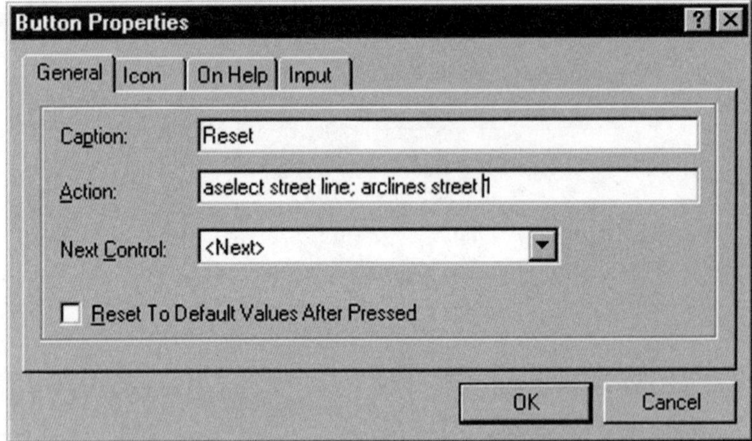

Button Properties menu.

Now add another button just below the first. This button will be used to quit the form. Once it has been added, double click on it to change its properties. In the Caption box, type *Quit*, and in the Action box, type *&stop*. The *&stop* AML directive ends execution of a form. Now your form should look like that shown in the following illustration.

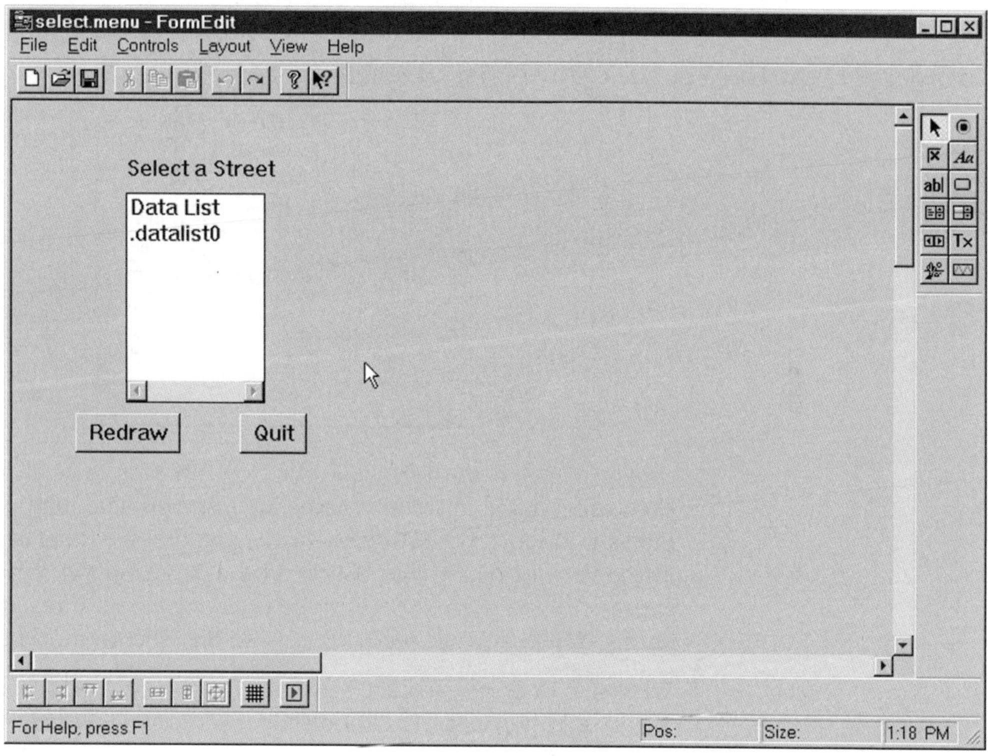

Form with two buttons added.

You can get a peek at what your menu will look like once it is running. To preview or "test" your form, select Layout ➥ Test. A window pops up showing your form as it would look to a user.

Finally, select File ➥ Save to save the menu to the workspace in which you have copied the sample data for this chapter. Name your newly created menu *select.menu*.

✓ **TIP:** *Always use the* .menu *extension when naming your forms.*

Step 3: Using the Select Menu

To test your menu, first start Arcplot. At the prompt, type the command *&menu select.menu*. Your new menu should appear on the screen and look like that shown in the following illustration.

Displaying SELECT.MENU form.

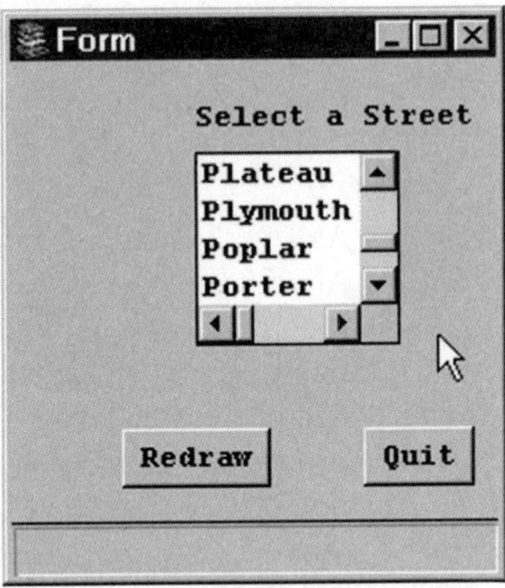

You can see in the scrollable window an alphabetical listing for each street name in the city. Selecting a street name with the mouse draws it in red on the ArcPlot canvas and lists the street's attributes in the ArcPlot dialog box. Click on the Reset button to return the street colors to white. Spend some time enjoying the results of your efforts. When you are ready to quit, select the Quit button.

What if it does not work? Welcome to the world of programming! You would proceed to methodically track down the error. Carefully check your work at each step of this exercise.

For other programs or menus you write, talking to other programmers can be a big help. ESRI even offers LISTSERVs to which you can post questions. You will get several replies to your programming question. It is protocol to summarize and post those responses so that others in the group can benefit.

NOTE: *For more information on LISTSERVs, and for ArcInfo information in general, visit the* www.esri.com *website.*

Careful documentation can also aid error fixing in your programs. You will know "what is supposed to be happening where" in your program. Use the *&pause* and related functions to pause the programs and check the values of variables. Be sure to comment out the *&pauses* and related functions before releasing your program for general use.

This section introduced the basics of creating menus. In the exercise you learned menu creation, including how to position controls on the screen, how to tie actions to each control, and how to add text explanations to the menu. Use what you learned here to create new forms that meet your specific needs. Forms are a great means of helping out occasional users of ArcInfo. They can also streamline daily operations.

Summary

AML provides real power to users with its ability to customize your ArcInfo sessions. You can create executable AMLs or create pop-up menu systems. As your knowledge of ArcInfo increases, use AML to make repetitive tasks go more quickly. Consider customizing ArcInfo so that you have a full-featured menu system suited to your projects. Experienced programmers will learn AML quickly, and newcomers can begin using AML to write simple scripts such as those in exercises 10-1 and 10-2. You will quickly find that using AML brings a new level of power to your GIS activities.

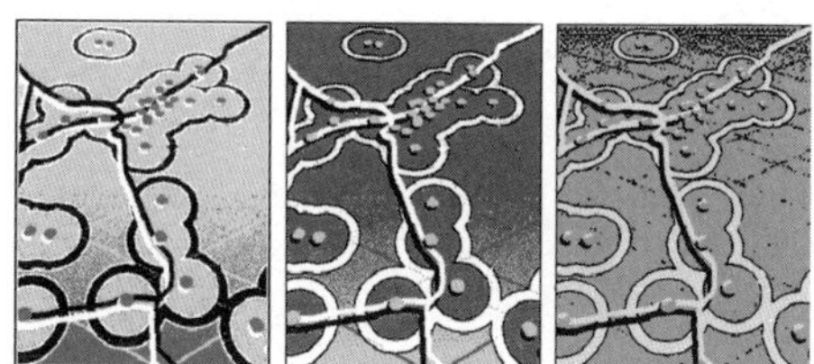

Appendix A

Internet Data Sources

The growth of the Internet has been a boon to the GIS community. Users previously had to digitize maps themselves or purchase digital data from third parties. Data collection has always been the most expensive cost of a GIS operation.

Such is not always the case today, because more data is available free of charge or at a minimal cost from sites on the Internet. Free data sources range from government agencies to educational institutions. Some private companies also support free data as a means of helping users get started with the companies' software.

The following sections contain lists of Internet data sources and links. This appendix is far from exhaustive. It is intended, rather, to familiarize you with the types of sources available, and as a starting point for each. Some sites may include a combination of free and per-charge data. Neither the authors nor the publisher make any endorsement of any site, and the individual lists are certainly not all-inclusive. Included are federal agency sources, state sources, international sources, private sources, and collections/link sources.

Federal Agencies

The following Internet sites are made available by agencies and departments of the federal government. Although much of the spatial data is available free of charge, some sites do carry charges for their data.

- U.S. Fish and Wildlife Service:

  http://www.fws.gov/data/datafws.html
- U.S. Geological Survey EROS Data Center:

  http://edcwww.cr.usgs.gov/dsprod/prod.html
- U.S. Geological Survey National Mapping Center:

  http://mapping.usgs.gov/

- U.S. Geological Survey US Geodata:
  http://edcwww.cr.usgs.gov/doc/edchome/ndcdb/ndcdb.html
- National Park Service GIS Office:
  http://www.nps.gov/gis/index.html
- U.S. EPA Spatial Data Clearinghouse:
  http://nsdi.epa.gov/nsdi/
- Bureau of Land Management Geospatial Information Center:
  http://www.blm.gov/gis/
- Bureau of Land Management National Applied Resource Science Center:
  http://www-a.blm.gov/gis/narsc/index.html
- U.S. Bureau of the Census TIGER Center:
  http://www.census.gov/geo/www/tiger/
- NASA Global Change Master Directory:
  http://gcmd.gsfc.nasa.gov/
- NOAA National Geophysical Data Center:
  http://www.ngdc.noaa.gov/
- NOAA Mapfinder:
  http://mapfinder.nos.noaa.gov/
- U.S. Department of Transportation Geographic Information Services:
  http://www.bts.gov/programs/gis/
- National Imagery and Mapping Agency:
  http://www.nima.mil/
- Federal Emergency Management Agency Map Service Center:
  http://www.fema.gov/msc/index.htm
- National Resources Conservation Center:
  http://www.ncg.nrcs.usda.gov/nsdi_node.html

State Sources

The following sites maintain data pertinent to individual states. This list is representative of what is available and should be used as a guide to the types of data used by state governments.

- Radford University Geoserver, Virginia:
  http://www.runet.edu:8800/~geoserve
- California DEMs, CalState Northridge:
  http://130.166.124.2/ca_dems.htm
- West Virginia Department of Environmental Protection:
  http://www.dep.state.wv.us/metadata/index.html
- Environmental Data Center, University of Rhode Island:
  http://www.edc.uri.edu/gis/
- Idaho Geospatial Data Center:
  http://geolibrary.uidaho.edu/
- Pennsylvania Geospatial Data Clearinghouse:
  http://www.pasda.psu.edu/
- South Carolina Department of Natural Resources:
  http://water.dnr.state.sc.us/gisdata/

International Sources

The following Internet sites highlight data available for areas outside the United States.

- United Nations Environment Program:
  http://grid2.cr.usgs.gov/
- British Geological Survey:
  http://www.bgs.ac.uk/

Private Sources

There are a growing number of private firms who market data to the public. Some have a combination of free sample data that you may download. Although the following list contains just a few Internet sites, there are many more available that you will find of interest.

- ESRI Data Online:
  http://www.esri.com/data/index.html
- Microsoft's Terraserver:
  http://terraserver.microsoft.com/
- GIS Data Depot:
  http://www.gisdatadepot.com/

Collections/Link Sites

The following sites contain links to many other sites that host spatial data on the Internet. They are often a good place to begin a search for a particular spatial data source.

- Department of Geography, University of Texas, Austin:
  http://www.utexas.edu/depts/grg/virtdept/resources/data/data.htm
- Alexandria Digital Library:
  http://www.alexandria.ucsb.edu/adl.html
- Department of Geography, University of Edinburgh:
  http://www.geo.ed.ac.uk/home/giswww.html

Index

G

H

I

J

K

L

M

N

S

T

U

V

W

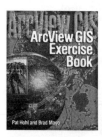

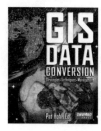

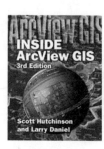

GIS Solutions in Natural Resource Management

Stan Morain

This book outlines the diverse uses of GIS in natural resource management and explores how various data sets are applied to specific areas of study. Case studies illustrate how social and life scientists combine efforts to solve social and political challenges, such as protecting endangered species, preventing famine, managing water and land use, transporting toxic materials, and even locating scenic trails.

Order number 1-56690-146-4

400 pages, 7″ x 9″ softcover

The GIS Book, 4th Edition

George B. Korte, P.E.

Proven through three highly praised editions, this completely revised and greatly expanded resource is for anyone who needs to understand what a geographic information system is, how it applies to their profession, and what it can do. New and updated topics include trends toward CAD/GIS convergence, the growing field of systems developers, and the latest changes in the GIS landscape.

Order number 1-56690-127-8

440 pages, 7″ x 9″ softcover

Raster Imagery in Geographic Information Systems

Stan Morain and
Shirley López Baros, editors

This book describes raster data structures and applications. It is a practical guide to how raster imagery is collected, processed, incorporated, and analyzed in vector GIS applications, and includes over 50 case studies using raster imagery in diverse activities.

Order number 1-56690-097-2

560 pages, 7″ x 9″ softcover

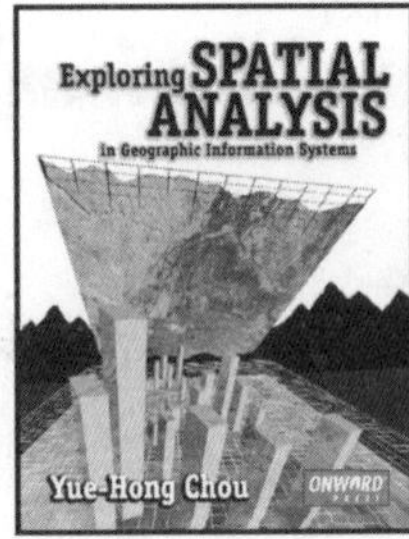

Exploring Spatial Analysis

Yue-Hong Chou

Written for geographic information systems (GIS) professionals and students, this book provides an introduction to spatial analysis concepts and applications. It includes numerous examples, exercises, and illustrations.

Order number 1-56690-119-7

496 pages, 7″ x 9″ softcover

Processing Digital Images in GIS: A Tutorial Featuring ArcView and ARC/INFO

David L. Verbyla
and Kang-tsung (Karl) Chang

This book is a tutorial on becoming proficient with the use of image data in projects using geographical information systems (GIS). The book's practical, hands-on approach facilitates rapid learning of how to process remotely sensed images, digital orthophotos, digital elevation models, and scanned maps, and how to integrate them with points, lines, and polygon themes. Includes companion CD-ROM.

Order number 1-56690-135-9

312 pages, 7″ x 9″ softcover

Focus on GIS Component Software: Featuring ESRI's MapObjects

Robert Hartman

This book explains what GIS component technology means for managers and developers. The first half is oriented toward decision makers and technical managers. The second half is oriented toward programmers, illustrated through hands-on tutorials using Visual Basic and ESRI's MapObject product. Includes companion CD-ROM.

Order number 1-56690-136-7

368 pages, 7″ x 9″ softcover

GIS: A Visual Approach

Bruce Davis

This is a comprehensive introduction to the application of GIS concepts. The book's unique layout provides clear, highly intuitive graphics and corresponding concept descriptions on the same or facing pages. It is an ideal general introduction to GIS concepts.

Order number 1-56690-098-0

400 pages, 7" x 9"

GIS: A Visual Approach Graphic Files

This set of 12 disks includes 137 graphic files in Adobe Acrobat, plus the Acrobat Reader. Corresponding with chapters in *GIS: A Visual Approach*, nearly 90% of the book's images are included. Available in Windows or Mac platforms. Ideal for instructors and organizations with a large GIS user base.

Order number 1-56690-120-0

Set of disks

Cartographic Design Using Arc-View GIS

Ed Madej

Both an effective primer on digital map design, as well as a classic software tutorial, this book makes particular reference to cartographic methods available in ArcView GIS fro m ESRI. Structured around fundamental concepts of map design, each standalone chapter introduces general map design theory followed by examples that apply these principles.

Order number1-56690-187-1

432 pages, 7-3/8 x 9-1/8".

Available June 2000.

FOR A COMPLETE LIST OF ONWORD PRESS BOOKS, VISIT OUR WEB SITE AT:

http://www.onwordpress.com

TO ORDER CALL: 800-347-7707

Your opinion matters! If you have a question, comment, or suggestion about OnWord Press or any of our books, please send email to *info@delmar.com*. Your feedback is important as we strive to produce the best how-to and reference books possible.